SECOND ASM
HEAT TREATMENT
AND
SURFACE ENGINEERING
CONFERENCE
IN EUROPE
Pt. 1

SECOND ASM HEAT TREATMENT AND SURFACE ENGINEERING CONFERENCE IN EUROPE

Dortmund, Germany, June 1st - 3rd, 1993

Conference Location :   Westfalenhallen

Conference Chairman :   R. Speri, Vide et Traitement SA(F)

Organizing Committee :   W. Moerdijk (Chair), Petrofer Benelux BV (NL)
G. Barten, Inco Alloys International Ltd. (D)
R. Gijswijt, Staps Industries (NL)
A. Hick, Wolfson Heat Treatment Centre (UK)
T. Khan, ONERA (F)
R. von Bergen, Houghton Vaughan plc. (UK)

Advisory Committee :   E.J. Mittemeijer (Chair), Delft University of Technology (NL)
T. Bell, University of Birmingham (UK)
O. Ericson, Brukens Nordic AB (S)
G. Frommeyer, Max-Planck-Inst. für Eisenforschung (D)
M. Gantois, Ecole des Mines (F)
G. Krauss, Colorado School of Mines (USA)
H. Kunst, Ingenieurbüro Dr. Kunst (D)
J. Roos, University of Leuven (B)
H. Veltrop, Hauzer Harding Centrum (NL)
P. von Czarnowski, Vereinte Aluminiumwerke (D)

List of Sponsors :   The Benelux Heat Treatment Society (VWT)
The Heat Treating Division Council of ASM
Industrial Heating Equipment Association (IHEA)
The International Federation of Heat Treatment (IFHT)
Metal Treating Institute (MTI)
Netherlands Society for Materials Science (BvM)
The Wolfson Heat Treatment Centre

Proceedings of the Second

ASM HEAT TREATMENT

AND

SURFACE ENGINEERING CONFERENCE

in Europe

Pt. 1

held June 1st - 3rd, 1993 in Dortmund, Germany

Editor

E.J. Mittemeijer

Laboratory of Materials Science
Delft University of Technology
Delft, The Netherlands

TRANS TECH PUBLICATIONS
Switzerland - Germany - UK - USA

Sep /ae
Chem

Distributed *in the Americas by*

Trans Tech Publications Ltd
LPS Distribution Center
52 LaBombard Rd. North
Lebanon, NH 03766
USA
Phone: (603) 448 0037
Fax: (603) 448 2576

and worldwide by

Trans Tech Publications Ltd
Hardstrasse 13
CH-4714 Aedermannsdorf
Switzerland
Fax: (++41) 62 74 10 58

Preface

THERMODYNAMICS, KINETICS AND PROCESS CONTROL

The study of equilibrium is crucial for identifying the state a system strives for. Thermodynamics provide the means to define equilibria from the condition of minimal energy. In this way phase diagrams can be established. Phase diagrams for binary systems for example show which phase or phases is or are to be expected at equilibrium as a function of temperature and composition at constant pressure, in accordance with Gibbs' phase rule. Heat treaters and surface engineers are frequent users of such tabulated phase diagrams as possible aids to understand the path followed by a material when subjected to a certain treatment in order to obtain a desired microstructure.

Reaching a state of equilibrium is not aimed for in general. As a matter of fact a process is often terminated in practice at a stage where the material concerned is far from equilibrium. Hence, the kinetics of material processes should be known to be able to arrive at a full control over the microstructure (e.g. governing the occurring phase transformations).

The present state of affairs in the field of heat treatment and surface engineering may be summarized as follows:

(i) the fundamentals of thermodynamics for binary systems are, more or less, well understood, but for ternary and in particular many component systems only more or less phenomenological approaches exist[1], and

(ii) the knowledge of kinetics is fragmentary and in many cases from phenomenological origin (e.g. the Johnson-Mehl-Avrami equation for phase transformations)[2,3].

Is is easy to predict that the emphasis in future research will be on the kinetics of the occurring processes, thereby pursuing to establish a nonobvious link with the corresponding thermodynamics. An example of the type of surprises that may then lie ahead of us is the following. It has often been thought that at a transformation front near "local equilibrium" prevails, i.e. the compositions at both sides of the interface are practically as prescribed by the equilibrium to which the system tends (e.g. see phase diagram). However, in contrast with the commonly adopted view, recent results indicate that this does not appear to hold in general for transformations in iron-carbon alloys. On the other hand, the kinetics of thermochemical processes as carburizing and nitriding may be described by imposing the "local equilibrium" assumption at the surface of the workpiece and layer/substrate interfaces.

Command of kinetics requires control of process parameters. An illustrative example of what can be achieved is provided by the gaseous carburizing process. By measuring the oxygen partial pressure in the furnace atmosphere by using a solid-state electrolyt ("oxygen sensor") on-line control of the developing carbon-concentration/depth profile in the workpiece is possible. The commercial success of this approach stimulates similar developments for other treatments. Thus a break-through is anticipated in the field of gaseous nitriding and nitrocarburizing. The nitriding potential can be controlled via the partial pressure ratio p_{H_2O}/p_{H_2} (through the oxygen partial pressure as measured by the "oxygen sensor"), recognizing that a small amount of H_2O is always present in ammonia gas employed in heat treatment practice (or a small amount of H_2O, or O_2, could be added deliberately). The development is discussed at length in issues of the journal Härterei-Technische Mitteilungen in 1993 and 1994. This progress has a peculiar and instructive facet for those performing research at university laboratories (which includes this author): they usually put a lot of effort in *avoiding* the presence of oxygen in nitriding atmospheres, which may well have been an important reason for the relatively late moment of time of this achievement.

The present-day computational force has certainly enhanced greatly the interest and activity in modelling of the thermodynamics and kinetics of heat treatment and surface engineering. Computer-aided predictions of residual stress profiles, hardenability and distortion are presented on the basis of some model for the occurring phase transformations. Progress is limited by the lack of numerical data ("material constants") to put into the models for application to materials used in practice. "Simply" measuring the character traits of practical materials, as exhibited by, for example, transfer and diffusion coefficients, is in order.

These proceedings provide a frozen picture of the current knowledge and technology on heat treatment and surface engineering. On comparing the proceedings of this second, "Dortmund" conference, with those of the first, "Amsterdam" conference[4], it becomes apparent that progress in this field is fast: for example compare the numbers and contents of the papers on "modelling". Therefore I do hope that these proceedings will be considered as useful by those active or interested in this classical field of materials science and engineering.

E.J. Mittemeijer

January 1994

Delft, The Netherlands

[1] L. Kaufman and H. Bernstein, Computer Calculations of Phase Diagrams, 1970, Academic Press, New York.

[2] J.W. Christian, The Theory of Transformations in Metals and Alloys, 2nd edition, part I, 1975, Pergamon Press, Oxford.

[3] E.J. Mittemeijer, Analysis of the kinetics of phase transformations. J. Mater. Sci. 27, 1992, 3977-3987.

[4] Proceedings First ASM Heat Treatment and Surface Engineering Conference, Mat. Sci. Forum, 102-104, 1992.

Editorial Notes

As compared to the proceedings of the first Heat Treatment and Surface Engineering Conference many more contributions, as submitted, satisfied the issued "notes of authors" and I felt it unnecessary to indicate this time in the Table of Contents which contributions were modified by me. However, in rare cases authors believed it to be justified to exceed the allotted number of pages with even a factor of two and more. I tried to truncate these manuscripts as well as possible or, if I considered this impossible, I rejected the contribution. Keynote and invited papers were allowed to occupy more space, as the authors concerned knew in advance.

The subdivision in chapters more or less resembles the one of the previous proceedings. The papers have been arranged in each chapter in an order primarily on the basis of logic of the sequence, unavoidably biased by my personal preferences.

E.J. Mittemeijer

Table of Contents

i   **Preface:** Thermodynamics, kinetics and process control

ii   **Editorial Notes**

Part 1

I.   **OVERVIEWS**

J. Ågren   *(invited lecture)*   3
Thermodynamics and Heat Treatment

G. Krauss   *(keynote lecture)*   15
State-of-the-Art Developments in the Heat Treatment of
Ferrous Alloys

M. Gantois   *(keynote lecture)*   37
Mechanisms and Modelling of Mass Transfer During Gas-
Solid Thermochemical Surface Treatment. New Process
and New Process Control for Carburizing and Nitriding

II.   **HEAT TREATMENT OF FERROUS ALLOYS**

Th.M. Hoogendoorn and A.J. van den Hoogen*(invited lecture)*   51
Heat Treatment in Hot Strip Rolling of Steel

M. Onink, C.M. Brakman, J.H. Root, F.D. Tichelaar,
E.J. Mittemeijer and S. van der Zwaag   63
Redistribution of Carbon during the Austenite to Ferrite
Transformation in Pure Fe-C Alloys

I.A. Wierszyllowski   69
The Influence of Thermal Path on the Beginning of
Austenite and Residual Austenite Transformation

T. Araki, K. Shibata and H. Nakajima   75
Metallurgical Aspects of Microstructures in Advanced Non-
Heat-Treated Type High Strength Low-C Steels

M. Isac   81
The Electron Microscope Analysis used to Study the
Martensite Morphology in High Strenght Low Alloy Steels

M. Przylecka, M. Kulka and W. Gestwa   87
The Activity of Carbon in the Two-Phase Fields of the Fe-
Cr-C, Fe-Mn-C and Fe-Si-C Alloys at 1173 K

P.F. Nizhnikovskaja   93
Thermal Treatment of Deformable White Irons Having
Plasticity Induced by Carbide Transformation (PICT-Irons)

M. Kocsis Baán   99
    Boron Hardenability Effect in Case Hardening ZF-Steels

E. Gariboldi, W. Nicodemi, G. Silva and M. Vedani   107
    Effect of Tempering Level on Mechanical Properties and
    Toughness of Spring Steels

J. Asensio, B. Fernández, J.I. Verdeja and J.A. Pero-Sanz   115
    The Relationship between Processing, Microstructure and
    Properties in TLP-50-AZ Steel for Off-shore Platforms

G.P. Rodríguez Donoso, A.J. Vázquez Vaamonde and J.J. de
    Damborenea Gonzalez   133
    Steel Heat Treatment with Fresnel Lenses

R.T. von Bergen   139
    The Effects of Quenchant Media Selection and Control on
    the Distortion of Engineered Steel Parts

D.L. Moore and S. Crawley   151
    Applications of "Standard" Quenchant Cooling Curve
    Analysis

III.   HEAT TREATMENT OF NON-FERROUS ALLOYS

L. Wagner and J.K. Gregory   *(invited lecture)*   159
    Thermomechanical Surface Treatment of Titanium Alloys

J. Mironi   173
    Thermomechanical Treatment of AA6061

G. Caironi, E. Gariboldi, G. Silva, M. Vedani   181
    Influence of Preliminary Heat Treatments on
    Microstructure, Mechanical Properties and Creep
    Behaviour of a 50Cr-50Ni Niobium Containing Alloy

C.W. Yeung, W.D. Hopfe, J.E. Morral and A.D. Romig Jr.   189
    Interdiffusion in High Temperature Two Phase, Ni-Cr-Al
    Coating Alloys

IV.   CARBURIZING AND CARBONITRIDING

V. Woimbée, J. Dulcy and M. Gantois   197
    Over-Carburizing Kinetics and Microstructure of
    Z38CDV5.3 Tool Steel

B.M. Khusid, E.M. Khusid and B.B. Khina   203
    Carburization of High-Chromium Steels

A.Ç. Can   209
    High Temperature (1050 °C) Carburizing of SAE 1020
    Steel and its Effect on Fatigue Failure

M. Kulka, M. Przylecka, W. Gestwa, M. Piwecki   215
    The Influence of Carburizing and Heat Treatment on the
    Mechanical Properties of LH15 Grade Bearing Steel

F. Schnatbaum and A. Melber   221
Pulse Plasma Carburizing of Steel with High Pressure Gas
Quenching

D.E. Goodman and S.H. Verhoff   227
Ion Carburizing-Applications in Industry

M.S. Yahia, Ph. Bilger, J. Dulcy and M. Gantois   233
Use of Thermogravimetry for the Study of Carbonitriding
Treatment of Plain Carbon Steel and Low Alloy Steel

M. Przylecka, W. Gestwa and M. Kulka   239
The Influence of Carbonitriding and Heat Treatment on the
Properties of Low- and High-Carbon Steels

V.   NITRIDING AND NITROCARBURIZING

Chr. Ortlieb, S. Böhmer and S. Pietzsch   247
Establishment of Quasi Equilibrium States in Technical
Systems of Iron and Nitriding Gas

S. Pietzsch, S. Böhmer and Chr. Ortlieb   253
Structure of Porous Areas of Nitride-Containing Compound
Layers

S. Pietzsch and S. Böhmer   259
Growth of Nitride Cover Layers on Compound Layers of Iron

M. Boniardi and G.F. Tosi   265
The Effect of Surface Finishing on Mechanical and
Microstructural Properties of Some Nitrided Steels

N. Geerlofs, C.M. Brakman, P.F. Colijn and S. van der Zwaag   273
The Kinetics of Internal Nitriding of an Fe-7at.% Al Alloy

J. Ratajski, J. Ignaciuk, J. Kwiatkowski and R. Olik   279
The Kinetics of Nitriding Layer Growth on Fe-Cr and Fe-Ti
Alloys

R. Russev and S. Malinov   285
Some More Data for the Eutectoid Equilibrium of the
Fe-N-C System

S. Deshayes, P. Jacquot, E. Denisse   (invited lecture)   293
The Sulphonitrocarburising Process Sulf Ionic®

G. Wahl   309
Present State of Salt Bath Nitrocarburizing for Treating
Automotive Components

VI.   BORIDING

B. Formanek   317
The Diffusion Boronizing Process of Reactive Atmospheres
Containing Boron Fluorides

A. Pertek   323
Gas Boriding Conditions for the Iron Borides Layers
Formation

L. Çapan and B. Alnipak   329
Erosion-Corrosion Resistance of Boronised Steels

Y. Özmen and A.Ç. Can   335
An Effective Method to Obtain Reasonably Hard Surface on
the Steel X210Cr12(D3): Boriding

H.J. Hunger and G. Trute   341
Successful Boronizing of Nickel-Based Alloys

VII.   INDUCTION AND FLAME HARDENING

H. Liao, Y. Wang and C. Coddet   349
Surface Hardening of Steels and Cast Irons by a DC Plasma
Torch

P.K. Braisch   355
On the Effect of an Additional Rolling on the Fatique
Behaviour of Surface Induction Hardened Components

Ch. Ch. Lal   367
Quality Control of Induction on Hardened Cast Iron
Cylinder Head Valve Seats

VIII.   LASER-BEAM TREATMENT

H.W. Bergmann, D. Müller, T. Endres, R. Damascheck,
J. Domes and A.S. Bransden   (invited lecture)   377
Industrial Applications of Surface Treatments with High
Power Lasers

J. Noordhuis and J.Th. M. De Hosson   405
Mechanical Properties and Microstructure of Laser Treated
AL Alloys

A. Zambon, E. Ramous, M. Bianco and C. Rivela   411
Surface Alloying of Ti6A14V Alloy by Pulsed Laser

K. Komvopoulos   417
Effect of Process Parameters on the Microstructure,
Geometry and Microhardness of Laser-Clad Coating
Materials

J.E. Flinkfeldt and Th.F. Pedersen   423
Laser Cladding in Pre-Made Grooves

Part 2

IX. ION-BEAM TREATMENT

J.P. Rivière   *(invited lecture)*   431
    Surface Treatments Using Ion Beam

J. Jedlinski, A. Nikiel, G. Borchardt and B. Rajchel   449
    The Effect of Composition, Heat Treatment and Surface
    Modification by Ion Implantation on the Oxidation
    Behaviour of Alloys from Fe-Cr-Al System

X. SHOT PEENING

C.P. Diepart   457
    Modelling of Shot Peening Residual Stresses - Practical
    Application

A. Zambon, A. Tiziani, L. Giordano, F. Bonollo and A. Molinari   465
    Shot Peening Induced Phase Transformations and Residual
    Stresses Measurements on Austempered Cast Irons by XRD

G. Poli, R. Andreetta and L. Baraldi   471
    Mechanical Treatments on Die-Surfaced Obtained through
    Electro-Discharge Machining: Role of Shot-Peening and
    Sandblasting

XI. CHEMICAL AND PHYSICAL VAPOUR DEPOSITION

R. S. Bonetti   *(invited lecture)*   479
    Protective CVD Coatings - Going from High Temperature
    Processes to Plasma Activation

A. Matthews and A. Leyland   *(invited lecture)*   497
    Plasma Processing for Enhanced Wear and Corrosion
    Performance

D. Drees, E. Vancoille, J.P. Celis and J.R. Roos *(invited lecture)*   509
    Structural Characteristics of PVD Ti-Based Coatings in
    Relation to their Wear and Corrosion Behaviour

J. Vetter, W. Burgmer, H.G. Dederichs and A.J. Perry   527
    The Architecture and Performance of Compositionally
    Gradient and Multi-Layer PVD Coatings

J. He, F. Feng, W. Li and Ch. Bai   533
    Thermal Shock Resistance of TiN Coated Material

W. Precht, W. Kacalak and A. Czyzniewski   539
    Wear and Lifetime of PVD TiN Coated Tools

S. Legutko 545
Wear and Tool Life of Taps with Hard Coats in Gray Cast Iron Machining

D.P. Monaghan, T. Hirst, D. Cross and R.D. Arnell 551
Mechanical and Electrical Properties of Co-Deposited and Multilayer Copper-Tungsten Films

XII.   THERMAL SPRAYING

H.-D. Steffens, M. Gramlich and K. Nassenstein*(invited lecture)* 559
Thermal Spraying

Th. F. Weber 573
High-Velocity Oxy-Fuel Spraying

G. Matthäus and W. Rother 579
Ceramic, Hard Metallic and Metallic Coatings Produced by High Speed Flame Spraying

K. Ebert, C. Verpoort and Ch. Karsten 587
Plasma Spraying - an Economical Process for Improving Wear - and Corrosion Properties of Machinery Parts

R. Pötzl 595
Laser Remelting of HVOF Coatings

J. Tehver and H. Sui 603
Porous Coating Parameters and Boiling Heat Transfer Enhancement

XIII.   COATINGS; MISCELLANEOUS

R.C. Jongbloed *(invited lecture)* 611
Chromizing

L. Swadzba 619
The Influence of Silicon on the Structure and Properties of Diffusion Aluminide Coatings on Nickel Base Superalloys

S.M.M. Khoee, A. Ata, A. E. Geckinli and N. Bozkurt 627
VC & NhC Coating on DIN 115CrV3 Steel in Molten Borax Bath

M. El-Sharif, X. Wang, C.U. Chisholm, A. Waston, A. Vertes and E. Kuzmann 633
Effect of Heat Treatment on Microstructure of Cr-Ni-Fe Coatings Prepared by Electrodeposition

XIV. MODELLING

XIVa. Modelling: Phase transformations

A. Thuvander and A. Melander   *(invited lecture)*   641
Prediction of Distortion and Residual Stresses of
Engineering Steels Due to Heat Treatment

M. Gergely, Sz. Somogyi and R. Kohlhéb   *(invited lecture)*   657
Computerized Property Prediction and Process Planning in
Heat Treatment of Steels

R. Kohlhéb, G. Buza, T. Réti and M. Gergely   667
Comparative Analysis of Kinetic Models Used for
Description of Non-Isothermal Austenite Transformation

T. Reti, T. Bell, Y. Sun and A. Bloyce   673
Computer Prediction of Austenite Transformation under
Non-Isothermal Conditions

XIVb. Modelling: Hardenability

K. Andersson, S. Kivivuori and A.S. Korhonen   683
Calculation of the hardness distribution in cooled steel
products

E.A. Geary, W.T. Cook and K.A.G. Lane   689
Calculated Hardenability - the Way Towards Improved
Consistency of Engineering Steels and Component
Properties

XIVc. Modelling: Thermochemical surface treatments

P. Brünner and K.H. Weissohn   *(invited lecture)*   699
Computer Simulation and Control of Carburization- and
Nitriding Processes

L. Torchane, Ph. Bilger, J. Dulcy and M. Gantois   707
Application of a Mathematical Model of Iron-Nitride Layer
Growth During Gas Phase Nitriding

W. Daves and F.D. Fischer   713
Finite Element Simulation of the Development of Residual
Stresses During Nitriding under Consideration of the
Micromechanical and Metallurgical Processes

H. Du, L. Sproge and J. Ågren   719
Modelling of Nitrocarburizing

A. Engström, L. Höglund and J. Ågren   725
Computer Simulation of Carburization in Multiphase
Systems

B.M. Khusid and A.V. Luikov   731
    Modelling of Diffusion Mass Transfer and Phase
    Transformations in a Ternary Alloy with Dispersed
    Particles

XIVd. Modelling: Heat flow

A. Sala, P. Gruszka, J. Kabata and A. Wozniak   739
    Minimizing of Energy Consumption in Heat Treatment
    Processes

D. Holoubek   745
    Thermodynamic Balance of Metallurgical Furnace

G. le Gouefflec   751
    Mastering Heat Transfer in Continuous Furnaces by
    Computer Modelling

S. Ovcharova and D. Stavrev   757
    Numerical Determination of Heat Flow in Plasma-ARC
    Surface Hardening

XV. FURNACE TECHNOLOGY

B. Edenhofer   (invited lecture)   765
    Progress in Automation and Control of Atmosphere - and
    Plasmacarburising Batch Chamber Furnaces

S.J. Sikirica, K.H. Hemsath and S.K. Panahi   777
    Development of Gas-Fired Vacuum Furnaces

P.F. Masters   783
    A User's View of Fluidised Bed Heat Treatment

I.   OVERVIEWS

Materials Science Forum Vols. 163-165 (1994) pp. 3 -14
© 1994 Trans Tech Publications, Switzerland

THERMODYNAMICS AND HEAT TREATMENT

J. Ågren

Division of Physical Metallurgy, Department of Materials Science and Engineering,
Royal Institute of Technology, S-100 44 Stockholm, Sweden

ABSTRACT

Over the last decade computational thermodynamics has become a powerful tool in materials and process development. Calculations by means of the Calphad technique may be applied to practical problems in alloy design and heat treatment. Moreover, the possibility of combining thermodynamic data with kinetic information allows the simulation of microstructural development during heat treating.

A short overview of the field and the underlying ideas will be given. Examples concerning design and heat treatment of complex steels, CVD of Ti(C,N) and nitrocarburizing will be discussed. Simulation of transformation of austenite upon cooling as well as carburizing of alloys will be discussed.

1. INTRODUCTION

When thermodynamics was born more than a century ago it was really as an engineering tool applicable to a number of practical problems. Newtonian mechanics, as used in those days, was simply not enough to answer questions about the efficiency of steam engines or why chemical reactions occur. Later, by the work of Gibbs and others, thermodynamics was developed in a more stringent way but unfortunately it also turned into a rather abstract science almost on the borderline of philosophy. Although useful in principle and always taught to the students, it was not much of an engineering tool, at least not in materials engineering. The main reasons were that the practical problems to which one would like to apply thermodynamics are far more complex than what could be managed by the simple methods of textbooks and, moreover, almost all data needed were missing.

The extensive development during the last two decades has brought thermodynamics back to engineering and it is now a most powerful tool for solving practical problems in heat treatment and materials design. This new field of the engineering sciences should now be called *computational thermodynamics*.

The properties of materials depend on their structure and the structure is controlled by composition and processing. In order to understand and fully control the processing it is thus necessary to have detailed information on phase equilibria and phase transformations involved in the structural changes. In this paper some of the recent achievements in thermodynamic calculation of multicomponent phase diagrams as well as simulation of structural changes by combining thermodynamic and kinetic data will be reviewed.

2. CALPHAD

In textbooks on heat treatment of steel it is customary to start with the Fe-C phase diagram and relate the structure (and the properties) to the various parts of the diagram, see ref. 1. Such a diagram might be useful in order to predict the structure of a steel after a heat treatment. In practice the diagram is not very useful because real steels contain a number of alloy elements in addition to carbon, which have been added to improve the performance, and a useful diagram is necessarily multidimensional. Although some attempts have been made to map the 3-dimensional phase diagrams of ternary alloys experimentally we can never expect mapping of higher-order systems experimentally as a general procedure. It would simply require too much work.

Evidently, some method of interpolating and extrapolating the sparse experimental information is needed in order to obtain useful multicomponent phase diagrams. The CALPHAD technique was initiated by the pioneering work of Kaufman and Bernstein (2) and by Hillert (3). The underlying idea is to use the fact that the Gibbs energy G for a material of a given composition, temperature and pressure is minimized at equilibrium. If the Gibbs energy of the different states of a material is known one thus can calculate the equilibrium by a minimization procedure.

In the CALPHAD technique mathematical expressions are chosen to represent the Gibbs energy of the individual phases as functions of composition, temperature and pressure and eventually internal order parameters. Usually the expressions are based on statistical mechanics and account is taken for various physical effects like positional and magnetic disorder. In a quite tedious procedure, usually partly computerized, the parameters in the expressions are adjusted to fit the experimental information available. A great deal of expertise is needed in order to obtain a useful result.

Examples of more recent work based on the CALPHAD technique are given in refs. 4-6.

A basic requirement for CALPHAD calculations is a general software for thermodynamic calculations. Thermo-Calc (7) is well established and the examples shown in this paper are all made by means of Thermo-Calc. Fig. 1, depicts a calculated isopleth (vertical section) through the Fe-Cr-Ni-C system at 18 mass% Cr and 8 mass% Ni from the recent work by Qiu (8).

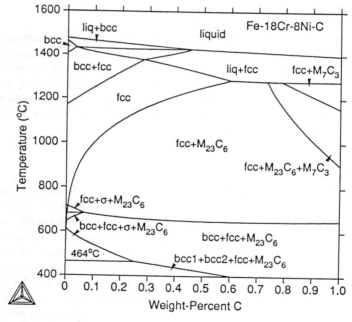

Fig. 1: Calculated isopleth (vertical section) through the Fe-Cr-Ni-C system at 18 mass% Cr and 8 mass% Ni from the work by Qiu (8).

2. COMBINED THERMODYNAMIC AND KINETIC CALCULATIONS

Of course the rate of structural changes is a critical parameter and the phase diagram must be supplemented by experimental TTT or CCT diagrams, one for each particular steel. Already 1946, in his classical paper, Zener (9) analyzed the decomposition of austenite upon cooling and assumed that thermodynamic equilibrium is established locally at phase interfaces. Moreover, Zener emphasized the physical significance of the extension of phase boundaries into the metastable regions of the phase diagram. He also recognized the necessity of extrapolating diffusivity data into the metastable regions where no experimental data is available. Actually, Hultgren already 1920 (10) extrapolated the $\gamma/\gamma + \alpha$ and $\gamma/\gamma +$ cementite phase boundaries into the region of undercooled austenite when discussing the transformation of austenite. In the late 50:ies Kirkaldy (11) outlined a general analytical procedure to solve the multicomponent diffusion equations for an advancing two-phase interface. His method involved the determination of the operating tie-line in the extrapolated ternary phase diagram by the solution of an equation system.

In addition to thermodynamic data, the approach of Zener and Kirkaldy requires the knowledge of diffusion coefficients. As in the case of thermodynamic data the experimental information on diffusion coefficients should be represented by simple models that take into account the physical behavior of the system rather than just power-series expansions fitted to the experimental data. Such an approach was outlined by Andersson and Ågren (12) and is the basis for Jönssons (13) assessments of diffusivity data in various systems.

The above approach, to predict the development of a phase transformation by combining two different types of information (thermodynamics and diffusion), has a large predictive power. The two types of information may be evaluated independently of each other from different types of experiments, but taken together they make it possible to calculate the rate of phase transformations. A software based on the above ideas should thus contain a module for calculating the local equilibrium at a phase interface and a module for solving diffusion equations.

3. PROJECT COSMOS

The above ideas form the framework for project COSMOS (COmputer Supported MOdelling of Steel transformations), which is a joint effort between the Max Planck Institut fur Eisenforschung in Dusseldorf, the Institut fur Werkstoffe at the Ruhr-Universität Bochum and the Royal Institute of Technology in Stockholm. The major goal has been to develop mathematical models and computer software to simulate the structural changes during heat treatment of steel from fundamental thermodynamic and kinetic data. A general software, given the name DICTRA (14), has been developed. Inden et al. (15) recently suggested a procedure for treating the growth (or shrinkage) of several particles by treating each particle in a computational cell and then treat the exchange of material between the cells. It is thus possible to simulate the evolution of a distribution of of particles. This procedure is now included in the DICTRA software.

When simulating the transformation of austenite upon cooling one starts by constructing an initial austenite grain structure. The procedure developed within COSMOS and first applied by Inden et al. (15) is to consider a volume element repeated periodically in space by cyclic boundary conditions. Within the volume element a finite number of points are generated, e.g. in random, and these are surrounded by Voronoi polyhedra. These polyhedra are taken as the austenite grains. In this grain structure a number of ferrite nucleation sites are defined, usually at some corners of the polyhedra, and a second generation of Voronoi polyhedra is constructed around the nucleation sites. This second generation of polyhedra defines the calculational cells which are further simplified with spheres having the same volume as the corresponding polyhedra. All the calculational cells are coupled by the requirements that the total mass of a component must remain constant during the simulation and that no driving force is dissipated by exchange of material between the cells (the latter requirement may be modified).

Fig. 2 shows a comparison between calculated and experimental volume fraction of allotriomorphic ferrite obtained by Inden et al. (15) for isothermal transformation of austenite at 749 °C having 0.51 mass% C. Figs. 3 a and b, taken from ref. 15, show the experimental microstructure of ferrite and martensite and the calculated one.

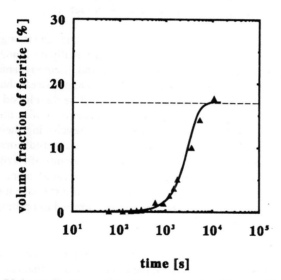

Fig. 2: Volume fraction of ferrite as a function of annealing time at 749 °C and 0.51 mass% C. Solid line: calculated, symbols: metallography by serial sectioning and dashed line: calculated equilibrium fraction. From ref. 15.

4. MATERIALS BY DESIGN
New materials are usually developed by trial and error. This is particularly true, and is probably still the most efficient method, for new classes of materials where very little is known a priori. Such was the situation for the high-temperature superconductors when they entered the stage some years ago. However, as the amount of knowledge grows

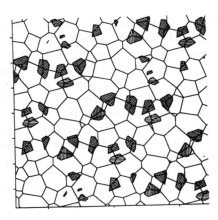

Fig. 3: a) Experimental and b) simulated microstructure (section) for same alloy as Fig. 2 after 1h annealing. The shaded areas represent the ferrite. From ref. 15.

and passes a critical level it certainly becomes advantageous to change strategy and apply a more fundamental approach in which the materials are actually designed from the knowledge about structure property relationships as well as the thermodynamic and kinetic behavior. As pointed out by Olson (16) more is known about steels than any other material and thus one may expect that new steels should be developed more efficiently by the fundamental approach rather than by trial and error. In fact, this was demonstrated in practice by the Sandvik Steel company when developing their new duplex stainless steels SAF 2304 and SAF 2507 (17). The required corrosion properties put some restrictions of the alloy composition: the amount of ferrite and austenite should be roughly the same in order to have the optimal mechanical properties and, moreover, the tendency for formation of the brittle σ phase must be minimized. Calculations with Thermo-Calc was successfully applied to optimize the chemical compositions of the new steels.

In a recent study Jansson et al. (18) applied materials design concepts to the development of a new class of bearing steels where the target is to combine high wear resistance, rolling contact fatigue resistance, toughness and hot hardness and for one steel, high corrosion resistance. These requirements are in conflict and an optimum must be found. The wear resistance is governed by rather large and very hard primary carbides like MC and M_6C as found in high-speed steels. If the Cr content is increased the primary carbides will be the softer M_7C_3 and $M_{23}C_6$. Toughness is controlled by the character and dispersion of second phase particles. Hot hardness may be obtained by a secondary hardening during tempering of martensite. The requirement is that at austenitizing one must have a sufficient amount carbide forming elements like Cr, W, Mo and V so that upon tempering at 500 - 650 °C the coarse cementite dispersion is replaced by a fine dispersion of MC and M_2C. In order to increase the toughness it is important that all the cementite dissolves. The primary-steel design criteria applied by Jansson et al. was to minimize the amount of primary carbides in the core of a component and to obtain a hard and wear resistant case by carburizing or induction hardening. In order to increase the wear resistance of the carburized case one should have a high content of V or Mo. For various reasons Mo is less good and V was chosen. However, the V content cannot be too high because then not all MC in the core will dissolve during austenitizing.

In order to avoid formation of ferrite during austenitizing the Ni content was increased. However, if both Ni and V are increased hardenability will decrease due to a too low M_s temperature. This may be compensated for by adding Co, the only alloy element that increases the M_s temperature. In this way one may reason and try different compositions until the design criteria are fulfilled. Since the whole procedure is done on the computer one rapidly covers a wide range of possible combinations and may sort out the most promising conditions which should be subjected for further experimental tests. Fig. 4, taken from the study by Jansson et al. shows an isopleth for Fe-4Cr-4Mo-2V-4.5Ni-2.5Co and shows the part of the phase diagram relevant for austenitizing the core. From this diagram we find that an alloy with 0.11 %C must be austenitized above 1090 °C in order to dissolve all the MC carbides.

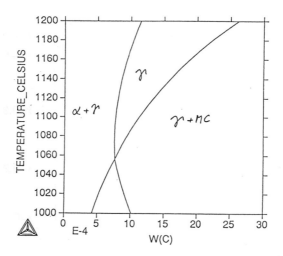

Fig. 4: Calculated isopleth for Fe-4Cr-4Mo-2V-4.5Ni-2.5Co. Horizontal axis is mass fraction of carbon.

5. CHEMICAL VAPOR DEPOSITION OF TITANIUM CARBO-NITRIDE
Cemented carbide tools are often coated with titanium carbo-nitride by chemical vapor deposition (CVD). The carbo nitride has the FCC structure and may be regarded as a solid solution between C and N on the interstitial lattice sites. Some of the sites may be vacant, i.e. the Ti content may be higher than expected from stoichiometry. As the properties of the coating depend on its composition it is interesting to investigate in

some detail the factors that determine the composition. We may denote the carbo nitride as TiC_xN_y, where $y = 1 - x$ for stoichiometric composition but $y < 1 - x$ in general. At a given temperature and pressure each combination of x and y corresponds to a combination of carbon and nitrogen potential which may be calculated from the thermodynamic properties of the carbo-nitride phase. This system was recently analyzed by Jonsson (19) and the present calculation is based on his thermodynamic parameters. In CVD one often uses gas mixtures based on $TiCl_4$, CH_4 and N_2 and the composition of the gas is used to control the C and N potentials and the composition of the carbo nitride. The constitution of a gas at internal equilibrium may be calculated using thermodynamic software and one may thus predict the relation between carbo-nitride composition and gas composition entirely by thermodynamic calculations provided that the coating may be approximated as in equilibrium with the gas. Such an example is shown in Fig. 5 calculated by Jansson (20). The diagram is calculated for 1050 °C and a total pressure of 1 atm and a $P_{N_2} = 0.05$ atm. Each solid line corresponds to a given x value and above the dashed line there is a tendency for formation of graphite. It is interesting to notice that the lines will be almost horizontal for P_{TiCl_4} larger than 0.20 atm and we may then control the x value by controlling P_{CH_4}. It should be noticed that the y value is not shown in this diagram but in the present case it is close 1 - x.

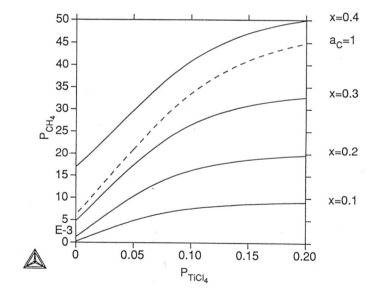

Fig. 5: Calculated composition of TiC_xN_y as function of gas composition during CVD with $TiCl_4$, CH_4 and N_2 at 1050 °C, 1 atm and $P_{N_2} = 0.05$ atm. Each solid line corresponds to a constant x value. Above the dashed line there is a tendency for graphite formation.

6. SIMULATION OF NITROCARBURZING

In nitrocarburzing C and N are added simultaneously to a steel surface at low temperatures ($\approx$ 575 °C) and a compound layer forms. The compound layer consists of a solid solution phase ε, having HCP structure, or γ' with the composition Fe_4N dissolving only minute amounts of C or a mixture of the two phases. In addition there may a varying degree of porosity. The properties of the compound layer depends on its constitution, i.e. the amounts of the different phases and the porosity. In gas nitrocarburizing this constitution will be controlled by the C and N potentials of the gas, i.e. by the composition of the gas, and by the processing time. In order to fully understand and control the process one should thus make combined thermodynamic and diffusion calculations. Du (21) recently assessed the Fe-C-N system and Fig. 6 shows a part of her calculated isothermal section at 580 °C. The symbols denote experimental diffusion-paths obtained during nitriding of Fe-C alloys by Somers et al. (22). At present DICTRA is applied to simulate the evolution of the compound layer during nitrocarburizing.

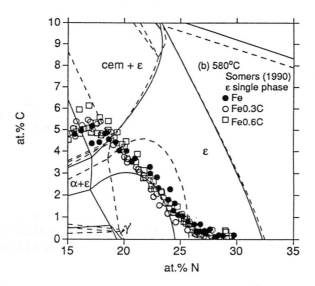

Fig. 6: Part of calculated isothermal section at 580 °C, From ref. 21. Symbols: experimental diffusion-paths obtained during nitriding of Fe-C alloys by Somers et al. (22).

7. SIMULATION OF CARBURIZING OF NI-BASE ALLOYS

Many authors have treated theoretically the carburization during case hardening of steel, see for example refs. 23-25. Since the carburization is performed at a high temperature where the steel is austenitic the process is almost ideal for simulations and involves only the solution of one diffusion equation with relatively simple boundary conditions. If secondary phases precipitate as a result of the carburization the situation becomes much more complex and a satisfactory treatment must be based on the thermodynamic properties of the system. Bongartz et al. (26) developed a model for carburization of high-temperature Ni-base alloys which made use of both thermodynamic and kinetic data. Recently Engström et al. (27) presented a method which only applied thermodynamic data and multicomponent diffusivities and did not include any adjustable parameters. Fig. 7 is taken from Engström et al. (28) showing a) the over-all C concentration profile after carburization of Ni-Cr alloys for 1000 h at 850 °C and b) the amount of carbides in one of the alloys. The symbols are experimental data from refs. 29-30

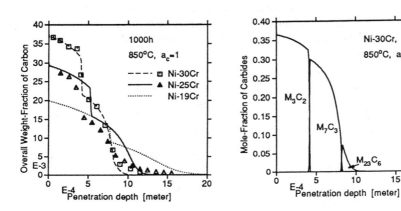

Fig. 7a: (from Engström et al. (28)) the over-all C concentration profile after carburization of Ni-Cr alloys for 1000 h at 850 °C. Sloid lines: calculated in ref. 28. Symbols: experimental data from refs. 29-30. 7b: the amount of carbides in one of the alloys.

8. SUMMARY

The predictive power of computational thermodynamics has been demonstrated with a few examples. Easy access to flexible software for thermodynamic calculations and combined thermodynamic and kinetic calculations, as well as assessed high quality data, have turned thermodynamics into a powerful tool in materials engineering. Reliable thermodynamic data is now available for steels and a several other types of materials but most of the kinetic data is still lacking. International collaboration is needed in order to assess the kinetic data, and possibly other type of data, required to use the full potential of computational thermodynamics.

REFERENCES

1. K.-E. Thelning: Steel and its Heat Treatment, Butterworths, London 1984, p4
2. L. Kaufman and H. Bernstein: Computer Calculation of Phase Diagram, Academic Press, New York 1970
3. M. Hiller: in ASM seminar on Phase Transformations 1968, Metals Park, Ohio 1970 pp 181-218
4. M. Hillert and C. Qiu: Metall. Trans. A Vol. 21A (1990) pp 1673-1680
5. K. Frisk: Metall. Trans. A Vol. 21A (1990) pp 2477-2488
6. W. Huang: Z. Metallkde. Bd 82 (1991) H.5 pp 391-401
7. B. Sundman, B. Jansson and J-O Andersson: CALPHAD Vol. 9 (1985) pp 153-190
8. C. Qiu: Thermodynamic Study of Carbon and Nitrogen in Stainless Steels, Dr thesis, Division of Physical Metallurgy, Rooyal Inst of Techn S-100 44 Stockholm, SWEDEN.
9. C. Zener: Trans. AIME Vol. 167 (1946) p 550
10. A. Hultgren: A metallographic study on tungsten steels, John Wiley & Sons, London 1920, (see e.g. p 29 and 30)
11. J.S. Kirkaldy: Can. J. Phys. Vol. 36 (1958) p 907
12. J.-O. Andersson and J. Ågren: J. Appl. Phys. Vol. 72 (1992) p 1350
13. B. Jönsson: in COMMP'93, International Conference on Computer-Assisted Materials Design and Process Development.
14. J.-O. Andersson, L. Höglund, B. Jönsson and J. Ågren: In *Fundamentals and Applications of Ternary Diffusion* Ed. G.R. Purdy, Pergamon Press 1990, pp 153-163
15. G. Inden, P. Francke and U. Knoop: in *Computer Aided Innovation of New Materials II*, Eds: M. Doyama, J. Kihara, M. Tanaka and R. Yamamoto, Elsevier Science Publishers 1993
16. G. B. Olson: in *Innovations in Ultrahigh-Strength Steel Technology* , Eds. G. B. Olson, M. Azrin and E. S. Wright, Sagamore Army Materials Research Conference Proceedings, pp 3-66
17. H. Widmark: Scand. J. Metall. Vol. 20 (1991) pp72-78
18. B. Jansson, S. Haglund, J. Ågren and A. Melander: to be published
19. S. Jonsson: TRITA-MAC 506, Internal report 1992, Royal Inst. Techn. SDept. of Materials Science and Engineering, S-100 44 Stockholm, SWEDEN
20. B. Jansson: unpublished work
21. H. Du: TRITA-MAC 511, Internal report 1993, Royal Inst. Techn. Dept. of Materials Science and Engineering, S-100 44 Stockholm, SWEDEN
22. M.A.J. Somers, P. F. Colijn, W.G. Sloof and E. J. Mittemeijer: Z. Metallkde. Vol. 81 (1990) pp 33-43
23. R. Coolin, S. Gunnarsson and D. Thulin: J Iron and Steel Institue Vol. 210 (1972) pp 777-784
24. H. J. Grabke and G. Tauber: Arch. Eisenhuttenwes. Vol. 46 (1975) pp 215-222
25. L. Sproge and J. Ågren: J. Heat Treat. Vol. 6 (1988) pp 9-19
26. K. Bongartz, D. F. Lupton and H. Schuster: Metall. Trans. A, Vol. 11A (1980) pp 1883-1893
27. A. Engström, L. Höglund and J. Ågren: TRITA-MAC-0515, Internal report 1993, Royal Inst. Techn. Dept. of Materials Science and Engineering, S-100 44 Stockholm, SWEDEN

28. A. Engström, L. Höglund and J. Ågren: This conference.

29. K. Bongartz, R. Schulten, W. J. Quadakkers, and H. Nickel, Corrosion vol 42 (1986) pp 390-97

30. W. J. Quadakkers, R. Schulten, K. Bongartz, and H. Nickel: 10:th Int Conf. on Metal Corrosion, Madras, India, Nov. 1987 pp 1737-1746

Materials Science Forum Vols. 163-165 (1994) pp.15 -36
© *1994 Trans Tech Publications, Switzerland*

STATE-OF-THE-ART DEVELOPMENTS IN THE HEAT TREATMENT OF FERROUS ALLOYS

G. Krauss

Advanced Steel Processing and Products Research Center,
Colorado School of Mines, Golden, Colorado, USA

ABSTRACT

This paper reviews selected developments in steels and heat treatment processing of steels as reported primarily in 1992 literature. Emphasis is placed on processing-microstructure-performance of through hardened quench and tempered steels, new bainitic steels, ultrahigh strength steels, thermochemical processing of steels by carburizing, nitriding, and nitrocarburizing, quenching, and modeling of microstructure and performance of quenched steels.

INTRODUCTION

Heat treatment of steels continues to be of major importance to manufacturing industries throughout the world. This importance is emphasized by the number of conferences which are annually devoted to topics related to the heat treatment and surface modification of steels. It has been my privilege to participate in several of these conferences in 1992, and my review of new developments is based on this direct but necessarily limited participation in the total of world-wide technology transfer. Considerable effort has been expended by the organizers and authors in each conference, and it is important to bring their efforts to the attention of broader audiences. Typically the conferences have attendance on the order of 200 participants more or less, and therefore each event has relatively small direct participation. However, the conferences also have published proceedings which are part of the permanent record available to those unable to attend a given conference. One of my purposes for this paper is to bring selected 1992 conference proceedings to the attention of the heat treatment and manufacturing community.

Specifically, the developments I will discuss are primarily from the following conferences and associated proceedings: The International Conference on Tooling Materials, Interlaken, Switzerland, September 7-9, 1992 (1), the First International Conference on Quenching and Control of Distortion, Chicago, Illinois, USA, September 11-15, 1992 (2), the Gilbert R. Speich Symposium on the Fundamentals of Aging and Tempering in Bainitic and Martensitic Steel Products, Montreal, Quebec, Canada, October 25-28, 1992

(3), and the 8th International Congress on Heat Treatment of Materials, Heat and Surface '92, Kyoto, Japan, November 17-20, 1992 (4). Also in 1992, the Proceedings of the First ASM Heat Treatment and Surface Engineering Conference in Europe were published (5).

A total of 341 papers totalling 2,755 pages of information are published in the proceedings volumes from the above list of conferences. This sample reflects a very impressive international technical effort in a single year related to the heat and associated fields. My review will be directed to subjects where physical metallurgy, i.e. the relationship of processing to microstructure and properties, is emphasized. Continued attention to processing-microstructure-property relationships should result in the most cost effective, optimum performance for heat treated materials. This review is limited to developments in steels and conventional heat treatment processing of steels, such as tempering of through hardened steels and thermochemical surface modification of steels. Plating and coating of steels by techniques such as physical vapor deposition, although of considerable interest (1,4-6), are outside of the scope of this paper.

NEW STEELS AND TEMPERING

Through hardening of steels, either by quenching to form martensite, or by air cooling to produce bainitic microstructures, remains a major approach to produce high strengths and fatigue resistance with moderate levels of toughness. The as-cooled microstructures are invariably tempered, primarily to increase toughness, but sometimes to produce additional strength by precipitation hardening. This section describes new understanding of the tempering of low alloy hardenable steels and new ultrahigh strength and bainitic steels.

Microstructural Changes During Tempering

The aging of quenched martensite in Fe-C and Fe-N alloys begins immediately at room temperature. Carbon and nitrogen atoms have sufficient room temperature mobility to diffuse to new configurations. Based on a review of evidence from x-ray diffraction and other experimental techniques, Böttger and Mittemeijer (7) show that some of the C and N concentrate at dislocations, a conclusion supported by direct atom probe observations, as reviewed by Smith (8). Larger scale concentration of interstitial atoms in martensite also occurs during low temperature aging, but takes quite different forms in Fe-C and Fe-N alloys (7,9). In Fe-N martensite the N atoms order in a coherent α''-type structure, while in Fe-C martensite carbon atoms tend to cluster in randomly distributed C-type octahedral sites.

At higher tempering temperatures, 150 to 200°C, transition carbides precipitate in the martensite of low-alloy carbon steels. The transition carbides have been identified as epsilon-carbide with a hexagonal crystal structure or as eta-carbide with an orthorhombic crystal structure. There is still controversy about the exact structure and morphology of the carbide in various alloy steels (10). Nevertheless, the morphology of the transition carbides has been confirmed to consist of rows of particles, each about 2 nm in size, within larger martensite laths of medium carbon steels (11). Also, the precipitation of the more stable theta-carbide or cementite was shown to nucleate at the transition carbide clusters and grow away from the clusters to establish the characteristic arrays of cementite on {110} planes of martensite tempered at intermediate temperatures.

The formation of the coarser cementite particles, either as a result of intralath formation as described above, or as a result of the transformation of interlath retained austenite, is associated with tempered martensite embrittlement (TME) which develops in hardened steels tempered around 300°C (12). Usually the TME is established by CVN impact testing. In 1992, evidence for TME during uniaxial tensile testing has been reported

(13). Figure 1 shows engineering stress-strain curves for a martensitic 4140 steel which has tempered at 200°C, 300°C, and 400°C. Despite the decreased flow stresses at 300 and 400°C, the total elongation does not increase as would be expected for the reduced strengths. All fracture was ductile, but the fine cementite particles produced during tempering at 300 and 400°C provided a higher density of sites for microvoid nucleation compared to the steel tempered at 200°C.

Embrittlement similar to TME also occurs in more highly alloyed medium carbon tool steels such as H11/H13 and roll steels which contain 5 pct Cr and significant amounts of Mo and V. Two 1992 papers evaluate impact toughness in such steels (14,15). Because of high alloy content, the stability of retained austenite in tool steels is increased relative to that in low alloy steels. Therefore the formation of detrimental arrays of carbides from retained austenite does not occur until after tempering at high temperatures. Figure 2, a schematic diagram from Branco and Krauss (14), shows the displacement of the impact toughness minimum to higher temperatures relative to those associated with tempering of lower alloy steels. The silicon modified 4340 type steel has an intermediate temperature minimum because Si retards the formation of cementite. The impact minimum in the H11/H13 steels is associated with cleavage fracture induced by coarse interlath carbides and a very high strength martensitic matrix produced by secondary hardening in the form of very fine alloy carbides (14).

Ultrahigh Strength Steels

There is continuing effort to expand the envelope of strength and toughness of ultrahigh strength steels. There are two approaches to designing microstructures with very high strengths. The one uses low alloy steels with medium carbon content, such as 4340, quenched to martensite and tempered at low temperatures, between 150 and 200°C. The other alloy design approach uses high alloy steels with low carbon content, quenched or air cooled to martensite and tempered at high temperatures, around 500°C. Steels produced by the latter approach are currently under most active development and evaluation (16-19). Figure 3 shows that the best combinations of strength and toughness are associated with steels such as AerMet 100, AF1410, and HY180 (16).

Tables 1 and 2, list respectively, the chemistries and mechanical properties of three high alloy ultrahigh strength steels (17). The steels all contain very high concentrations of Ni and Co, which together with other alloying elements produce high hardenability even with low carbon content. Strengthening of the low carbon martensite is accomplished by the precipitation of fine M_2C carbides at secondary hardening temperatures. Cobalt apparently retards the recovery of the martensitic dislocation structure during tempering, and thereby contributes to the fine M_2C distribution by providing a high density of nucleation sites for precipitation.

Bainitic Steels

The evaluation of steels with bainitic microstructures is now an extremely active area of development. There are at least three types of bainitic steels under consideration: low alloy steels with very low carbon content for flat rolled products (20-22), more highly alloyed steels with higher hardenability and copper additions for precipitation hardening (23-25), and medium carbon steels for direct cooled automotive components (26). Many of these steels are designed to produce bainite directly after hot work. As a result, reheating and quenching operations are not required and the costs of heat treatment and energy are eliminated.

The low-carbon bainitic steels are under development in order to improve combinations of strength and toughness or ductility. All of the steels transform to bainitic structures over a wide range of cooling rates between those which form martensite or mixtures of polygonal ferrite and pearlite. Also, the temperatures of which the bainitic

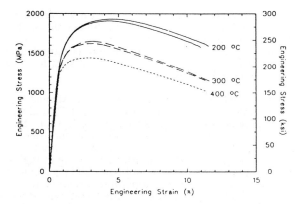

Figure 1.   Engineering stress-strain curves for 4140 martensitic steel tempered at 200°C, 300°C, and 400°C. Reference 13.

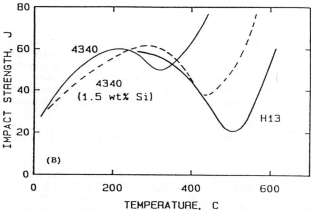

Figure 2.   Impact toughness minima as a function of tempering temperature for various steels. Reference 14.

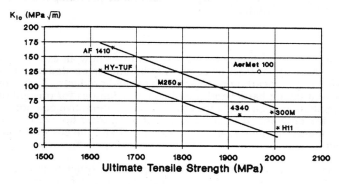

Figure 3.   Fracture toughness as a function of ultimate tensile strength for various ultrahigh strength steels. Reference 16.

TABLE 1 - COMPOSITION OF THREE NI-CO STEELS WITH HIGH
STRENGTH AND TOUGHNESS (17)

Element, Wt. Pct.	Alloy		
	AerMet 100	AF1410	HY180
C	0.21-0.27	0.13-0.17	0.10-0.14
Mn	0.10 max	0.10 max.	0.005-0.20
P	0.003 max.	0.008 max	0.010 max.
S	0.002 max.	0.005 max.	0.006 max.
Si	0.10 max.	0.10 max.	0.10 max.
Ni	11.00-12.00	9.50-10.50	9.50-10.50
Cr	2.5-3.3	1.80-2.20	1.80-2.20
Mo	1.0-1.3	0.90-1.10	0.90-1.10
Co	13.3-13.5	13.50-14.50	7.50-8.50
Al	0.01 max.	0.015 max.	0.020 max.
N	0.001 max.	0.0015 max.	0.0075 max.
0	0.001 max.	0.0020 max.	0.0025 max.
Ti	0.01 max.	0.015 max.	0.040 max.

TABLE 2 - TYPICAL ROOM TEMPERATURE MECHANICAL PROPERTIES OF
SEVERAL HIGH STRENGTH STEELS (17)

Property	AerMet 100[1]	AF1410[1]	HY180[1]
0.2% Yield Strength, ksi (MPa)	250 (1724)	220 (1517)	195 (1344)
Ultimate Tensile Strength, ksi (MPa)	285 (1965)	240 (1655)	205 (1413)
Elongation, %	14	15	16
Reduction in Area, %	65	68	75
Hardness, HRC	53.5	49	43
Longitudinal Charpy V-Notch Impact ft-lbs. (J)	30 (41)	45 (61)	140 (190)
Fracture Toughness, K_{IC}, ksi in. (MPa√m)	115 (126)	150 (165)	185 (203)

[1]Carpenter Technology, internal data

*Heat Treatment for properties shown:
  AerMet 100   1625(lh)/AC/-100°F(1h)AW/900°F(5h)AC
  AF1410   1525°F(1h)/OQ/-100°F(1h)AW/950°F(5h)AC
  HY180   1650°F(1h)/WQ/1550°F(1h)WQ/950°F(5h)AC

microstructures begin to form are intermediate to those at which martensite and polygonal ferrite begin to form, leading Japanese researchers to the adoption of the German term "zwischen" to characterize the bainitic microstructures (22). Figure 4 shows examples of CCT curves for low-carbon bainitic, Cu-strengthened steels and medium-carbon direct cooled forging steels (27). Compositions of the steels are given in Table 3.

TABLE 3 - COMPOSITION (WEIGHT PERCENT) OF ALLOYS FOR WHICH CCT DIAGRAMS ARE GIVEN IN FIGURE 4

Alloy	C	Mn	Si	Mo	Cr	Ni	Other
A (A710/HSLA-80)	0.05	0.50	0.28	0.20	0.71	0.88	1.12 Cu, 0.035 Nb
B (A710 Modified)	0.06	1.45	0.35	0.42	0.72	0.97	1.25 Cu, 0.040 Nb
C (HSLA-80/100)	0.05	1.00	0.34	0.50	0.72	1.77	1.25 Cu, 0.040 Nb
D (HSLA-100)	0.06	0.83	0.37	0.59	0.58	3.48	1.66 Cu, 0.028 Nb
E (0.24C-Mn-Mo-V)	0.24	1.67	0.39	0.22	0.17	0.14	0.11 V
F (0.35C-Mn-Mo-V)	0.35	1.40	0.76	0.19	0.07	0.06	0.14 V

The low carbon and alloying additions, primarily manganese, chromium and/or molybdenum, of the new bainitic steels produce microstructures quite different from those typically formed in medium-carbon steels (12). In the latter steels, the transformed bainitic microstructures consist of ferrite and cementite. In upper bainite the cementite forms between laths of ferrite and in lower bainite the cementite is present as very fine carbides within the ferrite. However, in the newer low-carbon steels, the bainitic microstructures do not contain cementite. Instead austenite is retained between ferrite of various morphologies formed at intermediate transformation temperatures. Sometimes on further cooling, the austenite transforms to martensite. As a result the new bainitic steel microstructures consist largely of ferrite and finer regions of martensite and austenite. The latter regions are often referred to as the martensite/austenite, M/A, constituent of the microstructure. The size and distribution of the M/A constituent influences mechanical properties, as demonstrated in the 1992 paper by Iwama et al. (26).

In view of the non-classical microstructures which form in the new bainitic steels, new terminologies are being developed to describe the microstructures (22). The characterization of the microstructures, the mechanisms by which the austenite transforms, and the relationship of microstructure to properties in the new bainitic steels will continue to be important future research and development activities.

THERMOCHEMICAL SURFACE MODIFICATION TREATMENTS

Thermochemical surface modification techniques consist of the exposure of steels to gaseous or liquid atmospheres containing chemical species such as C or N, the transfer of C and N from the atmosphere into the steel, and the time-temperature dependent diffusion of the C and N into the interior of the steel. The extensive technological base associated with classical thermochemical processing to traditional steels has encouraged development of vacuum and plasma thermochemical processing techniques and the application of the techniques to a broader range of steels. Parallel to the technological

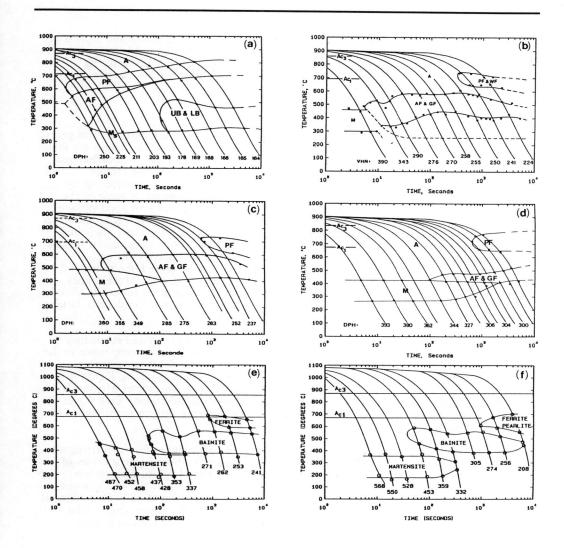

Figure 4.   Continuous cooling transformation diagrams for (a) A710, (b) A710 modified, (c) HSLA 80/100, (d) HSLA 100, (e) 0.24C-Mn-Mo-V, and (f) 0.35C-Mn-Mo-V steels.

developments, has come increased interest in understanding the fundamental aspects of thermochemical treatments, a process driven by the need for better and better performance of heat treatment components in today's competitive manufacturing climate. The above strategies have been intensively applied for the last decade, and have received significant attention in 1992.

Carburizing

Traditional gas carburizing of alloy steels is widely used for machine parts such as gears and shafts which require high resistance to bending and contact fatigue. There are many processing and microstructural parameters which affect the performance of carburized steels. When one measure of performance of carburized steels, bending fatigue endurance limit, is selected, the measured values of endurance limits reported in the literature vary widely as shown on the left in Figure 5 (28). However, if some of the processing variables are removed by the selection of test specimens with good surface finish, two microstructurally dependent ranges of fatigue resistance emerge, as shown on the right in Figure 5. In steels with typical phosphorus contents and direct quenched and tempered after carburizing, fatigue resistance is dependent on intergranular fatigue crack nucleation. Phosphorus segregates to austenite grain boundaries during carburizing, and subsequently cementite forms on the austenite grain boundaries, lowering the fracture strength of the prior austenite grain boundaries. Much higher fatigue endurance limits, Figure 5, develop when carburized steels are reheated, shot peened, or are produced with very fine austenite grain sizes. In such conditions, fatigue crack initiation appears to occur by transgranular mechanisms.

A major cause of the variability of fatigue performance of gas carburized steels may be the intergranular oxidation which occurs as a result of oxygen-bearing compounds in the gas atmosphere. Namiki and Hatano (29) have shown that the depth of surface oxidation is related to the Si, Mn, and Cr contents of the carburized steel, Figure 6. Therefore, they have designed a series of alloy steels, Table 4, which uses Mo and Ni additions which do not oxidize, and minimizes Si, Cr, and Mn additions. Low P and O contents, and shot peening also were recommended to improve fatigue performance of carburized steels.

An innovative study of new processing and new alloying approaches for carburizing was reported in 1992. Kimura and Namiki (30) applied plasma carburizing to a series of chromium steels, ranging in Cr content from 1 to 12 percent. High surface carbon contents, between 2 and 4 pct, were achieved and the case microstructures contained high densities of carbides. Figure 7 shows C content as a function of alloy Cr content at various distances from the surfaces of the carburized steels. The processing, referred to as super carburizing, was shown to produce improved bending and contact fatigue resistance compared to a conventionally carburized alloy steel.

Machine and automotive components are major applications for carburizing. However, a critical application for carburized components is in helicopter transmission. Wells et al. (31) review the status of steels for such applications. Traditionally, steels such as 9310 have been used for aircraft applications. However, with ever increasing transmission speeds and loads, operating temperatures increase to levels which cause softening of low alloy steels such as 9310. As a result, more highly alloyed secondary hardening steels which have higher hardness and which resist softening on heating have been developed. Table 5 lists nominal compositions for these high performance carburizing steels and Figure 8 compares the tempering resistance of a 9310 steel with two secondary hardening grades of carburizing alloy steels (31).

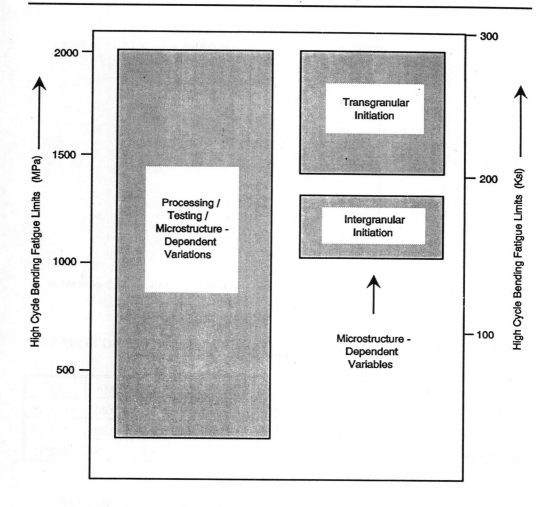

Ranges of Reported Bending Fatigue
Limits in Carburized Steels

Figure 5.   Ranges of reported bending fatigue endurance limits in carburized steels.
Reference 28.

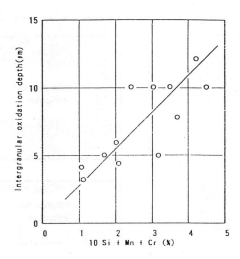

Figure 6.   Depth of intergranular oxidation as a function of Si, Mn, and Cr content of carburized steel. Reference 29.

TABLE 4 - CHARACTERISTICS OF NEW GRADES OF CARBURIZING STEELS. MASS PCT., REFERENCE 29

Grade	C	Si	Mn	P	S	Ni	Cr	Mo	O
0.4Mo	0.20	<0.15	0.70	<0.015	0.015	---	1.00	0.40	<0.0015
1Ni-0.30Mo	0.20	<0.15	0.60	<0.015	0.015	1.00	0.80	0.30	<0.0015
2Ni-0.75Mo	0.20	<0.15	0.30	<0.015	0.015	2.00	0.30	0.75	<0.0015

TABLE 5 - NOMINAL COMPOSITIONS OF GEAR AND BEARING STEELS

Steel	Alloy Composition (wt. pct.)								
	C	Mn	Si	Ni	Cr	Mo	V	W	Cu
AISI 9310	0.10	0.55	0.25	3.25	1.2	0.1	--	--	--
Vasco X-2M	0.15	0.3	1.0	--	5.0	1.4	0.5	1.35	--
Pyrowear 53	0.1	0.3	1.0	2.0	1.0	3.0	0.1	--	2.0
M50NiL	0.12	0.3	0.3	3.0	4.5	4.0	1.25	--	--
M50	0.8	0.25	0.15	--	4.0	4.25	1.0	--	--

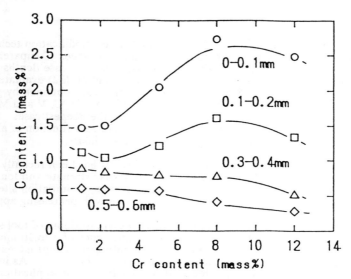

Figure 7.   Carbon content as a function of depth and Cr content of steels. Plasma carburized at 920°C for 10.8 ks. Reference 30.

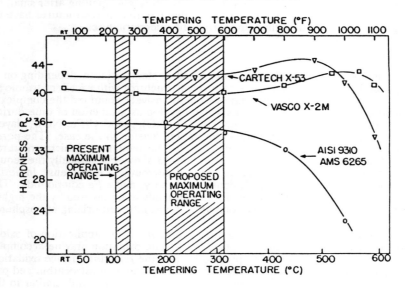

Figure 8.   Hardness as a function of tempering temperature for various alloy steels tempered for 2 plus 2 hours. Reference 33.

Nitriding

Gas and plasma or ion nitriding are well established surface modification techniques which significantly enhance surface strength and fatigue resistance. Compared to carburizing, nitriding produces parts with very low distortion, but case depths require much longer treatment times. In a 1992 paper, Kobayashi et al. (32) evaluated the effects of alloying elements on the rate at which case depths are obtained by gas nitriding in the temperature range between 570 to 650°C. At 570°C, V and Mo produce deep cases and at higher temperatures V and Al produce the deepest cases. Figure 9 shows effective nitrided case depth to a hardness of HV420 as a function of alloy content for various alloy additions for steels nitrided at 570°C for 9 hours.

The rotating bending fatigue behavior of gas nitrided steels has been recently described in detail by Spies et al. (33). The effects of case depth, case strength and toughness and core hardness and applied stress state were considered. Fatigue initiation sites undergo a transition from surface initiation to subsurface initiation with decreasing applied stress.

Haberling and Rasche (34) discussed the benefits of plasma nitriding of tool steels. They point out that dimensional changes, typically on the order of 5 to 10 microns are caused by nitriding tool steels. The increased surface nitrogen content increases volume, and causes the changes in dimension shown schematically in Figure 10. As in all surface engineering, the processing and properties must be matched to the application. Haberling and Rasche demonstrated that the best performance of taps correlates with shallow nitrided cases without compound layers. For cutting and punching tools, surface hardness must not be excessive in order to avoid fracture and spalling after small numbers of load applications. Figure 11 shows that bainitic microstructures have much lower toughness after nitriding than do martensitic microstructures.

Nitrocarburizing

The simultaneous addition of nitrogen and carbon to steel surfaces, depending on processing condition, produces a range of surface structures with excellent tribological properties. Mittemeijer and his colleagues have intensively examined the complex microstructural changes and mechanisms which develop as a function of time during nitrocarburizing (35,36). Cementite (Fe_3C), ϵ_2 $(Fe_2(NC)_{1-x})$ and $\gamma'(Fe_4(NC)_{1-x})$ layers form and eventually porosity and channel formation develop in the case. The pores form as the ϵ and γ' dissociate to produce N_2 gas. The sequence of microstructural change is a function of processing. Figures 12 and 13 show schematically the sequence of changes in salt-bath nitrocarburized and gas nitrocarburized specimens. Important differences are related to the C and N contents of the γ' and ϵ Fe compounds. The ϵ has higher C contents, and therefore its formation tends to be favored by the higher C content of nitrocarburizing salt baths, compared to gas nitrocarburizing atmospheres.

Another major area of activity regarding nitrocarburizing is the application of oxidation treatments to nitrocarburized components (37-40). The oxidation may be accomplished by either salt bath or gas atmosphere processing, and the resulting surface oxidation produces significant improvements in the corrosion resistance of nitrocarburized parts. Iron oxides form on the compound layer of nitrocarburized parts, and, similar to the surface changes produced by nitrocarburizing, a complex sequence of phase and structural changes occurs (37,38). The improved wear and corrosion resistance of oxidized parts may be due to the filling of the pores and channels in nitrocarburized layers by iron oxides rather than the incorporation of oxygen into the Fe-C-N compounds which are produced during nitrocarburizing.

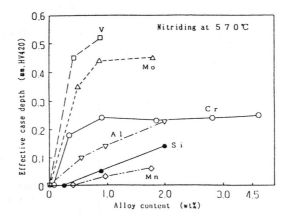

Figure 9.   Effective case depth as a function of alloy content for steels nitrided at
              570°C for 9 hours. Reference 32.

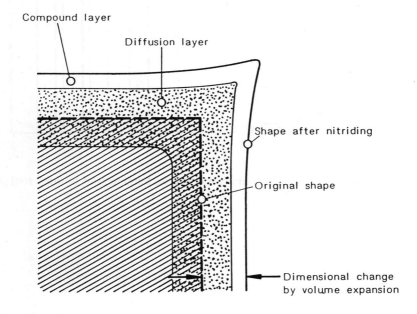

Figure 10.   Schematic of shape changes produced by nitriding at the edges of tools.
              Reference 35.

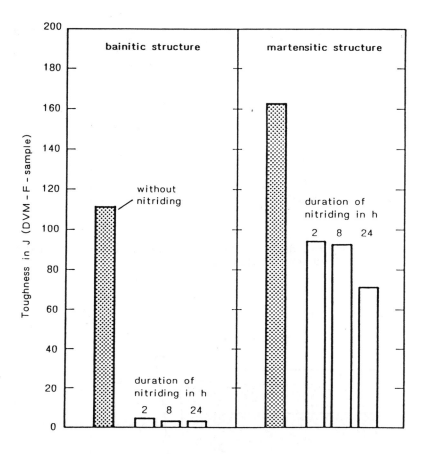

Figure 11.   Effect of substrate microstructure on toughness of nitrided tool steels.
Reference 35.

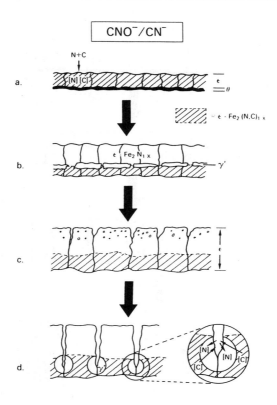

Figure 12. "Compound-layer formation during (commercial) salt-bath nitrocarburizing: (a) initially cementite (Θ) nucleates at the surface (tentatively drawn too thick), followed by nucleation of ϵ on top of Θ; (b) gradually higher nitrogen uptake and lower carbon uptake at the outer surface leads to the sequence of phases (bottom to top); ϵ (carbon rich) /γ'/ϵ by considerable carbon uptake; this leads to an almost monophase ϵ layer with a carbon accumulation near the substrate; dissolution of the layer along grain boundaries leads to channel development; pore formation by N_2 precipitation occurs too; (d) penetration of the salt into the channels allows exchange of carbon against nitrogen in the bottom part of the compound layer, leading to the reappearance of γ''. Reference 35.

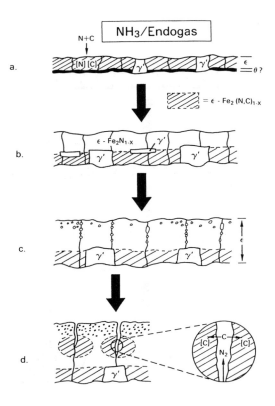

Figure 13.   "Carbon uptake during (commercial) gaseous nitrocarburizing: (a) nucleation of cementite (presumed; tentatively drawn too thick) followed by the formation of carbon rich ϵ on top; (b) gradually more nitrogen uptake and less carbon uptake at the outer surface leads to the sequence of phases (bottom to top): ϵ (carbon rich) /γ'/ϵ (nitrogen rich); c) pore formation in nitrogen rich ϵ by precipitation of N_2-gas; (d) carbon uptake through the channel walls establishes a (second) carbon enrichment in the near surface region of the compound layer, that is not connected with the (first) carbon accumulation in the bottom part of the layer (see a) and (b) above". Reference 35.

QUENCHING AND PERFORMANCE

The need for better control of the properties of through hardened and carburized steels has led to intense interest in quenching and in modeling the cooling and transformation phenomena which occur during quenching (2,41). There are many reasons for this interest. Quenching, in particular the characteristics of the quench media and the heat transfer at the quenchant-workpiece interface, establishes the cooling rates within the workpiece, which together with the austenitic structure and composition of steel, controls the transformation of austenite to mixtures of ferrite and cementite or martensite and therefore the final mechanical properties of a quenched part. Also, the temperature and transformation gradients established during quenching influence the development of residual stresses, distortion and cracking. Thus the understanding of quenching and associated phenomena is essential not only for the production of steel parts with desirable mechanical properties but also to minimize manufacturing difficulties associated with quench products.

Much of the coordinated effort to evaluate the quenching process has come from the "Scientific and Technological Aspects of Quenching" technical committee of the International Federation of Heat Treatment and Surface Engineering (IFHT). This activity has resulted in the 1992 publication of the handbook, Theory and Technology of Quenching, in which 17 authors address all aspects of quenching (42).

Important aspects of quenching include the characterization of the cooling power of quenching media and steel response to cooling. With respect to the former aspect, laboratory tests with several types of probes, primarily manufactured with INCONEL Alloy 600 or silver, with imbedded thermocouples, are now well established (42,43). Figure 14 shows temperature and cooling rate as a function of time during a typical oil or water quench. As shown there, are three stages of the cooling process: vapor blanket formation, boiling, and convection. The most rapid temperature decreases and cooling rates are associated with the boiling stage, and this stage together with the duration of the vapor blanket stage determine if the austenite in a given steel transforms to martensite or nonmartensitic products. Following the boiling stage, cooling rate drops significantly during the convection stage. It is during this phase of cooling where residual stresses, cracking and distortion may develop. Quenchants which provide low convection cooling rates are therefore desirable for sensitive steels and complex part geometries. Every commercial quenching oil has its characteristic cooling capacity, and Bodin and Segerberg (43) outline a method to compare the hardening power (HP) of various oils based on parameters associated with the 3 stages of cooling. For unalloyed steels they have developed the following regression equation.

$$HP = 91.5 + 1.34\,T_{vp} + 10.88\,CR - 3.85\,T_{cp}$$

Figure 15 defines the parameters, and shows results applied to a number of quenching oils.

The computer modeling of phase transformations, microstructural evolution, mechanical properties, residual stresses and distortion during quenching is also currently under intensive development (2,4). Figure 16, from Buchmayr and Kirkaldy (44), shows schematically the many phenomena which must be quantitatively coupled in order to predict microstructure and mechanical properties which develop during quenching. There has been great progress in the development of computer modeling for the various phases of quenching. However, the complexity of the quenching process and steel response, in particular the need for the temperature-dependent flow properties of many microstructures which are required by the models, insures that much future effort will be necessary for the development of accurate predictive models.

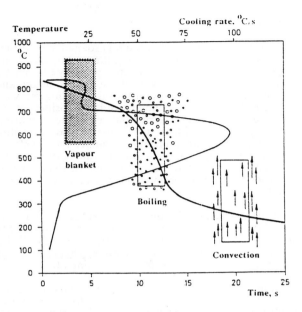

Figure 14.   Typical cooling curve and plot of cooling rate versus time for quenching. The three stages of cooling of a part are shown. Reference 43.

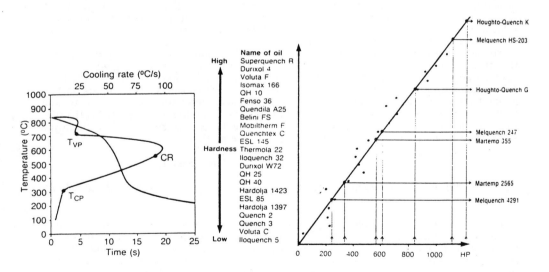

Figure 15.   Cooling rate parameters used to calculate the hardening, power, HP, of a quenchant, left, and HP for various oil quenchants, right.

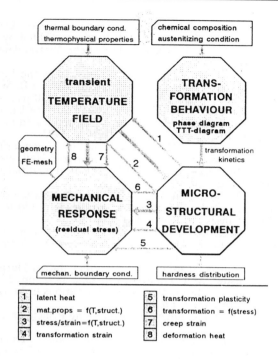

Figure 16.   Model development to predict microstructure and mechanical properties of quenched steels. Reference 44.

SUMMARY

The above review has presented a view through a selected window of the recent activity devoted to improving steels and heat treatment processing of steels. Most of the literature cited has come from 1992 contributions to the literature. This approach highlights current activity, but lacks the depth of a review which traces the evolution of a given type of steel or heat treatment processing. That evolution is covered in part in the reference lists of the papers cited in this paper, and the reader is urged to seek out the original papers if he or she has an interest in a given topic. Despite the above reservations, this brief review reflects a very vigorous world-wide effort devoted to heat treatment processing. This effort continues in the present conference and will continue into the future as the need for high quality and economy in manufacturing intensifies.

REFERENCES

1.   New Materials, Processes, and Experiences for Tooling, Edited by H. Berns, M. Hofmann, L.A. Norström, K. Rasche, and A.-M. Schindler, Published by MAT SEARCH, CH-1015 Pully, Switzerland 585 pages, 1992.
2.   Quenching and Distortion Control, Edited by G.E. Totten, Published by ASM International, Materials Park, Ohio, USA, 341 pages, 1992.

3.   Gilbert R. Speich Symposium Proceedings, Fundamentals of Aging and Tempering in Bainitic and Martensitic Steel Products, Edited by G. Krauss and P.E. Repas, Published by the Iron and Steel Society, Warrendale, Pennsylvania, USA, 267 pages, 1992.

4.   Heat and Surface '92, Edited by I. Tamura, Published by the Japan Technical Information Service, Tokyo, Japan (Available from Robert B. Wood, Secretary, IFHT, 55 Church Road, Richmond, UK, TW10 6LX), 630 pages, 1992.

5.   Proceedings of the First ASM Heat Treatment and Surface Engineering Conference in Europe, Edited by E.J. Mittemeijer, published in Materials Science Forum, Volumes 102-104, Parts 1 and 2, 902 pages, 1992.

6.   Advanced Surface Coatings: A Handbook of Surface Engineering, Edited by D.S. Rickerby and A. Matthews, published by Chapman and Hall, New York, 1991.

7.   A. Böttger and E.J. Mittemeijer: "Redistribution of Interstitial Atoms on Aging of Iron Based Martensites", in Reference 3, pp. 5-17.

8.   L. Chang, S.J. Barnard, and G.D.W. Smith: "The Segregation of Carbon Atoms to Dislocations in Low-Carbon Martensites: Studies by Field Ion Microscopy and Atom Probe Microanalysis", in Reference 3, pp. 19-27.

9.   M.J. van Genderen, A. Böttger, and E.J. Mittemeijer: "Ternary Iron-Carbon-Nitrogen Martensitic Alloys: As Quenched State and Development of Transition Precipitates", in Reference 3, pp. 29-37.

10.   Y. Ohmori and I. Tamura: "Epsilon Carbide Precipitation During Tempering of Plain Carbon Martensite", Met. Trans. A, vol. 23A, 1992, pp. 2737-2751.

11.   H.-C. Lee and G. Krauss: Intralath Carbide Transitions in Martensitic Medium-Carbon Steels", in Reference 3, pp. 39-43.

12.   G. Krauss: Steels: Heat Treatment and Processing Principles, ASM International, 1990, pp. 229-255.

13.   J.A. Sanders, P. Purtscher, D.K. Matlock, and G. Krauss: "Ductile Fracture and Tempered Martensite Embrittlement of 4140 Steel", in Reference 3, pp. 67-76.

14.   J.R.T. Branco and G. Krauss: "Toughness of H11/H13 Hot Work Die Steels", in Reference 1, pp. 121-134.

15.   M. Lin, R.L. Bodnar, and S.S. Hansen: "Tempering Behavior of High-Carbon Roll Steels Containing 3 to 5 pct Chromium", in Reference 3, pp. 77-91.

16.   M.L. Schmidt: "A Closed Die Forging Study of Computer AerMet 100 Alloy", 34th Mechanical Working and Steel Processing Conference Proceedings, ISS, Warrendale, PA, 1992, pp. 297-308.

17.   P.M. Novotny: "An Aging Study of Carpenter AerMet 100 Alloy", in Reference 3, pp. 215-230.

18.   J.S. Montgomery and G.B. Olson: "M_2C Carbide Precipitation of AF 1410", in Reference 3, pp. 177-214.

19.   W.M. Garrison, Jr., J.L. Maloney, and A.L. Wojcieszynski: "Influence of Tempering and Second Phase Particle Distribution on the Fracture Behavior of HY180 Steel", in Reference 3, pp. 237.

20.   T. Araki, M. Enomoto, and K. Shibarta: "Microstructural Identification and Control of Very Low Carbon Modern High Strength Low Alloy Steels", in Reference 5, pp. 3-10.

21.   T. Araki, K. Shibata, and H. Nakajima: "Self Tempering and Transformation Mode in Bainitic Intermediate Microstructures of Continuous-Cooled Low-C HSLA Steels", in Reference 3, pp. 113-118.

22.   Atlas for Bainitic Microstructures, Vol. 1, Edited by T. Araki, The Iron and Steel Institute of Japan, Tokyo 100, Japan.

23.   S.J. Mikalac, A.V. Brandemarte, L.M. Brown, M.G. Vassilaros, and M.E. Natishan: "Thermal Cycling in the Austenite Temperature Region for Grain Size Control in Steels", in Reference 5, pp. 11-30.

24.   A.K. Lis, M. Mujahid, C.I. Garcia, and A.J. DeArdo: "The Role of Copper in the Aging Behavior of HSLA-100 Plate Steel", in Reference 3, pp. 129-138.

25. R.P. Foley and M.E. Fine: "The Effect of Tempering on the Tensile Behavior of a Low Carbon Steel at 77, 188, and 298K: Normal and Intercritical Tempering Effects", in Reference 3, pp. 139-154.

26. N. Iwama, T. Tsuzaki, and T. Maki: "Effect of Alloying Elements in Microstructure and Mechanical Properties of Low-Carbon Bainitic Steels", in Reference 4, pp. 49-52.

27. S.W. Thompson and G. Krauss: "Structure and Properties of Continuously Cooled Bainitic Ferrite-Austenite-Martensite Microstructures", in Mechanical Working and Steel Processing Proceedings, ISS, 1989, pp. 467-481.

28. G. Krauss: "Advanced Performance of Steel Surfaces Modified by Carburizing", in Reference 4, pp. 7-12.

29. N. Namiki and A. Hatano: "High Toughness and High Fatigue Strength Carburizing Steel", Reference 4, pp. 361-364.

30. T. Kimuara and K. Namiki: "Plasma Super Carburizing Characteristics and Fatigue Properties of Chromium Bearing Steels", in Reference 4, pp. 519-522.

31. M.G.H. Wells, J.S. Montgomery, and E.B. Kula: "High Strength Steels: Army Research and Applications", in Reference 3, pp. 165-175.

32. K. Kobayashi, K. Hosoda, and K. Tsubota: "Effects of Alloying Elements on Gas Nitriding Properties", in Reference 4, pp. 377-380.

33. H.-J. Spies, M. Scharf, and N.D. Tan: "Einfluss des Nitrierens auf die Zeitfestigkeit von Stählen", Härterei-Technische Mitteilungen, vol. 46, 1991, pp. 288-293.

34. E. Haberling and K. Rasche: "Plasma Nitriding of Tool Steels", in Reference 1, pp. 369-392.

35. M.A.J. Somers and E.J. Mittemeijer: "Verbindungsschichtbildung Während des Gasnitrierens und des Gas-und Salzbadnitrocaburierens", Härterei-Technische Mitteilungen, vol. 47, 1992, pp. 5-12.

36. M.A.J. Somers and E.J. Mittemeijer: "Model Description of Iron-Carbonitride Compound Layer Formation During Gaseous and Salt-Bath Nitrocarburizing", in Reference 5, pp. 223-228.

37. M.A.J. Somers and E.J. Mittemeijer: "Phase Transformations and Stress Relaxation in γ'-Fe_4N_{1-x} Surface Layers During Oxidation", Metallurgical Transactions A, vol. 21A, 1990, pp 901-910.

38. M.A.J. Somers, B.J. Koui, W.G. Sloof, and E.J. Mittemeijer: "On the Oxidation of γ'-Fe_4N_{1-x} Layers; Redistribution of Nitrogen", Surface and Interface Analysis, vol. 19, 1992, pp. 633-637.

39. C. Dawes: "Nitrocarburizing and its Influence on Design in the Automotive Sector", in Surface Engineering and Heat Treatment, Edited by P.H. Morton, The Institute of Metals, London, 1991, pp. 80-120.

40. G. Wahl: "Nitrocarburizing Plus Post Oxidation on Economical Surface Engineering Process for a Wide Range of Applications", in Reference 4, pp. 401-404.

41. Quenching and Carburizing, A Seminar Sponsored by the International Federation for Heat Treatment and Surface Engineering and the Institute of Metals and Materials Australia, Melbourne, Australia, September 1991.

42. Theory and Technology of Quenching, Edited by B. Liscic, H.M. Tensi, and W. Luty, Springer-Verlag, 1992.

43. J. Bodin and S. Segerberg: "Measurement and Evaluation of the Quenching Power of Quenching Media for Hardening", in Reference 2, pp. 1-12.

44. B. Buchmayer and J.S. Kirkaldy: "A Fundamental Based Microstructural Model for the Optimization of Heat Treatment Processes", in Reference 2, pp. 221-227.

Materials Science Forum Vols. 163-165 (1994) pp. 37-48

MECHANISMS AND MODELLING OF MASS TRANSFER DURING GAS-SOLID THERMOCHEMICAL SURFACE TREATMENT: NEW PROCESS AND NEW PROCESS CONTROL FOR CARBURIZING AND NITRIDING

M. Gantois

Laboratoire de Science et Genie des Surfaces (URA CNRS 1402),
I.N.P.L.-Ecole des Mines, Parc de Saurupt, F-54042 Nancy, Cédex, France

ABSTRACT

Using methods allowing basic mass transfer mechanisms analysis and modelling, mass transfer monitoring with thermogravimetric instrumentation and gas mixture analysis, metallurgical microstructure checking, we show that the first stage of the gas solid reaction is the step which controls the mass transfer mechanisms at the gas/solid interface and consequently the speed of the treatment and the metallurgical superficial microstructures. Two examples are described : steel carburizing to obtain minimum treatment time and steel nitriding to control nucleation and growth of nitride compound layer's thanks to a solid state diffusion model. Industrial developments need a comprehensive analysis of the gaseous reactive species, the design of "a perfectly agitated" reactor and the monitoring of the gas flow rate inlet.

INTRODUCTION

Improvement and development of new thermochimical surface treatments by means of a gas/solid reaction should consider two aims :

- *a better productivity*, meaning treated pieces with zero defect and a lower treatment time,

- *environmental protection* by lowering all the pollution sources : reduction of polluting gases rejected in the atmosphere and diminution of the energy cost.

To reach these aims, one needs to understand :

- The nature, formation mechanisms and flowing modes of chemical active species in the gas.

- The mechanisms of phase formation (carbiɔes, nitrides, oxides) and mass transfer at the gas/solid interface.

- The diffusion mechanisms in the solid state that monitor the surface hardness.

This last point was largely studied and the relation diffusion mechanisms/phase transformation/hardening is well known by now, in particular in the steel carburizing and nitriding field.

On the contrary, chemical active species formation and flow in the reactor and mass transfer at the gas/solid interface are usually less understood.

Therefore process optimizing and monitoring are limited and the development of new treatments is difficult.

All the studies we are conducting on gas/solid reactions in atmospheres where reactive species are obtained by chemical or physical ways, imply an analysis of the mass transfer mechanisms in the gas, at the gas/solid interface and in the solid.

This paper will be limited to steel carburizing, carbonitriding and nitriding. We are going to show that the mass transfer at the gas/solid interface monitors the nature of the formed compounds and the elaboration speed of the concentration gradients [1,2,3].

Monitoring the surface metallurgical microstructure and the maximum treatment speed supposes to monitor the interface phenomena from the beginning of the treatment.

GENERAL   APPROACH

The method used here is the following :

- basic mass transfer mechanisms analysis,

- modelling of the mass transfer in the solid and of the metallurgical microstructures,

- mass transfer monitoring from the data given by the model using thermogravimetric instrumentation. Metallurgical microstructure checking.

- operating in an industrial reactor.

Metallurgical microstructures and surface physico-chemical study used X-ray diffraction, electronic microscopy, Auger spectroscopy, electronic microprobe analysis and glow discharge optical spectrometry.

The study of the flowing active species used specific technics :

- *Gas chromatography* for the gas/solid reactions realized at atmospheric pressure for molecular species.

- *Optical spectrometry* for species formed at low pressure with physical methods (electrostatic or microwave discharges or post-discharges).

Thermogravimetry allows to test several hypothesis and mass transfer models.

A quenching device allows to determine the physico-chemical and structural states of the samples. The set-up is the following :

- *a thermobalance* including an alumina or high temperature resistant stainless steel cylindrical reactor with a special device for oil or water quenching. The flow study showed that this reactor is a piston-reactor,

- *chromatographes* allowing chemical species studies in the gas,

- *mass flow controllers* allowing to adjust the mass flow of each constituant of the gas in order to monitor the mass transfer in the sample according to the model laws,

- *a computer* allowing on the one hand data acquisition and on the other hand, computation of mass transfer thanks to the model and computation of the superficial concentration gradients in the sample. In particular, we can compute the mass balance of the reaction.

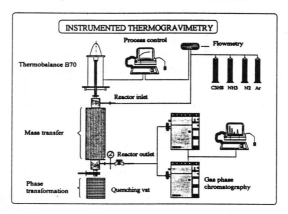

Figure 1

- an *industrial "furnace-reactor"* should be a perfectly mixed reactor. That's controlled by a gas flow characterization.

I - CARBON AND NITROGEN TRANSFER IN AUSTENITIC PHASE WITH FIXED INTERFACE CONDITIONS

I.1 - Carbon transfer. Development of a new steel carburizing treatment.

Usually steel carburizing is realized in a carbon oxide (CO) atmosphere. CO split on the steel surface and carbon diffuses in the austenitic phase, whereas oxygen is first adsorbed on the steel surface and then transfered in the gas phase (as O_2, CO_2 or H_2O).

So, for a 0.2% carbon steel placed in an atmosphere corresponding to a 1.3 carbon potential, the transfer resistance resulting from the adsorbed oxygen prevents the superficial carbon concentration to reach rapidly the carbon potential value. Then, it exists a transitory period the duration of which depends on several parameters (atmosphere composition, furnace hydrodynamic, etc.).

This period, where the superficial carbon concentration grows and for which the interface condition is not known and corresponds to a superficial concentration below the potential, does not allow a precise control of the treatment, nor a maximum treatment speed.

In order to avoid a transfer resistance at the gas/solid interface, resulting from adsorbed oxygen, we use a reactive gas mixture which includes one single hydrocarbon diluted in an inert gas. For example, one could use a proper propane-nitrogen mixture in a 800°C-1050°C temperature range. Carbon transfer in the austenitic phase is monitored by the diffusion phenomena in the solid state, only. **Figure 2** clearly shows that we can impose the interface conditions :

- carbon transfer with a constant mass flow (*curves 1,2 and 3*),
- carbon transfer with a constant concentration (*curve 4*).

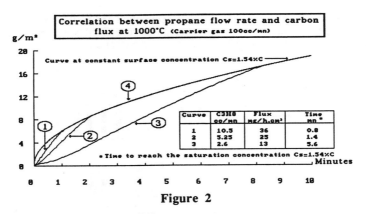

Figure 2

We verify that we can realize a carbon concentration profile using both boundary conditions :

- *constant mass flow* : this condition is achieved with injection of a constant hydrocarbon mass flow during precise time slots. During these slots, carbon is transfered in the sample. They are followed by diffusion periods. This alternation of enrichment and diffusion periods allows to have a superficial carbon concentration varying between a high limit (carbon saturation in the austenitic phase) and a low limit which is freely chosen.

- *constant concentration* : this condition is achieved with injection in a continuous way of a variable hydrocarbon mass flow which decreases with time. This enrichment period is followed by a single diffusion period. If the superficial concentration, which is constant during the enrichment period, is equal to the carbon saturation concentration in the austenitic phase, the carbon concentration gradient is obtained with the highest possible speed at the considered temperature.

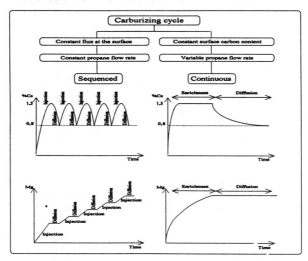

Figure 3

Figure 3 schematizes this two possibilities. **Figures 4** and **5** show the carbon concentration profiles realized at 980°C and 870°C on a XC10 steel, maintaining the superficial carbon concentration equal to the saturation concentration during the enrichment period.

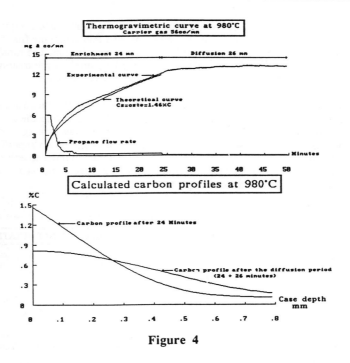

Figure 4

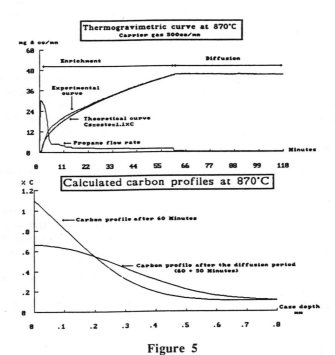

Figure 5

In the particular case where the carbon concentration at the interface is equal to the saturation carbon concentration in the austenitic phase, we can point out a thin cementite film formed on the steel surface. This thin cementite film fixes the interface carbon concentration. The rapid formation of this thin cementite layer allows to fix and to monitor, from the beginning of the treatment, the condition which ensures the maximum transfer speed of carbon in the austenitic phase.

In short, using a proper gas mixture which avoids any transfer resistance at the interface allows to choose the transfer conditions of carbon at the interface and, knowing the steel composition, to calculate the carbon concentration profile for each instant. So, we can construct a two phases concentration profile : an enrichment period with a constant and maximum superficial concentration (saturation concentration), a diffusion period with a zero flow rate in order to adjust the profile shape and, in particular, the final superficial carbon concentration. **Figure 6** shows as an example, a cycle with the corresponding carbon concentration profile.

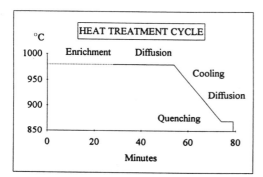 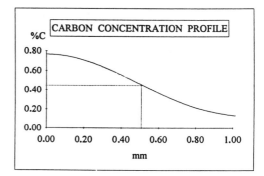

Figure 6

The same principles are used to operate an industrial reactor, but then we need to monitor the process in order to fix the interface condition. This interface condition should be achieved on all the surfaces to be treated. Obviously, in order to reduce the treatment cycle length, we have to search for a superficial concentration equal to the saturation carbon concentration during the enrichment period. This condition can be achieved by forming on the steel surface a thin Fe_3C cementite layer (around 1 μm) which is dissolved during the diffusion period.

For a temperature higher than 850°C, this cementite layer forms in a very short time in an hydrocarbon gas mixture obtained from propane decomposition.

Since the carbon enrichment time is very short, carbon concentration has to be uniform on all the treated parts, that means that temperature has to be homogeneous (about ± 5°C) and that the reactor has to be "perfectly mixed". Its an example, on effective depth of 1.2 mm can be achieved after a 54 minutes enrichment at 950°C and a 174 minutes total time before quenching at 850°C.

Considering the gas flow rates used, the CO_2 rejected in the atmosphere is about 50 times lower than for a conventional carburizing process. Associated with a pressurized gas quenching, this new process sensibly reduces the treatment time, lowers the pieces deformation and is quite totally pollution free

I.2 - Nitrogen and carbon transfer. Realization of a new carbonitriding treatment.

This new carburizing treatment is characterized by a diffusion period with a zero carbon flow rate at the gas/solid interface.

This period can be used to realize a nitrogen transfer, the concentration profile of which can be adjusted in order to optimize the surface properties.

In order to monitor the nitrogen transfer in a carbon concentration gradient, it is necessary :

- to analyse the carbon-nitrogen interaction in the austenitic phase,
- to study the nitrogen transfer mechanisms at the interface and its diffusion mechanisms in the solid.

Finally, a complementary metallurgical study is necessary to determine the carbon and nitrogen concentrations which allow, in particular, to avoid the nitrides precipitation.

A geometrical exclusion model applied to the Fe-C-N ternary solid solution shows that nitrogen increases the carbon activity in the austenitic phase and that carbon increases the nitrogen activity in the austenitic phase (**Figure 7**).

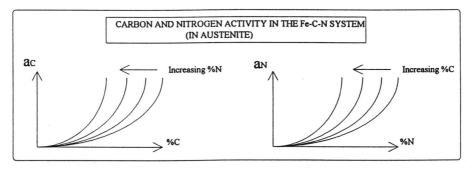

Figure 7

Thermogravimetry clearly shows (**Figure 8**) these interactions. For a same ammonia partial pressure, the nitrogen dissolving in the austenitic phase is greater for a 0.1% carbon steel than for a 0.7% carbon steel. Since carbon increases nitrogen activity and reciprocally, computation shows the effect of carbon concentration gradient onto nitrogen chemical potential gradient : thermogravimetry allows to verify that a carbon concentration gradient considerably increases the nitrogen diffusion speed in the austenitic phase.

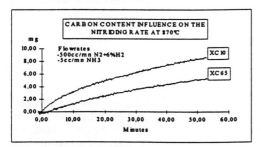
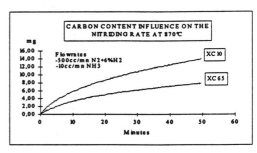

Figure 8

Then, thanks to a precise mathematical model, we can monitor a carbonitriding treatment, as shown on **Figure 9**.

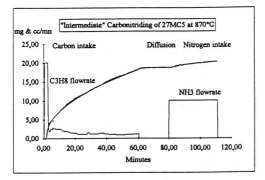

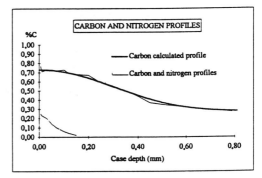

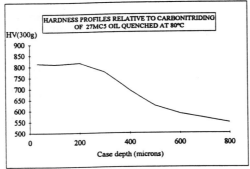

Figure 9

II - NITROGEN TRANSFER IN FERRITIC PHASE - APPLICATION TO NITRIDING

A similar method can be applied for the study of steel nitriding in gaseous phase. Nitrogen transfer can be obtained in an ammonia gas mixture or in a gas mixture containing excited species obtained with a electrostatic or electromagnetic field (microwaves for example). So, we want to develop a mathematical model in order to compute the nitrogen transfer rate in the solid, the nitrides layers growth rate and the nitrogen concentration profiles, and then we want to compare theses results with thermogravimetry experiments and metallurgical analysis and observation.

Considering the industrial interest and in order to simplify this paper, we only show here :

- *for a Fe-N binary system :*
    * nitrogen transfer in the α solid solution α-Fe,
    * nitrogen transfer in the two phases system α - Fe + γ' - Fe$_4$N.
- *for a Fe-C-N ternary system :*
    * nitrogen transfer in the two phases system α - Fe + ϵ - Fe$_{2\text{-}3}$ (C,N).

The mathematical model is established in semi-infinite system, considering that the nitrogen diffusion coefficient is independent of the nitrogen concentration gradient, that the carbon concentration is constant in the ϵ phase and that there is no carbon transfer. On the other hand, we assume that the nitrogen concentrations at the interfaces are constant and equal to the equilibrium values and that the C_N^S nitrogen concentration at the gas/solid interface does not vary during the treatment.

In theses conditions, we can solve the diffusion equations from the boundary conditions. For example, in the $\gamma' - \alpha$ system, we obtain :

- The nitrogen concentration profiles in the γ' and α phases :

$$C_{\gamma'}(x,t) = C_N^S + (C_{1\gamma'\alpha} - C_N^S) \frac{\text{erf}\left(\dfrac{x}{2\sqrt{D_{\gamma'}\cdot t}}\right)}{\text{erf}\left(\dfrac{b_{\gamma'\alpha}}{2\sqrt{D_{\gamma'}}}\right)} \qquad \text{for } 0 \le x \le \lambda_{\gamma'\alpha}$$

$$C_{\alpha}(x,t) = C_{2\gamma'\alpha} \frac{1-\text{erf}\left(\dfrac{x}{2\sqrt{D_{\alpha}\cdot t}}\right)}{1-\text{erf}\left(\dfrac{b_{\gamma'\alpha}}{2\sqrt{D_{\alpha}}}\right)} \qquad \text{for } \lambda_{\gamma'\alpha} \ge x \ge \infty$$

- The evolution of the γ' layer thickness $\lambda_{\gamma'\alpha}$ and the interface displacement speed $V_{\gamma/\alpha}$:

$$V_{\gamma/\alpha} = \frac{d\lambda_{\gamma\alpha}}{dt} \ .$$

- The transferred nitrogen mass as a fonction of time :

$$M(\alpha,\gamma')(t) = \frac{2\, S_L \cdot (C_N^S - C_{1\gamma'\alpha})}{\text{erf}\left(\dfrac{b_{\gamma'\alpha}}{2\sqrt{D_{\gamma'}}}\right)} \cdot \sqrt{\frac{D_{\gamma'}\cdot t}{\pi}}$$

All these values can be determined experimentally in order to verify the model. The treatment is conducted while adjusting the gas flow rates in order to follow the transferred mass computed from the model. The γ layer thickness and the nitrogen concentration gradients evolution are measured for different treatment time.

For the ε layer growth on a solid solution, it is necessary to know the nitrogen diffusion coefficient in the ε phase. Since this coefficient is not available, we determined it thanks to the application of the model to the iron-nitrogen binary system in the case where the $\varepsilon/\gamma'/\alpha$ configuration is formed.

Indeed, knowing the nitrogen diffusion coefficient in the γ' phase and the relative evolution of the ε and γ' layer thickness when operating with a $\sqrt{t}$ law growth, allow, after measuring the C_N^S nitrogen concentration of the formed ε carbonitride, to determine the kinetic constants (from the measurement of the layers thickness).

Then the model allows to determine the diffusion coefficient value.

$$D_{\varepsilon} = 2.1 \cdot 10^{-8} \exp\left(\frac{-93\,517}{RT}\right)$$

This value is constant for a nitrogen + carbon composition of the ε carbonitride between 7 Wt% and 9.5 Wt%. It is possible to follow a law m = f(t) by adjusting the ammonia flow rate for a chosen layer structure : ε and γ' layer thickness and nitrogen concentration (i).

The growth model so validated can be applied to the three configurations α ; γ'/α ; ε/α using for the latest case an ammonia and propane gas mixture.

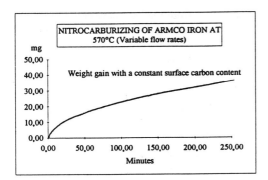 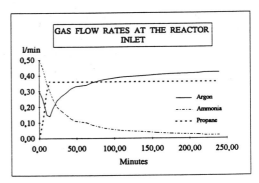

Figure 10

Figure 10 shows the thermogravimetric curve corresponding to the formation of ε carbonitride onto α solution. X rays show that ε carbonitride was formed without γ′ nitride and its analysis shows that it contains 8.0% of nitrogen and 1.0% of carbon, wanted values introduced in the theoretical model.

We can see that the reactive gas flow rate, introduced in the reactor, is regularly decreasing : very high at the beginning of the treatment in order to immediately form ε carbonitride, and rapidly decreasing to be low at the end of the treatment.

If the reactive gas flow rate is chosen constant during all the treatment, two cases can occur :

- The constant ammonia flow rate is at least equal to the initial flow rate needed to form ε **(Figure 11)**.

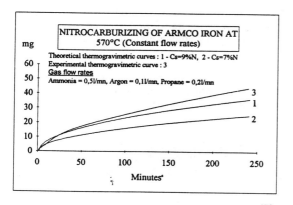 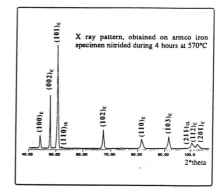

Figure 11

We can see that the nitrogen concentration of the ε carbonitride increases with time, the transferred mass increases with the formation of porosities.

- The constant ammonia flow rate is lower (5 times less) than the initial flow rate needed to form ε (**Figure 12**).

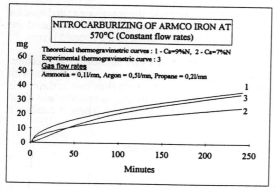
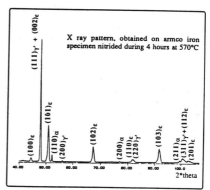

Figure 12

A (γ+ε) two-phase layer is formed, the nitrogen concentration and relative thicknesses of which keep on evoluting.

So, the reactive species flow rate is, with temperature, the fundamental parameter which monitors the treatment. The initial stage of the treatment governs, in particular, the nature of the formed layers. **Figure 13** represents this conclusion in the case where reactive species are obtained in a microwave post-discharge.

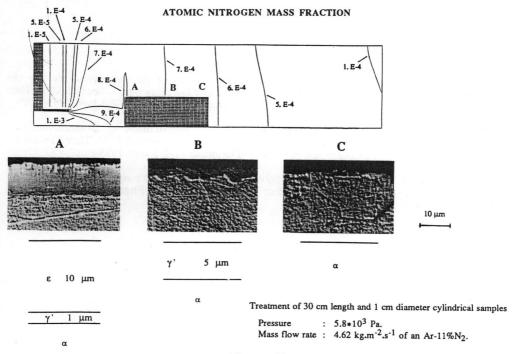

Figure 13

Is it possible to monitor the reactive species flow rates in industrial conditions, in particular in the initial phase of the treatment ?

This is easy with the ionic bombardment nitriding process, because the reactive excited species density can be monitored with the gas mixture composition and because they are formed around the metal surface. This is why this process allows to precisely monitor the nature of the formed layers.

Our studies show similar possibilities for a gas/solid reaction with an ammonia gas mixture. However, all our experiments were conducted in an alumina reactor behaving as a piston-reactor with a few centimeter heated area. In these conditions, without sample, the ammonia flow rate at the exit is equal to the entrance flow rate. We showed that this is not the case if the reactor is metallic and if the residence time of the ammonia molecules at the high temperature is increased. Then, a catalysed decomposition on the reactor walls is observed.

These phenomena should be studied before planning an industrial set-up.

CONCLUSIONS

The initial stage of a gas/solid thermochemical surface treatment is the step which controls the mass transfer at the gas/solid interface.

The mass transfer mechanisms during the initial stage of the treatment governs the nature of the superfical layer which controls the treatment's speed (steel carburizing) or the metallurgical microstructures (steel nitriding).

Monitoring the first step of a treatment is fundamental for a treatment's maximum speed and an effectiv control of the metallurgical microstructures.

REFERENCES

1) **M.S. Yahia, Ph. Bilger, J. Dulcy, M. Gantois** - *"Use of thermogravimetry for the study of carbonitriding treatment of plain carbon steel and low alloy steel"* - Second ASM "Heat Treatment and Surface Engineering Conference in Europe"- Dortmund - 1-2-3 June 1993

2) **L. Torchane, Ph. Bilger, J. Dulcy, M. Gantois** - *"Application of a mathematical model of iron nitride layer growth during gas phase nitriding"* - Second ASM "Heat Treatment and Surface Engineering Conference in Europe"- Dortmund - 1-2-3 June 1993

3) **V. Woimbée, J. Dulcy, M. Gantois** -*"Over-carburizing kinetics and microstructure of Z38CVD5.3 tool steel"* - Second ASM "Heat Treatment and Surface Engineering Conference in Europe" - Dortmund - 1-2-3 June 1993

II. HEAT TREATMENT OF FERROUS ALLOYS

Materials Science Forum Vols. 163-165 (1994) pp. 51-62

HEAT TREATMENT IN HOT STRIP ROLLING OF STEEL

Th.M. Hoogendoorn and A.J. van den Hoogen

HOOGOVENS GROEP BV, Applied Physical Metallurgy Dept.,
IJmuiden, The Netherlands

ABSTRACT

Heat treatment aims at optimal properties by control of metallurgical mechanisms during a controlled temperature-time cycle. This can be combined with deformation during this cycle.

During hot rolling of steel in a hot strip mill all the metallurgical aspects play an important role, often in such a way that they even influence the control of the process of rolling and cooling itself.

The installation is designed for bulk production. Due to market development however the specifications of the final product change very frequently. This makes control of process and material properties very difficult.

To obtain a good process control all the metallurgical processes have to be known and quantified during the specific rolling cycle, including the interactions with the process.

To solve this problem we have chosen at Hoogovens an approach with physical mathematical models, where the metallurgical models are integrated with the process models. In such a way a practical and economical solution can be found for the production of high quality coils.

This approach will be dealt with and illustrated with results from rolling experiments in the Hot Strip Mill.

1   INTRODUCTION

Hot strip rolling of steel is a very complicated thermo-mechanical treatment. Slabs are heated up to high temperatures, then rolled during cooling in the austenite. After that the steel is water cooled to the coiling temperature. In most cases the austenite-ferrite transformation takes place in this area. Subsequently the coil cools down to room temperature in still air.

This process is designed for bulk production of hot rolled steel coils, the base for further cold rolling and annealing but also for direct application. Bulk production means the production of large amounts with the same material specifications. Market developments however have caused that the lot size - the number of coils which are successively rolled, with the same specifications decreased from about 20 in the sixties to about 1 in the nineties. This has large implications for the way of control. Instead of a control system based on statistically learning and feed back, now a predictive control and feed forward systems are necessary.

To predict what will happen during processing and what settings will give the desired product in terms of dimensions and mechanical properties, a profound knowledge of the process and material behaviour is needed. This knowledge has to be quantified for the actual conditions during the process.

To be able to do so physical-mathematical models were built at Hoogovens which describe the metallurgical mechanisms. These are combined with models which describe the processing conditions, like e.g. temperature evolution. From these off-line models simplified on-line models can be derived for actual control of the mill.

This paper will deal with some of the metallurgical off-line models, and how we make use of them for the design of heat treatment aiming at optimal properties after hot rolling.

2   HOT STRIP MILL AND METALLURGICAL MECHANISMS

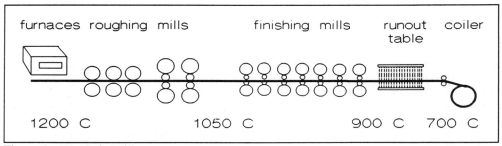

Figure 1 - Hot Strip Mill.

2.1   Furnace

Starting point of the hot strip mill is the slab furnace. In this furnace slabs are reheated. The maximum temperature is limited by the formation of liquid oxides which are difficult to remove before the actual rolling. The minimum temperature is determined by the temperature losses during rolling, as the finish rolling temperature has to stay in the austenitic phase of the steel. This means that thin end gauges require higher furnace temperatures than thicker ones, as they lose more temperature during rolling.

During heating of the slab precipitating elements in solution Al, N, Ti, V, Nb etc. will precipitate completely at first. As the temperature rises high enough the precipitates can dissolve again. Depending on the type and amounts of precipitating elements this dissolution can be complete or incomplete at a certain temperature.

During heating the phase transformation from ferrite

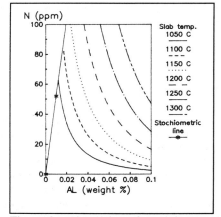

Figure 2 - Dissolution of precipitates.

and carbides into austenite takes place. The austenite grain structure will coarsen at higher temperatures, specially when the precipitates which inhibit grain growth will coarsen and dissolve.

In the soaking section temperature inhomogeneities have to even out and a homogeneous distribution of most of the alloying elements is formed. due to the construction of the furnace in real practice, like way of firing and way of support of the slabs, quite considerable inhomogeneities at the exit of the furnace have to be taken into account. This means in terms of heat treatment, that quite large safety margins have to be maintained.

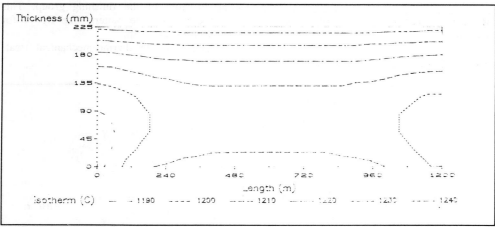

Figure 3 - Slab temperature distribution.

2.2   Rolling

After the exit of the furnace, the slab starts cooling. During this cooling rolling takes place. This rolling is subdivided in two stages: roughing and finish rolling. During roughing the slab is in one stand at a time, while during finish rolling material from one slab is rolled in all the six or seven finishing stands at the same time.

2.2.1   Roughing
During roughing the slab is rolled from 225 mm down to 37 mm in a temperature region from some 100°C under the furnace exit temperature, say 1150°C to 1050°C.

The function of roughing is obtaining a homogeneous grain structure over the slab. This is realized by subsequent vertical deformations of the material followed by recrystallization.

At the same time the slab width can be reduced to the finish rolling width by horizontal reductions.

As the slab is only in one stand at a time, the interpass times are large, and the next deformation takes always place on fully recrystallized material.

In unalloyed and micro alloyed low carbon steels no other relevant metallurgical mechanisms than recrystallization take place.

2.2.2   Finishing
During finishing the material is rolled at Hoogovens in all seven finishing stands at the same time. This means, that, as the material is reduced in thickness in every pass, that the next

deformation has to take place at a higher speed. So rolling speed increases at every pass.
As the end of the slab has to wait longer before it is finish rolled than the front of the slab, the
tail end loses more temperature in waiting than the front end.

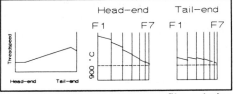

Figure 4 - Temperature profiles during finishing.

To compensate for this temperature inhomogeneity over the length, an extra acceleration during finish rolling is used. So apart from the "normal" acceleration for compensating for the thickness reduction, an extra speed up is used for compensating temperature losses during waiting in front of the finishing group. This leads to a temperature pattern like fig. 4 for head and tail of the strip.

This means that the thermo-mechanical treatment varies over the length of the strip.

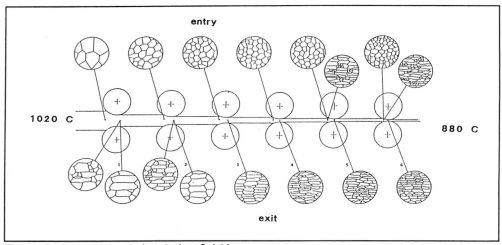

Figure 5 - Structure evolution during finishing.

The reductions in the first or first two finishing stands can give rise to dynamic recrystallization during rolling. In the following passes, this does not occur any more, but only static recrystallization can take place. Depending on steel quality, draft and temperature there can be complete, incomplete or no static recrystallization. See fig. 5. This is of great influence on the deformation resistance. Some precipitations can take place in this stage (Nb, Ti) which inhibit static recrystallization.

This highly complicated processing step gives the incoming material for the next phase in hot strip rolling - the cooling on the run out table.

2.3   Cooling

2.3.1   <u>Run out table cooling</u>
The rolled strip enters the run out table at finish rolling temperature. At this temperature the material is as mentioned earlier still austenitic. the steel can be non-, partly-, or fully recrystallized.

The run out table consists of three parts, as can be seen in fig. 1.

In the first short part there is cooling in air. This part is used for equalizing the temperature through thickness and measuring the finish rolling temperature and the thickness of the strip.

In the second part the steel is water cooled by means of sprays or mostly laminar flow nozzles at the top and bottom side of the strip. The density of the cooling pattern must be selected before rolling. The increase in speed from head end to tail end is compensated by an increase in cooling length, so by using more cooling sections of the same density.

The water cool section is followed by an air cool zone. The strip cools down to the coiling temperature in air. Sometimes a fourth section is used, a so called trimming section, which cools away too large deviations from the coiling temperature. The coiling temperature is considered one of the most important process factors determining mechanical properties.

During the run out table cooling the transformation from austenite to ferrite and carbides takes place. The structure of the hot rolled steel formed here, is one of the main factors determining mechanical properties.

During this transformation heat is released.

Where on the run out table the transformation takes place depends on austenitic grain size, degree of recrystallization of the austenite, steel composition and cooling speed. Very difficult to predict, but very important to know.

The efficiency of the water cooling is determined for a great deal by the surface temperature. The higher this temperature, the greater the cooling efficiency. The evolution of transformation heat at low temperatures costs more water to cool down the steel to the same coiling temperature. To cool a strip of a St41 from finishing temperature to coiling temperature takes about 30-50 % more water, than for cooling the same strip dimensions through the same temperature region than for a St52!

As the time the steel is on the run out table is limited and the temperature drops very fast, precipitations do not play an important role in this area.

2.3.2 Coil cooling

After leaving the run out table the strip is cooled at a temperature which varies normally between about 500°C and 650°C. From this temperature the coil cools to room temperature in still air at a very low cooling rate. during this coil cooling precipitation is an important factor.

As we are concerned about the homogeneity of properties over length and width of the trip it is important to realize that cooling of a coil is a typical batch process with an inhomogeneous character.

The ends and sides of the coil cool in a different way from the bulk. Fig. 6 illustrates this by giving the cooling speeds at different positions in the coil.

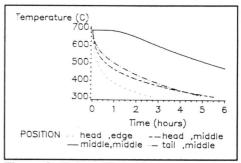

Figure 6 - Coil cooling.

3   INTEGRATED MODEL

From the overview of the successive processes which together form the hot strip rolling process, it will be clear that this is a very complicated process. Each part of it has connections and interaction with the other parts. To be successful has to be controlled very accurately.

As explained before the feed back system with trial and error system and learning from the mistakes is no longer fit for purpose. Market developments together with the increase in coil weight have reduced the number of coils which are rolled successively with the same specifications to a too low number for doing so.

A system based on feed forward with predictive power is needed. However it will never be possible to exclude feed back altogether.

To be able to make use of feed forward control, one has to know in advance what will happen when a specific strip will be rolled under specific conditions.

Influence of all relevant parameters, temperature, strain, strain rate, interpass time has to be quantitatively known for every composition and for every thermo-mechanical history.

At Hoogovens we tried to describe the various processes in the hot strip mill by mathematical models based on physics. As in the hot strip rolling all processes are linked, the various models which describe the processes have to be linked either. So we came to the integrated metallurgical model for the process, which we called CASH, Computer Aided Simulation of Hot rolling.

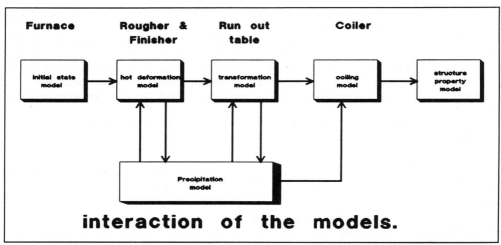

Figure 7 - CASH.

The model has a modular structure and its architecture reflects the process parts of the hot strip mill. It is a so called off-line model for scientific use. In the different modules metallurgical calculations are done for actual processing conditions. This means, that relevant parts of the processing line, e.g. interstand distances, and relevant parts of the processing models, e.g. work roll flattening, are integrated parts of the modules. From this off-line model, simplified versions, specialized on a specific part of the installation can be derived for actual control.

For investigating a specific influence, relevant modules can be selected to calculate the total effect. Irrelevant parts of the process do not need to be taken into account for the calculation. New developments can easily be introduced in the modular structure. The programme is written in Pascal and runs on a PC.

In fig. 7 the general lay out of CASH is given. Above the blocks the most relevant connections to the installation are given.

The first model defines the initial state for the rolling process. The last model gives the

Table I - OVERVIEW OF C A S H (Computer Aided Simulation of Hot rolling).

Model	Part of installation	Specific installations & process settings	Input	Metallurgical aspects	Output
Initial State	Furnace	Lay-out Rougher section for recalculating temperature at the exit of the furnace from Rougher 6 temperature. Reheating Cycle	Chemical composition (Al, N, B, V, Nb, Ti, C). Temperature measurement at Rougher 6	Connected equilibrium equations	Amount of elements in solution from head-end to tail-end of the slab. Grain size
Hot Deformation	Rougher & Finisher	Distance between stands. Elasticity of work rolls. Coefficient of friction per stand	Reduction schedule. Temperature schedule. Initial grain size. Chemical composition. Thread speed. Undeformed roll radius. Width	Hardening. Recovery. Dynamic & static recrystallisation. Grain growth	Deformation resistance. Grain size. Fraction recrystallized. Dislocation density. Slip factor. Roll force
Transformation	Run out table	Position of cooling banks. Length of run out table. Extended formulations of heat transfer of water. Actual flows of cooling banks	For each position of the strip: - Time-speed schedule - Cooling pattern - Flows of the cooling banks - Temperatures Finishing Coiling 3 Extra measurements in wet section - Temperature of the water - Thickness and width - State of austenite after rolling - Chemical composition	Phase transformation	Temperature-time curve at each position (width and thickness) of the strip. Fraction, ferrite, pearlite, bainite and martensite at each position of the strip. Heat released during transformation
Coiling	Coiler	Inner radius of coiler	Length, width & thickness of the strip. Measured coiling temperature (SCOAP) as a function of width & length of the strip. Crown. Distance between wraps. Room temperature. Time of contact with coiler. Heat exchange factor with coiler. Cooling time. Fraction transformed		Temperature-time curves at each place of the coil
Structure & Properties	Product		Mean transformation temperature. Chemical composition		Mechanical properties
Precipitation	Finisher Run out table Coiler		Amount of elements in solution	Interaction between precipitation and recrystallisation during deformation, and Interaction between precipitation and transformation is under development. AlN precipitation during cooling at the coil	Fraction precipitated. Coherency

structure and the mechanical properties of the hot rolled product. The models in between are arranged according the processing sequence. The precipitation model runs in parallel with this line. These phenomena can play an important role in nearly all the other models, so this is calculated separately. The connections/interactions of the modules are indicated with the arrows.

The initial state model takes chemical composition and temperature of the slabs as input. It gives grain sizes as output and the state of precipitation, the amount of precipitating elements in solid solution, based on mainly thermo-dynamic calculations.

The initial state is an input to the next model, the hot deformation model. This block has a close connection with the roughers and the finishing train in the mill. It will be clear that the rolling programme and the temperature during rolling are also used as input.

Based on the production and annihilation of dislocations, strengthening by the rolling deformation and softening by recrystallization, and grain growth are quantified.

The average state of the grains in a certain roll gap gives the deformation resistance. From this the roll force can be calculated. The model gives the state of the austenite prior to transformation as output. The state of precipitations is calculated again by the precipitation model. In case of interactions with the recrystallization the precipitation results can be fed back to the hot deformation model.

The transformation model has a close relationship with the run out table, where the hot strip is water cooled from the final rolling temperature to the coiling temperature. Based on state of the austenite, composition, cooling pattern and thickness, the transformation is calculated together with the transformation heat and the resulting structures quantified. The precipitation model can if necessary calculate the precipitation.

The coiling model uses this state of the ferrite together with the coiling temperature, and the cooling pattern over the coil as input and calculates together with the precipitation model the precipitations and eventually the precipitation hardening over the length and the width of the strip.

All structure aspects will be combined in the structure-property model which will give the mechanical properties for every part of the strip.

Since all this must seem very complicated in the next chapter some illustrations of the actual use of CASH are given. Table I is a summary of the various components of the CASH model.

4   EXAMPLES

The aim of heat treating is influencing structure and properties of the material. This process has to take place well controlled, so that in a reproducible way material with the desired properties can be produced. To demonstrate how the use of CASH can transform the normal hot strip rolling into a real heat treatment two examples are chosen:

1   The development of a quality without nitrogen in solid solution.
2   Phase transformation on the run-out table.

4.1   The development of a quality without nitrogen in solid solution

Many hot rolled coils are further processed by cold rolling and annealing. For deep drawing steels a soft quality with a favourable texture is demanded.

These properties are acquired after annealing. For the annealing two basic production principles are available. The oldest one is Batch Annealing. In this process complete coils are placed in a furnace.

During the annealing cycle, which takes days, the material recrystallizes.

The other process is Continuous Annealing. The coils are decoiled and the strip proceeds through the annealing furnace and is recoiled again after this. Now the cycle time is counted in minutes.

To produce soft qualities with the right texture the incoming hot strip material has to contain as little nitrogen in solid solution as possible.

To investigate the possibilities to produce this material, given the limitations of our installations, we used CASH extensively.

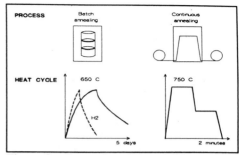

Figure 8 - Comparison BA and CA.

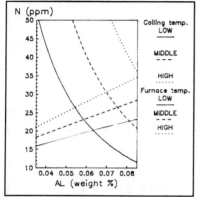

Figure 9 - Dissolution and precipitation.

With the initial state model we calculated the dissolved amounts of aluminium and nitrogen as a function of the slab temperature. Taking into account the maximum allowable aluminium content in term of surface quality, the minimum nitrogen level the steel mill can guarantee, we could determine the effect of lowering slab temperatures on nitrogen in solid solution. With the hot rolling model we calculated the effect of the lower rolling temperatures on the roll forces. In this way we could develop a rolling procedure. During rolling and run-out table cooling no AlN precipitation takes place. This means the nitrogen which is dissolved in the slab furnace, has to precipitate up to the allowable limit during coil cooling. Calculations of the precipitation were done for different compositions and coiling temperatures. The combined effect of slab temperature and coil temperature and composition gives fig. 9, from which we can derive the processing procedure.

We set the maximum allowable limit of nitrogen in solid solution at 5 ppm. A normal coiling temperature straight over the length of the strip gives a free nitrogen distribution like fig. 10. Due to the inhomogeneous cooling of a coil, head, tail and the edges show higher free nitrogen contents then the middle of the strip. By using higher coiling temperatures the rejectable edges diminished in size.

Although the coiling temperature in practice is bound to a maximum, it was possible to reduce the reject area at the edges so far that they fall in the normal side trimming scrap. For the head and tail end this was different. The reject parts here were

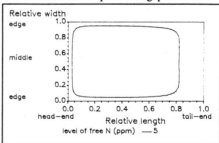

Figure 10 - Free N distribution - normal coiling.

far too large. Far higher coiling temperatures were needed here. This led to the development of a U shaped coiling temperature profile over the length of the strip - see fig. 11. The result for the nitrogen distribution is seen in fig. 12.

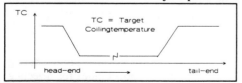

Figure 11 - U shape coiling.

The processing procedure developed in this way proved to be a sound base for regular production.

4.2   Phase transformation on the run-out table

The next example of the use of CASH in heat treatment is still more in the stage of development. The γ/α phase transformation of low carbon steels as we normally produce takes place mostly in the wet section of the run-out table. During this transformation the final structure of the hot rolled material is formed, which determines the properties for a great deal. So control of the transformation is very important. However, the means of control are limited. The final rolling temperature and the coiling temperature are the only measurements available. During transformation heat is released.

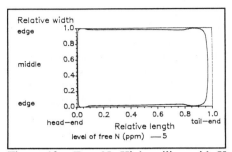

Figure 12 - Free N. High coiling with U shape.

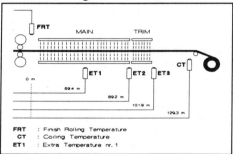

Figure 13 - Extra temperature measurements.

The place and thus the temperature at which this heat of transformation is released is of great influence on the efficiency of the water cooling. The surface temperature is the determining factor. Small variations e.g. in composition can lead to dramatic variations in the amount of water to cool down the steel to its coiling temperature. All this caused by varying transformation temperatures. Although one can try to calculate the transformation, these calculations have to be verified.

First of all we developed extra temperature measurements which were installed in the wet section of the run-out table, see fig. 13. The

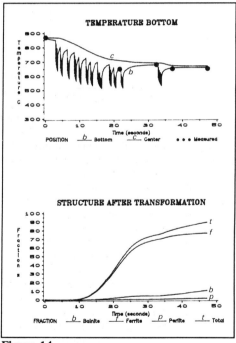

Figure 14
a) Temperature profiles during cooling.
b) Transformation products.

extra temperature measurements give extra information about the temperature profile during cooling, but linked with the transformation model they give also extra possibilities for detecting the transformations on-line, making use of thermal analysis. Fig. 14a gives the calculated temperature profiles - bottom-surface and mid-thickness - of the strip on the run-out table. The

temperature measurements are plotted in the profiles of the bottom side as black dots.
The agreement of calculations and measurements seems quite good. Linking these temperature profiles with the transformation model enables us to calculate the character and amount of the transformation products. Like in fig. 14b this opens many possibilities for improvement of processing conditions and product development.

5   FINAL DISCUSSION

In the previous chapters the Hoogovens' integrated metallurgical model CASH was discussed. The integration means in this case in the first place the integration of metallurgical models and processing models for an installation. The processing models define the conditions for the metallurgical process and contain the relevant information of the installation. In second place integration means the linkage of the several models for the various installation parts. In this way we can speak of an integrated model for our hot strip mill, describing the metallurgical processes from furnace to coiler.
It is an important tool for:
-       product and process development.
-       development of on-line process control models.

Internally, within our research group, we also consider it as a very efficient tool to document our knowledge efficiently, and also a tool which shows us which knowledge we lack, and has to be developed.
By restricting the application to our hot strip mill the model is limited in size and runs on a PC and is easy to run, not only by the people who developed the model. Application on other processes or installations is possible, but then the relevant parameters have to be changed in the model.
The physical base of the models is of prime importance for development work. Only on this base predictions can be made for not yet existing situations.
Another advantage of the limitation of the model to one specific processing line is, that processing data from the installation can be used as input for the model. A great help in the analysis of crisis situations.
The use of the model for solving problems in daily production and analyzing trials can be considered as standard practice. The introduction of the CASH model is a big step forward in the optimisation of the heat treatment of hot strip rolling. This does not mean, the work is finished. The model is not complete yet. Some parts still have to be formulated or extended. Keeping it up to date asks for continuous attention, but already now it contributes considerably to the success of our research. For us only the research results which are applied in practice count.

Materials Science Forum Vols. 163-165 (1994) pp. 63-68
© 1994 Trans Tech Publications, Switzerland

REDISTRIBUTION OF CARBON DURING THE AUSTENITE TO FERRITE TRANSFORMATION IN PURE Fe-C ALLOYS

M. Onink[1], C.M. Brakman[1], J.H. Root[2], F.D. Tichelaar[1],
E.J. Mittemeijer[1] and S. van der Zwaag[1]

[1] Laboratory of Materials Science, Delft University of Technology,
Rotterdamseweg 137, NL-2628 AL Delft, The Netherlands

[2] Neutron and Condensed Matter Science Branch, AECL Research,
Chalk River Laboratories, Chalk River, Ontario, K0J 1J0, Canada

ABSTRACT

Experimental evidence is presented on the carbon redistribution in Fe-C alloys at the austenite/ferrite interface during isothermal transformation in the temperature region 1000-1040 K. Combined optical microscopy, neutron diffraction and dilatometry showed that in the pure Fe-C alloys investigated carbon pile-up at the austenite/ferrite interface could lead to the precipitation of non-equilibrium carbides. This formation of carbides reduces the carbon concentration in the remaining austenite.

1. INTRODUCTION

Cooling of Fe-C alloys, containing 0.02- to 0.8 wt.% C, from a temperature in the austenite (γ) region to a temperature below its A₃ temperature, will lead to the formation of ferrite (α) [1]. Since the solubility of carbon in ferrite is very low, the formation of ferrite not only requires a transformation of the iron lattice from an fcc to a bcc crystal structure but also the diffusion of carbon away from the ferrite nucleus. This results in a carbon enrichment of the remaining austenite. The carbon concentration of the austenite at the interface is usually assumed to be equal to the concentration indicated by the (γ+α)/γ boundary in the Fe-C phase diagram [2,3,4], i.e. the "local equilibrium" concept. The mobility of the transformation interface is assumed to be controlled by the diffusion of carbon in the austenite [5].

Studying the redistribution of carbon during the austenite to ferrite transformation in high purity Fe-C alloys, results inconsistent with the local equilibrium concept were obtained. The observations suggest that in pure Fe-C alloys the carbon concentration in austenite at the interface is not restricted to the concentration given by the (α+γ)/γ phase boundary. The enrichment is such that carbides may precipitate at the interface. The carbide formation leads to a reduction of the average carbon concentration in the remaining austenite and affects the dimensional changes upon phase transformation.

2. EXPERIMENTAL

The iron-carbon alloys were prepared by arc-melting iron under an argon flow while adding pure graphite to obtain the carbon concentration required. The chemical composition after hot rolling was determined using Inductively Coupled Plasma Optical Emission Spectroscopy combined with Atomic Absorption Spectroscopy. The chemical composition of each alloy is given in Table 1. The total impurity level in the alloys is less than 0.02 wt.%, with the exception of the Fe-0.8 C alloy which contains 0.06 wt.% of contaminants.

TABLE 1 Chemical composition of the Fe-C alloys used in this study. All concentrations are given in wt.%. The uncertainty in the carbon concentration is 0.03 wt.%. The total impurity level does not include the carbon concentration. The alloys marked by * were used for neutron diffraction; the alloys marked by # were used for dilatometry.

Name	C	Cr	Cu	Mn	Mo	Ni	P	S	Sn	Total
Fe*	0.01	0.001	0.001	0.0007	0.002	0.003	< 0.001	< 0.005	0.002	< 0.016
Fe - 0.2 C#	0.17	0.001	0.007	0.0007	0.001	0.004	< 0.001	< 0.005	0.001	< 0.015
Fe - 0.3 C*	0.28	0.002	0.001	0.0002	< 0.002	< 0.001	< 0.004	< 0.005	< 0.004	< 0.019
Fe - 0.4 C#	0.36	0.001	0.015	0.0007	0.001	0.003	< 0.001	< 0.005	0.001	< 0.021
Fe - 0.4 C*	0.38	< 0.001	0.001	0.0001	< 0.002	< 0.001	< 0.003	< 0.005	< 0.004	< 0.016
Fe - 0.6 C*#	0.57	0.001	0.002	0.0009	0.002	0.001	< 0.001	< 0.005	0.001	< 0.014
Fe - 0.8 C*	0.80	0.0002	0.006	0.0005	< 0.002	0.003	0.038	< 0.005	0.003	< 0.056

Optical Microscopy was performed using a Jena Neophot 30 microscope with a 100 x objective corrected for oil immersion. Samples were etched for 10 seconds with a 1% Nital (HN03 in ethanol) solution.

Neutron diffraction experiments were performed using a high-throughput neutron diffractometer equipped with a 800 wire detector spanning an arc of 80° 2θ in steps of 0.1° 2θ [6]. The neutron wavelength was 0.15002 nm. The specimens were radiation heated in a vacuum of $3x10^{-8}$ bar using a graphite resistance heater. The specimens were heated to 1250 K and subsequently cooled from 1250 K to 850 K in steps of 20 K. Two diffraction patterns were collected at each temperature. The collection of a single diffraction pattern lasted about 20 minutes.

Dimensional changes during the austenite to ferrite transformation were measured using a Bähr dilatometer, type 805A. The tubular specimens had a length of 10 mm, an outer-diameter of 3 mm, and a wall-thickness of 0.25 mm. The cooling rate during quenching with Helium gas was approximately 200 K·s $^{-1}$. The temperature was measured with a Pt/Pt-10%Rh thermocouple, which was spot-welded to the specimen. Heating of the specimen in a vacuum of $3x10^{-8}$ bar occurred by inductive heating using a high frequency generator. Additional experiments were performed on an MMC dilatometer, showing exactly the same results as measured by the Bähr dilatometer [8].

Samples from the dilatometer experiments were also used for the optical microscopy studies because of their well-defined thermal history. The thermal treatment consisted of an austenitising step (50 K above the A$_3$-temperature for 10 minutes), quenching to the transformation temperature in the α/γ bi-phasic region, isothermal anneal at the transformation temperature, followed by quenching to room temperature. Both fully transformed and partially transformed samples were prepared.

3.   RESULTS AND DISCUSSION
3.1   Optical microscopy
Optical analysis of the microstructures after completed transformations in the bi-phasic region for the Fe-0.2 C, Fe-0.4 C and Fe-0.6 C alloys, showed the phases to be expected: ferrite (transformation product phase) and martensite (former austenite, formed during quenching to room temperature). The fraction of each phase in the fully transformed samples was in reasonable agreement with the phase diagram. However, samples of the Fe-0.4 C alloy isothermally transformed at 1005 K showed unusual precipitates at the majority of the ferrite/martensite interfaces. A typical example is shown in figure 1. The size of the precipitates was about 200 nm, which is close to the theoretical resolution of the optical miroscope. For samples with a higher carbon concentration, or samples transformed at a higher temperature, also signs of troostite were found at the ferrite/martensite interface, see figure 2.
Analysis of the microstructure after etching with Murakami etchant, which is selective for carbides, supports the hypothesis that the precipitates formed are carbides. If so, the presence of these carbides would decrease the carbon concentration of the remaining austenite.

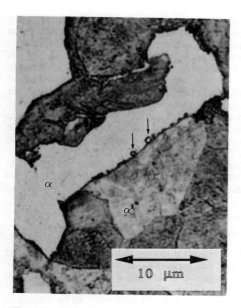

Figure 1. Optical micrograph (Bright Field; Oil-Immersion) of a ferrite(α)/martensite(α') interface marked with precipitates (arrows) (Fe-0.4 C, 1003 K, 30 s.)

Figure 2. Optical micrograph (Bright Field; Oil Immersion) of a ferrite (α) /troostite (t) interface marked with precipitates (arrows) (Fe-0.4 · C, 1025 K, 30 s)

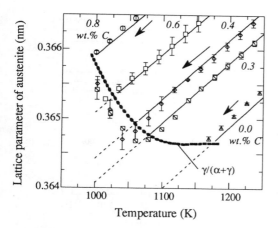

Figure 3 Lattice parameter of austenite as a function of temperature measured using neutron diffraction [6]. The numbers printed in italics indicate the carbon concentrations of the starting alloys. The closed arrows indicate that the measurements are performed on cooling. The heavy dashed line shows the expected lattice parameter of austenite based on the phase diagram.

3.2　Neutron diffraction experiments

The average carbon concentration of the austenite during transformation can be determined from in-situ neutron diffraction experiments, since the lattice parameter of austenite depends on the carbon concentration [6]. The lattice parameter of austenite for the Fe-C alloys is shown in figure 3 as a function of the temperature. For temperatures above the A_3 temperature a linear dependence of the lattice parameter on the temperature is observed. The carbon enrichment of austenite at temperatures in the bi-phasic region results in an increase of the lattice parameter with decreasing temperature. The heavy dashed line shows the temperature dependence of the austenite lattice parameter calculated using the carbon concentrations at the $\gamma/(\alpha+\gamma)$ phase boundary from the literature. Clearly, the lattice parameter of the austenite is smaller than expected. This effect can not be attributed to undercooling in view of the long annealing times at each temperature. To determine the carbon enrichment of the austenite as a function of annealing time, the Fe-0.3 C alloy was austenitised at 1100 K for one hour and cooled to a temperature of 1025 K. At both temperatures the lattice parameter of the austenite was measured as a function of time during the isothermal anneal. Lowering of the temperature to 1025 K resulted not only in an increase of the lattice parameter of the austenite, but also in an increase of the line width in early stages of the transformation (Full Width at Half Maximum=FWHM) as shown in figure 4. The line broadening is attributed to local variations in the carbon concentration in the austenite due to the transformation, with a higher concentration of carbon at the austenite/ferrite interface than in the bulk. The carbon concentration in the bulk of the austenite as calculated from the position of the maximum of the austenite line profile is shown in figure 5. An increase of the carbon concentration is observed during 30 minutes after cooling to 1025 K, attributed to the enrichment of the austenite. The carbon concentration in austenite at the end of the transformation is 0.55 wt.%, which is lower than the carbon concentration of 0.60 wt.% expected on the basis of the phase diagram [1].

To estimate the variation in the carbon concentration across the austenite grains at intermediate stages of the transformation, the broadening of the $\{113\}_\gamma$ line profile was analysed in more detail. Line broadening can be due to instrumental and structural causes. For Gaussian peaks it holds [7]:

$$FWHM_{structural} = \sqrt{(FWHM)^2_{measured} - (FWHM)^2_{instrumental}} \quad (1)$$

In the analysis the lowest recorded value of FWHM (i.e. at 1100 K after 1 h of annealing) is used as a measure of the instrumental line broadening. For each diffraction pattern the structural line broadening is calculated using equation 1. Assuming that the presence of a carbon concentration variation is the only cause for structural line-broadening and that the width of the peak at the base is twice that at half height, the maximum local carbon concentration in austenite can be estimated from the scattering angle $2\theta'=(2\theta\{113\} - FWHM_{structural})$. The results are also shown in figure 5. The maximum local carbon concentration at the start of the transformation is 0.64 wt.% C, whereas 0.60 wt.% C should have been obtained if local equilibrium prevails. The variation in the carbon concentration decreases with increasing transformation time.

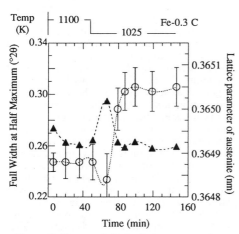

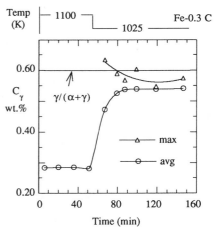

Figure 4 Change of the lattice parameter of a Fe-0.3 C alloy with time during isothermal anneal showing the carbon enrichment of the austenite (circles). Inhomogeneous distribution of the carbon causes an increase in the FWHM of the $\{113\}_\gamma$ reflection at early stages of the enrichment (triangles)

Figure 5 Average carbon concentration in wt.% in the bulk of the austenite (circles) as determined from the lattice parameter. The triangles represent the maximum carbon concentration deduced from the FWHM. The horizontal line indicates the equilibrium carbon concentration given by the Fe-C phase diagram.

3.3 Dimensional changes

The formation of non-equilibrium carbides, as deduced from optical microscopy, and the corresponding decrease in average carbon concentration of the austenite, as deduced from neutron diffraction, will lead to an overall volume change upon transformation different from that calculated for the austenite to ferrite and austenite transformation based on the data from the equilibrium phase diagram. Equation 2 gives the volume change for the decomposition of austenite into ferrite, carbon enriched austenite and an iron carbide, assuming the carbide to be cementite [8]. It is assumed that no carbon is dissolved in the ferrite.

$$\left[\frac{\Delta V}{V}\right]_{\gamma_{\chi}\to\alpha+\gamma_{\chi'}+\theta} = \frac{x_{\alpha}\frac{v_{\alpha}}{2}+\left(\left(1-x_{\alpha}\cdot\left(1+3\cdot f_{\theta}\cdot\frac{\chi}{1-\chi}\right)\right)-\chi\right)\frac{v_{\gamma_{\chi'}}}{4}+\left(3\cdot f_{\theta}\cdot x_{\alpha}\cdot\frac{\chi}{1-\chi}\right)\frac{v_{\theta}}{12}-(1-\chi)\frac{v_{\gamma_{\chi}}}{4}}{(1-\chi)\frac{v_{\gamma_{\chi}}}{4}} \quad (2)$$

In this equation x_{α} is the atom fraction of ferrite as calculated from the phase diagram, (since it is assumed that no carbon is dissolved in the ferrite this means the fraction of Fe atoms in ferrite of all atoms), χ is the atomic fraction of carbon in the starting alloy, v_i/N_i is the unit-cell volume of phase i divided by the number of Fe-atoms, N_i, per unit-cell. f_{θ} is the fraction of the redistributed carbon atoms on transformation that precipitates as cementite.

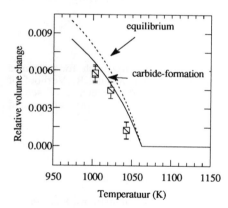

Figure 6 Calculated (lines) and experimentally (squares) determined volume changes as a function of the transformation temperature for the Fe-0.4 C alloy.

The relative volume changes upon transformation from the undercooled austenite to ferrite and carbon enriched austenite as measured for the Fe-0.4 C alloy are shown in figure 6. The dashed line in the figure indicates the calculated volume change as a function of temperature based on equilibrium conditions, i.e. f_{θ} =0 and the carbon concentration in the austenite as given by the phase diagram. The experimental volume changes are clearly smaller than predicted. The solid line indicates the calculated volume change as a function of temperature taking f_{θ} = 0.5 and the carbon concentration in the austenite as estimated from neutron diffraction experiments. Good agreement between the experimental volume changes and the calculated volume changes is observed.

The volume percentage of cementite corresponding to f_{θ} = 0.5 is about 1%, which is in agreement with the optical microscopy data.

It should be pointed out that the observed discrepancy between the experimental volume changes and the calculated volume changes using equilibrium conditions cannot be explained from decarburisation of the sample. Measurements of the carbon concentration on heat-treated samples showed that no loss of carbon had occurred. Furthermore, decarburisation would lead to even larger discrepancies.

Studies on isothermal γ/α transformations in pure Fe-C alloys were also reported in [9] and [10], but in these cases the transformations were interrupted at an early stage. Contrary to our results, no carbide precipitation above the A$_1$ temperature was reported. The carbon concentration at the γ/α interface (measured by Energy Dispersive Spectroscopy in a Transmission Electron Microscope at 10 nm intervals perpendicular to the interface) reported in [10] was shown to be consistent with the local equilibrium model [2]. This discrepancy is attributed to the fact that the local carbon concentration and the formation of the carbide particles depend on the degree of the transformation, or that the carbides were too small to observe at these low degrees of the transformation.

4.   CONCLUSIONS AND FINAL REMARKS
The microsopic, diffraction and dilatometric results obtained in this study are compatible with the following tentative description of the austenite-ferrite transformation above the A_1 temperature in pure Fe-C alloys:
1.   Carbon diffusion in austenite is not sufficiently fast to maintain a homogeneous carbon distribution within the austenite during the transformation process.
2.   Carbon enrichment in the austenite at the ferrite/austenite interface can even exceed the carbon solubility leading to the precipitation of carbides.
3.   These carbides pin the transformation interface.
4.   A (metastable) stage of completed transformation occurs with a carbon concentration of the austenite smaller than that of the $\gamma/(\alpha+\gamma)$ phase boundary in the accepted Fe-C phase diagram.

The non-equilibrium distribution of carbon resulting from the austenite to ferrite transformation observed for the present Fe-C alloys is attributed to a relatively fast migrating ferrite/austenite interface. It is suggested that in less pure steels such interface mobility is lower and carbide formation does not occur.

5.   ACKNOWLEDGEMENTS
The authors are grateful to mr. N.B. Konyer and mr. D. Tennant of AECL Research, for assistance with the neutron experiments, ir. Th. Hoogendoorn, ing. P. Marchal and dr. R. Verhoef of Hoogovens Research, IJmuiden, The Netherlands, for the use of the Bähr dilatometer, dr. Tesch of Thyssen Stahl AG, Duisburg, Germany, for experiments performed on the MMC dilatometer, and mr. P.F. Colijn for assistance with optical microscopy.

6.   REFERENCES
1) Kubaschewski, O.:Iron Binary Phase Diagrams, Springer, Berlin, 1982, 23.
2) Ågren, J.: Acta Metall., 1982, 30, 841.
3) Aaronson, H.I., Enomoto, M., Reynolds Jr., W.T.: Advances in phase transitions, Proc. Int. Symp., ed. J.D. Embury, G.R. Purdy, Pergamon Press, 1988, 20.
4) Enomoto, M.: ISIJ Intern., 1992, 32, 297.
5) Trivedi, R., Pound, G.M.: J. Appl. Phys., 1967, 38, 3569.
6) Onink, M., Brakman, C.M., Tichelaar, F.D., Mittemeijer, E.J., Zwaag, S. v.d., Root, J.H., Konyer, N.B.: submitted for publication in Scripta Metall.Mater.
7) Warren, B.E.: X-Ray diffraction, Addison-Wesley, 1969, p. 258.
8) Onink, M., Brakman, C.M., Tichelaar, F.D., Mittemeijer, E.J., Zwaag, S. v.d.: to be published.
9) Wunderlich, W., Foitzik, A.H., Heuer, A.H.: Ultramicrosc., 1993, 49, 220.
10) Crusius, S., Höglund, L., Knoop, U., Inden, G., Ågren, J.: Z. Metallkde., 1992, 83, 729.

Materials Science Forum Vols. 163-165 (1994) pp. 69-74

THE INFLUENCE OF THERMAL PATH ON THE BEGINNING OF AUSTENITE AND RESIDUAL AUSTENITE TRANSFORMATION

I.A. Wierszyllowski

Department of Metallurgy, Poznan Technical University,
Pl. Sklodowskiej - Curie 5, 60-965 Poznan, Poland

ABSTRACT

The influence of thermal path on transformation kinetics was determined by true isothermal transformation theory [1].This work shows experimental evidence that the theory is true for the beginning of austenite transformations which take place during cooling , and that which take place during heating (residual austenite).

1.0 INTRODUCTION

Determination of influence of the temperature change rate on the way in which the transformation proceeds is important from the theoretical and practical points of view because heating , and cooling to a transformation temperature are essential operations of heat treatment . Such an influence was determined with the help of the true isothermal transformation (T I T) theory [1]. The T I T depends only on the material considered and characterizes the transformation independently of the experiment's condition . The condition necessary to obtain the T I T diagram is an infinitely fast change of temperature from an initial (T_s) to an isothermal - (T_t) temperature.

There are two methods of deriving the T I T diagram , the first on the basis of isothermal experiments and the second on the basis of continuous temperature change experiments . From the T I T diagram one can derive any experimental TTT or CCT diagrams (1). The aim of this work was experimental verification of the theory based on an example of the austenite transformations which proceed during cooling , and based on the example of the residual austenite transformation which proceeds during heating.

2.0 EXPERIMENTAL

2.1 Choice of material

2.1.1 Material for studying austenite transformations during cooling .

In order to study all transformations of austenite during cooling hypoeutectoid steel was selected . In order to decrease the critical cooling rate steel should contain about 1.5 % Mn which in addition decreases a bit the amount of carbon in pearlite. The amount of carbon of about 0.2 % (together with Mn) is sufficient to produce easily detectable volume of pearlite and bainite. Therefore, commercial 510 steel was chosen for experiments.

2.1.2 Material for studying residual austenite transformations during tempering (heating experiments).

In order to study the residual austenite (R.A.) transformation one needs considerable amount of it (the order of 10 to 20 %).

The RA should be stable enough to avoid transformation during relatively slow heating rates and contain a sufficient amount of carbon to produce detectable volume increase during the transformation . 100Cr6 steel (about 1.0 % C and 1.5 %Cr \ ) was chosen for experiments .

2.2 Choice of research methods .

In order to obtain easily controlled and recorded temperature changes during heat treatment of specimens , dilatometry was chosen as the main research method. The method enables fast cooling and heating in a controlled way , and precise recording of length volume changes. Experiments were performed with the use of Theta and Adamel dilatometers. Small , thin tube like specimens were used. Microscopy , microhardness ,and magnetometry were applied as supplementary methods .

2.3 Heat treatment of specimens.

Heat treatment of specimens for studying of the austenite transformations during cooling was the following : the austenitizing time10 min. at 1000 C , and final cooling with constant rates varied from 0.00015 C/sec. to 242.0 C/sec. Austenitizing temperature and time are high enough to produce relatively coarse grains of austenite which is required for incubation time increase . Heat treatment was performed in the dilatometer in the vacuum , high cooling rates were possible with the use of the compressed helium jet . Heat treatment of specimens for studying of the residual austenite transformations during heating was the following : the austenitizing time 15 min. at 1000 C , quenching , afterwards tempering 30 min. at 130 C and finally heating with the constant rates varied from 0.08 C/sec. to 15.0 C/sec. Austenitizing temperature and time were high enough to produce a considerable amount of relatively stable residual austenite . Heat treatment was performed in the dilatometer in the vacuum , specimens were quenched with the use of the compressed helium jet .

3.0 Derivation

3.1 Derivation of TIT diagrams for the beginning of transformations.

The diagrams were derived according to a method based on continuous change of temperature experiments [1]. From this ,one can obtain a relationship between the particular rate C_p , necessary to start the transformation at a certain temperature, and the temperature T. Then , because

$$C_p(T) = f(T) = \int_{T_0}^{T_t} \frac{dT}{t_i(T)} \qquad (1)$$

taking T_t as a variable and differentiating C_p (T) with respect to T_t , one obtains the integrand function $1/t_i$(T) and , finally the necessary true transformation function t_i(T) . The derivations were verified at certain selected temperatures by the method based on isothermal experiments [1] .

3.2 Derivation of experimental TTT diagrams for the beginning of transformations.

The diagrams were derived according to a simplified equation for the actual transformation time t_{sx} [1] , for the constant rate of temperature change (C) to isothermal transformation temperature (T_t) .

$$t_{si}(T_t) = t_i(T_t) + \frac{\Delta T - I(T_t)}{C} \qquad (2)$$

where

$$\Delta T = (T_t - T_s)$$

and

$$I(T_t) = \int_{T_0}^{T_t} \frac{t_x(T_t)}{t_x(T)} dT$$

4.0 Results .
4.1 The austenite transformations during cooling .

As a result of dilatometric experiments with constant cooling rates ,and structure studies (light and scanning microscopy) CCT diagram of 510 steel was produced.The diagram is shown in Fig.1.

CCT diagram of 510 steel

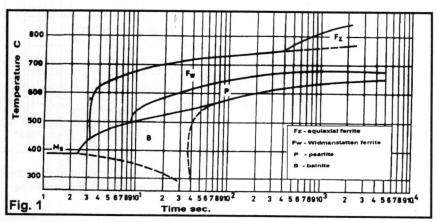

Fig. 1

One can see that at slow cooling rates the austenite transforms to equiaxial ferrite (Fz) and pearlite (P) , at faster cooling the austenite transforms into Widmannstatten ferrite (Fw) pearlite and bainite (B). When cooling rate increases the austenite transforms to bainite and martensite (M)
On the basis of the CCT diagram a TIT diagram for the beginning of the austenite transformation was produced (Fig.2).

TIT diagram of 510 steel

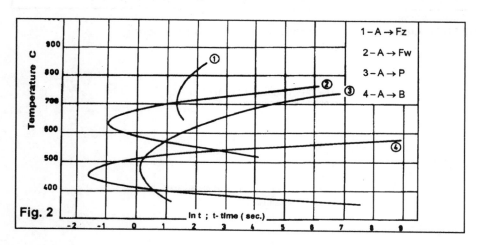

Fig. 2

On the basis of the T I T diagram an actual TTT diagram for the beginning of austenite transformation was derived (Fig.3).

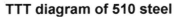

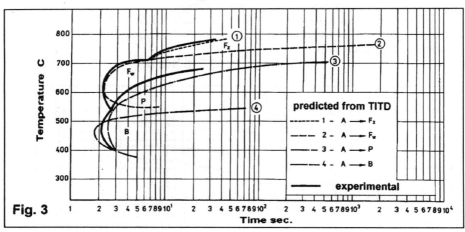

Fig. 3

The T T T diagram was derived for the assumed cooling rates presented in Table 1. The derivation were verified by experiments , experimental cooling rates are shown in Table1. The results are indicated in Fig.3 .

Table 1.

C[C/sec.]	T_i [C]	C[C/sec.]	T_i [C]	C[C/sec.]	T_i [C]
10.0 (10.2)	775	125.0(124.3)	700	264.7(266.0)	550
14.0(14.5)	760	162.5(162.8)	675	277.8(278.5)	500
42.0(42.3)	740	194.5(193.2)	650	343.0(343.8)	450
81.0(80.8)	725	235.5(236.0)	600	328.0(327.2)	410

(C)- experimental cooling rates,

4.2 The residual austenite transformations during tempering .

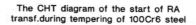

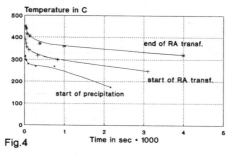

Fig.4

On the basis of dilatometric experiments with constant cooling rates the CHT(continuous heating transformation) diagram of the residual austenite of 100Cr6 steel was produced (Fig.4).

The TIT diagram for the beginning of residual austenite transformation was derived on the basis of the C H T diagram (Fig.5).

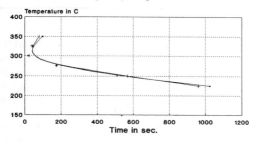

The TIT diagram of the start of RA transf. during tempering of 100Cr6 steel

Fig.5
— Series 1   —+ Series 2
derived from C.T.   derived from I.T.

On the basis of the T I T diagram an actual TTT diagram for the beginning of the residual austenite transformation was derived (Fig.6).

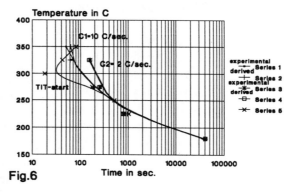

TTT diagrams of the start of RA transf. during tempering of 100Cr6 steel

Fig.6

The diagram was derived for the assumed heating rates shown in Table 2. The derivations were verified by experiments , experimental cooling rates are shown in Table 2 . The results are indicated in Fig.6 .

Table 2.

C2 [C/sec.]	C1 [C/sec.]	T_i [C]	C2 [C/sec.]	C1 [C/sec.]	T_i [C]
-	10.0 (10.3)	350	2.0 (2.1)	10.0 (10.01)	225.05
2.0 (1.96)	10.0 (9.6)	325	2.0 (2.03)	10.0 (9.8)	180.02
2.0 (2.02)	10.0 (10.2)	275	-	-	-

(C)- experimental heating rates,

5.0 Discussion and conclusion .
Experimental and derived TTT diagrams for the beginning of austenite transformations of 510 steel show that the differences are small. Agreement of both the results is really good for the transformation of the austenite in equiaxial and Widmannstatten ferrite.The differences appear for the transformation in pearlite and bainite. In case of pearlite the difference is caused by different accuracy of reading the beginning of the pearlite transformation from dilatometric diagrams. The beginning of the pearlite transformation when structure consist of austenite and ferrite can be read better from dilatometric diagrams of isothermal transformations than from continuous cooling ones. This can be corrected using different settings of the dilatometer. In case of bainite the difference is caused by difficulties with the choice of the best fit curve. These which fitted the best, were meaningless in respect to the nature of the transformation . However, even for the pearlite and bainite transformation the differences between the derived and experimental TTT diagrams are acceptable (see Fig. 3) . The comparison of the experimental and derived TTT diagrams for the beginning of the residual austenite transformation during tempering shows very small difference (see Fig.6). Thus, in conclusion one can say that the methods of derivation of the TIT and the TTT diagrams for established rates of temperature change (thermal path) are correct , and the theory agrees with experimental results .

Literature.
1.Wierszyłłowski, I.A.: Met.Trans. , 1991 , 22A p.993,

Materials Science Forum Vols. 163-165 (1994) pp. 75-80

METALLURGICAL ASPECTS OF MICROSTRUCTURES IN ADVANCED NON-HEAT-TREATED TYPE HIGH STRENGTH LOW-C STEELS

T. Araki[1], K. Shibata[2] and H. Nakajima[3]

[1] Kobe Steel Ltd., R & D, 1-8 Marunouchi, Tokyo 100, Japan

[2] University of Tokyo, Materials Science, Hongo 7-chome, Tokyo 113, Japan

[3] Formerly: National Research Institute for Metals, Nakameguro, Tokyo 153, Japan

ABSTRACT

Activitiy results of the present authors' Bainite Committee of the Iron & Steel Inst. Japan are introduced concerning the researches on microsructural problems of most modern (very) low carbon non-QT type high strength steels for various usages. Invesigating a large number of micrographs of these steels summerized in the book "Atlas for Bainitic Microstructures Vol.1" just published by the said Committee, new tentative classification method, concerned metallurgical disciplines and various influencing factors of the subject steels are described. Thus classified intermediate Zw matrix phases: α°_B (and granular α_B) and αq are explained with examples of composite structures, and also accompanying minor C-enriched film or island shaped constituents, are discussed in respect of aimed applications for line pipe, automobile, ship, carriers etc. and reqired strength levels and properties.

1. INTRODUCTION

The conventional quench-and-tempered type high strength low alloy (HSLA) steels have long been utilized for many applications of over 800 MPa strength level as the most reliable sructural materials when relatively fine and uniform structures (tempered martensite) are ensured. However since recent decade other categories of "non-QT" type low carbon low alloy steels are being developed to meet the better cost performance for the high strength usages by utilizing the novelty technologies such as TMCP (thermo-mechanical control process) and other new controlled cooling techniques [1].

The microstructures of these new steels necessarily consist of low temperature transformed microstructures acccording to required higher strength level. So these structures often involve a vaiety of (and composite of) the <u>intermediate</u> transformation products and <u>martensite</u>, all which phases likely have different level of carbon contents from each other and generally acompany carbon-enriched secondary phases, γ_{retain}, $\alpha_M{}'$ etc. Such complexity existing in these microstructures [1,2,3] is a <u>significant</u> problem for wider application in connection with the practical quality control and further improvement and development of these new steels.

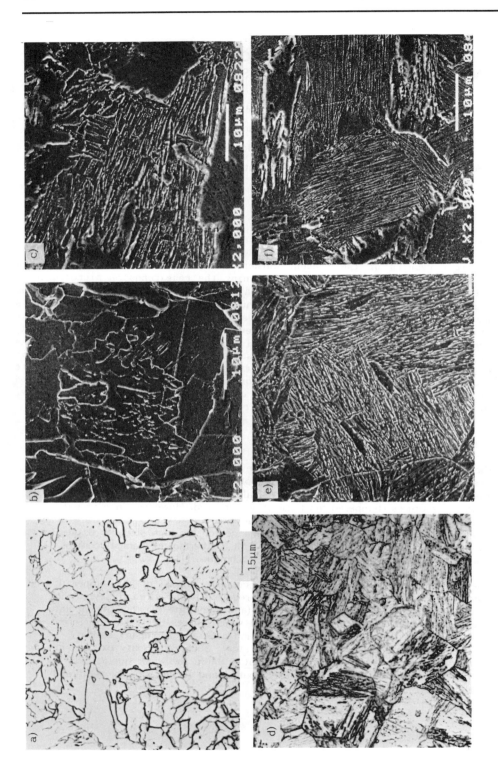

Photo. 1. Optical- a), d) and electron-(SEM) micrographs b), c), e), f) of 3%-Mn HSLA steels; a), b), c), f) =0.004% C; d), e) =0.04% C Cooling rate is: a), b), d), e) =40°C/s; c) =10³°C/s; f) =10⁴°C/s

The Bainite research Committee of the Iron and Steel Institue of Japan have been studying the above stated complicated metallographical problems of low carbon bainitic and similar intermediate microstructures. And as the first fruits of its collaborative study, we have just published an "Atlas book of bainitic microstructures" of low carbon HSLA steels. As the chair and the members of the said Committee and the editors of this Atlas book, the present authors have been analizing and summarizing the results of the concerned investigation and many times discussed on these problems with a number of researchers outside Japan. Hereafter some aspects of the metallography of the continuously cooled transformation structures of the subject steels from a metallurgical view are introduced.

2. PHASE CLASSIFICATION IN ACCORD WITH INFLUENCING FACTORS

The subject steels are, as stated in the Introduction, low-C higher-strength low alloy steels (with some microalloying elsments), which are continuously cooled from recrystallized or unrecrystallized austenite region. Some research notes on the metallografical identification of various bainitic and similar ferritic matrix-phases (abbr.: Zw) were already described in the preceding paper in 1991 [1] of the authors. After then the later investigation results were well illustrated in the "Atlas" book [4] of the continuously cooled microstructures as it has systematically arranged numerous micrographs by optical and electron microscopy with the data of the principal influencing factors.

The subject steels with 0.00x~0.1% carbon content consist of a variety of bcc matrix phases and often in composite state. Some examples are shown in Photo. 1. The steel specimen of a) in Photo. 1 had 0.004% C and showed quasi-polygonal (massive— α like)α as a main phase accompanying some granular Zw-α phase at continuous cooling rate ~40℃/s from 920 ℃. In this case the transformation (stagnant) temperature was ~ 650℃. Meanwhile the microstructure d) of the steel with 0.04% C and boron addition showed mainly bainitic α accompanied by C-enriched film-like MA constituents at the same cooling rate with transformation range around 600~500℃. The steel of a), 0.004%C, also showed a coarse bainitic α at a higher cooling rate up to 10³℃/s as can be seen in c) of the Photo., and furthermore at an extra-high cooling rate 10⁴℃/s, a fine bainitic α with some α' appears as shown in f) of the Photo. Thus the influences of carbon and cooling rate on the microstructure are observed very clearly in this range.

The past research papers studied on these low-C composite bcc phases have been generally using rather simple nomenclatures of classification such as M (α'), F (or polygonal α), and B (or BI) [5] in the metallographical terms and CCT diagrams in the practice.
However in the said Atlas [4], low-C martensite and bainitic/Zw bcc phases, other than polygonal ferrite, can be classified as shown in Table 1, where the intermediate phases are more divided than ever, and additionally the usually accompanying C-enriched secondary minor phases are also explained. Here, the abbreviations α, α'M, B and γ stand for, as usual, ferrite, martensite, bainite, and austenite respectively.
The representative micrographs for each category of this tentatively defined

Table 1   Tentative symbols and nomenclatures of main matrices phases and C-enriched phases observable in continuous-cooled very low carbon HSLA steels

SYMBOL	NOMENCLATURE	Characteristics
I o	Major Matrix-Phases:	--------
αp:	Polygonal Ferrite	equiaxed, polyhedral shaped, mostly recrystallized
αq:	Quasi-polygonal α	irregular changeful shape, formed at lower temperature crossing over γ-grain boundary; mostly recovered.
(αw:)	Widmanstätten α	characteristic lath/plate-like shape; not usual in very low-C steels; mostly recovered.
α_B:	Granular (bainitic) α	granular (bainitic) ferritic Zw structure; dislocated substructure but fairly recovered like "lath-less".
$\alpha^{\circ}{}_B$:	Bainitic Ferrite	sheaf-like with laths but no carbide; conserving the prior γ-grain boundary, generally enough dislocated.
α'm:	Dislocated cubic Martensite	lath type (massive) martensite; highly dislocated; conserving prior γ-gr. boundary.
II o	Minor Secondary Phases (C-enriched and scattered as islands or films):---	
γ_r:	Retained Austenite	highly carbon enriched; often film-like.
MA :	Martensite-Austenite constituent	M and A phases coexist in a crystallographic relationship; with (high) dislocation density.
α'_M:	Martensite	C-enriched; lath (dislocated) type at medium C; twinned type at higher C enrichment.
aTM:	Auto-tempered Martensite	slightly carbon enriched α' tempered; transformation being succeeded by autotempering process.
B :	BⅡ, B2: Upper Bainite	carbon enriched to some extent;
	Bu: Upper Bainite	carbon enriched
	B_L: Lower Bainite	more carbon enriched
P':	Degenerated Pearlite	
P :	Pearlite	eutectoidal reaction product

classification by means of optical and electron microscopy can be seen in the said Atlas [4] in accord with the influencing factors e.g. C-content, Mn and other equivalent alloying elements (Mneq.), cooling rate and the state of the mother austenite γ that is recrystallized (coarse or fine) or hot-deformed and unrecrystallized. The classified matrix-phases can be, in a qualitative way, judged of the approximate position in a map plotting Ceq/Mneq values in latitude and the cooling rates in longitude [4]. The hot-deformed and unrecrystallized γ show a trend of enhanced nucleation of higher temperature transformation products i.e. shifting to $\rightarrow \alpha_B \rightarrow \alpha q$ in comparison with the case of mainly α°_B from recrystallized γ of the same steel.

Otherwise, the significant role of the carbon on such microstructures with respect to the paraequilibrium solubility of carbon in α as well as the partitioning phenomena in local equilibrium and also the strong effects of Nb and boron to these microstructures were already explained elswhere by the authors [1,2,6].

3. STRNGTH LEVEL AIMED FOR APPLICATION AND MICROSTRUCTURE

The aimed applications of the subject new steels may be grouped in the following three:

a) Weldable construction HSLA seels with strength level of more than 700~750 MPa: mainly aimed for: -- line pipe materials (up to X80 grade), : -- hot-rolled plate steels (ship building, civil engineering etc. with higher strength)
b) Machine structural materials for more than 1 GPa strength level: machine parts for carriers and vehicles, (in particular higher level requirement from 1 up to 1.4 GPa with better cost performance for automobile parts in as rolled or as forged state).
c) Highly shapable sheet steels with strenth level over 750 MPa (mainly aimed for automobile).

The microstructure to apply for the above usage a), b), c) can be respectively summarized as the following A), B), C):

C level %	Matrix-phase	strength MPa	and required properties
A) 0.02/0.07	α°_B (main) $-\alpha'm,\ \alpha q,\ \alpha_B$.	700~750	weldability, toughness(low T)
B) 0.06/0.12	$\alpha^\circ_B + \alpha'm$ $-\alpha q,\ aTM$	over 1000 up to 1.4GPa	toughness, fatigue, machinability
C) 0.004/0.01	$\alpha^\circ_B,\ \alpha q$ $-(\alpha'm, \gamma_r,\ M)$	750~800	ductility /superior shapability, (spot welding)

4. ASPECTS OF BCC ZW-STRUCTURE CONCERNING STRENGTHENING

The knowledge of strengthening factors of the aforementioned microstructures was described to some extent in the preceding paper [1] on the basis of the microstructural strengthening mechanisms. Such concept could be applied for the simply classified structures by means of optical microscopy (e.g. [8]).

However, further consideration based on the above stated <u>finer</u> microstructures may approach, even by a step, to the final composite rules to analyze
and predict the various mechnical properties of the composite structures of
these continuous-cooled high strength steels in respect of the foregoing
chapter 3 as the followings:

α q: quasi-polygonal ferrite---consist of low-dislocation (soft) bcc-matrix
with high and <u>low</u> (mixed) angle boundaries. The strength level by single α q
phase: below ~ 700 MPa without other particular strenthening mechanisms even
if the grain size be very fine down to $\sim 2\mu$ m. For the ductility: act as a
favourable factor when composite with harder phases.
Toughness: can act as favourable factor due to its nature of growth crossing
over the prior grain boundary (lessening intergranular fracture mode).

$\alpha^{\circ}{}_{B}$: bainitic ferrite---highly dislocated and lath-packet substructure
similar to α 'm but coarser. Strengthened mainly by dislocation, but little
by low angle boundaries between laths and minor phases. Strength: may exeed
700 MPa. Ductility: moderate. Toughness: excellent when fine lath-
substrcture and minor constituents, but inferior by coarser and a bit
unified lath-formation depending on the transformation temperature/cooling
rate.
α_{B}: granular Zw-α ---granular dislocated α other than sheaf of lath substructure presumably by a different mode from bainitic transformation [7].;
sometimes it seems unified lath by recovering process during and just after
the transformation usually at much slower cooling rate and higher
temperature than $\alpha^{\circ}{}_{B}$.
When compared with $\alpha^{\circ}{}_{B}$, ductility: might be better, but toughness inferior.

In addition to above, it should be noticed the roles of the accompanying C-
enriched minor phases such as MA, γ ,, α '$_{M}$ etc. are also important as
strengthening and influencing factors with respect to their morphology,
properties, transformation behaviours.

REFERENCES

1] T.Araki,M.Enomoto,K.Shibata; Proc. ASM Heattr.Surf.Conf.'91, Materials
   Sci.Forum,(1992), <u>102</u>, p.3
2] T.Araki, K.Shibata, M.Enomoto; Procdgs. ICOMAT-9 Sydney, Mate. Sci. Forum,
   1990, <u>56</u>, p.275
3] D.Edmonds and R.C.Cochrane: Metall. Trans.A, 1990, 21A, p.1527
4] Bainite Research Committee ISIJ; Atlas f. Bainitic Microstructures,Vol.1,
   (Iron. Steel. Inst. Jpn.), 1992, pp.1-165
5] Y.Ohmori,H.Ohtani,T.Kunitake; Trans.ISIJ, 1972, <u>11</u>, p.25
6] T.Araki,M.Enomoto,K.Shibata; Materials Transac. JIM, Jpn Inst. Metal,
   1991, <u>8</u>, p.729
7] S.W.Thompson, D.J.Colvin, G.Krauss; Metall. Trans.A 1990, <u>21</u>, p.1493
8] U.Lotter, H.P.Hougardy; Praktische Metallogra.,(C.Hanser Verlag), 1992,
   <u>29</u>, p.151

Materials Science Forum Vols. 163-165 (1994) pp. 81-86

THE ELECTRON MICROSCOPE ANALYSIS USED TO STUDY THE MARTENSITE MORPHOLOGY IN HIGH STRENGTH LOW ALLOY STEELS

M. Isac

Polytechnic Intitute of Bucharest, P.O. Box 39-111, 73200 Bucharest, Romania

ABSTRACT

The class of HSLA steels have a great development due to the spread of quenched and tempered type steels.

The present paper has in view the research by means of electronic microscope analysis of structural characteristics of a HSLA steel after quenching and tempering.

The martensite-bainite structure is investigated by means of secondary electrons images, transmitted electrons images or by diffraction of electrons on selected areas.

Martensite has an aspect in slats ,it is made up of packets with a high dislocation density. The martensite packets are made up of subgrains with slight differences of orientation among them, having the form of a thin blade with depth of 0.1 μm.

At the tempering over 450 °C begin the loss of carbide coherence and the advanced decomposition of the qunching structure. Between 450 °C and 600 °C in the bainite cristal the ferrite cells become more evident. There appear subgrains with limits at the short angles. Gradually as the subgrains grow in the ferrite poliangled matrix appears limits for the big angles. There starts the matrix recrystallisation. In this stage these can be noticed a powerful growth of the alloy elements precipitates.

The electronic microscope analysis allowed the study of the evolution of morphologycal aspects of the martensite-bainite duplex structure leading to an efficiet use of this type of steel.

EXPERIMENTS AND RESULTS

The class of high strength low alloyed and microalloyed steels, called HSLA, is increasingly developing now; this development is also determined by a large use of quenched and tempered steels.

First of all, for using this type of steels, we have to learn about the connection between their structure and properties, in order to use efficiently the specific strenghtening mechanisms:

- solid solution hardening ,in austenite, in tempered martensite or in ferrite.
- hardening by increasing the density of the defects that consist in hardening by dislocations and hardening by grain boundaries.
- hardening by particles (precipitates) of alloying and microalloying elements in austenite, tempered martensite and ferrite.

The development of modern research means of structure by electonic microscopy, the comprehension of presence and dislocations movements effects during the alloying and microalloying of these steels, represent the necessary conditions of this type of steels development.

This paper displays the structural characteristics of C-Mn-Cr-Mo-B-V-steel (Table 1).

Table 1

The chemical composition of C-Mn-Cr-Mo-B-V steel

Chemical composition (wt. %)										
C	Mn	Si	S	P	Cr	Mo	B	Al	V	Ni
0.2	1.00	0.35	0.019	0.018	0.98	0.41	0.008	0.006	0.006	0.72

The steel structure after quenching in water at 920 °C is made of martensite and inferior bainite.

From a crystalo-morphologic point of view martensite has a slats like aspect (called martensite of slats) with CVC network and containing the packets of great density of dislocation (30-90 cm^{-2}).

The packets are of parallel slats ;each packet contains a matrix of slats with an identical orientation separate only by limits of sharp angles (<20 °) and sets of slats with different orientation, separated from matrix by limits of large angles.

The singular martensite slats-like crystals cannot be noticed by optical microscopy research means, using them, one can just notice packets of 20 µm, in which one can distinguish parallel adjacent strips. By electronic microscopy one can remark the structure of the packets formed by subgrains with a slight difference in orientation,in the shape of their slats from 0.1 µm to a few microns. Beside the slats-shaped martensite, after tempering in water of Cr-Mo-B-V steel there also appears the inferior bainite.

This duplex-structure gives a greater resilience to the steels as a completely martensite structure.

The structural determinations has been made with an electronic JEM C X JEOL microscope with an ASID scanning annex,with the resolution of the 2.04 Å, network, point- point 3.0 Å resolution and possibilities to magnify 70-250,000 times.

We have used samples of 3 mm electrolytically thinned in a 20% percloric acid solution in ethylic alcohol at 30 °C and 30 V working tension in a dynamic regime and 6 V static regime, in order to notice the foil by transmission and at 10-15 V tension, specific for determinations of electronic scanning microscopy.

Fig 1a and b represent secondary electrons images made by primary scanning electrons beam of the areas of steel samples, quenched at 920 °C in water. We have used a tension of accelerating the electrons of 60 kV. These images make evident the duplex inferior bainite-slats-shaped martensite.

Inside the needles of inferior bainite one can notice precipitates of rough carbides oriented at 60 ° according to placket axis cast in a ferritic matrix of dark color (Figure 1). Between the martensitic slats and their interior and on the bainite-martensite interfaces,as well we remark precipitates,thinner that those from bainite composition.

The morphologic type of slats-shaped martensite has been made evident by the transmission electronic microscope on thin foil, in light field,as well as,by images obtained by diffraction of electrons on selected areas (Figure 2).

In Figure 2a and b there are presented images of transmitted electrons magnified 20,000 times, as well as, images of electronic diffraction of selected areas, made on samples of quenched steel at 920 °C in water;these determinations have been made at a 120 kV electrons acceleration tension.

The images obtained with transmitted electrons point out the alternative distribution of the martensitic slats (Figure 2a), as well as, a great density of this type of dislocations specific for this martensite.

The circular images of electronic diffraction made of concentric rings are characteristic for martensite (Figure 2b).

The alloying and microalloying elements from the presented steel influence the kinetics of processes which take place during tempering, prolonging them so that the loose of carbide coherence and the great decomposition of the quenching structures by the formation of equilibrium Fe_3C carbide take place starting from 450 °C when also starting the processes of recrystalization in matrix.

The bainite decomposition also accurs at 450 °C temperature;in bainite cristals there appear cells with density of dislocations, law in the center, high at the edge, as an effect of the matrix polygonization [2].

The rising of the temperature between 450-600 °C, in the bainite crystal, the ferrite cells become more distinct,their dimensions enlarge (there are subgrains, free or almost free of dislocation).

While the subgrains increase themselves it is possible to appear limits at the big-ended in the polygonized ferrite matrix; this means the begining of the first recristalization of matrix.

In Figure 3 there are a secondary electrons image made of the samples of quenched steel at 920 °C and tempered at 700 °C/90'/air, at 100 kV tension of acceleration of electrons.

After tempering at 570 °C one can clearly see the precipitation of alloying and microalloying elements, precipitate oriented and represented by rough carbides distributed on the former austenite grains limits. The tempering at 700 °C points out the coalescence of precipitates, carbides and nitrides of B,V,Al,as well as,chrome carbide (Cr_7C_3, $Cr_{23}C_6$) and molybdenum carbide, $[(FeMo)_{23}C_6]$,[1], [2] .

In Figure 4 there are presented the following images (for the quenched steel at 930C/water and tempered at 700 °C/90'/air):

a - images of secondary electrons,obtained by scanning at accelerating tension of 100kV.

b - images of transmitted electrons at accelerating tension of 120 kV

c - images of electronic diffraction on a selected area which emphasize the ferritic phase representing the matrix, respectively $Cr_{23}C_6$ carbide.

After tempering at 700 °C, the coalescence phenomenon of carbides and recrystalization of ferrite are made evident (Figure 5a).

The image of trasmitted electrons (Figure 5b) shows a great density of precipitates distributed in ferritic matrix.

In the image of electrons diffraction, obtained on a selected area,with circular aspect, corresponding to ferrite, we have identified position in the image of solid solution diffraction due to the $M_{23}(B,C)_6$ carbide (Figure 5c).

The electronic microscope analysis permits the minute study of the martensite and bainite morphologic type in the C-Mn-Cr-Mo-B-V quenched steel structure, the study of sorbite structure obtained by tempering and the following of the precipitation processes of alloying and microalloying elements in bainite-martensitic or ferritic matrix.

BIBLIOGRAPHY

[1] ISAC M., Doctoral paper, Pol.Inst.of Bucharest, 1986.

[2] TILKIN M.A.,Structural Steel Structure, Metallurghia, Moskva, 1983.

[3] NISHIYAMA Z., Martensitic Transformation, Academic Press, New-York, 1978.

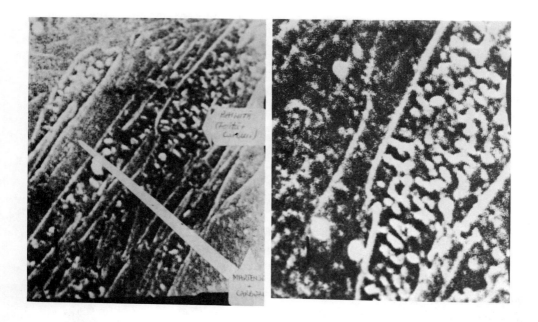

Figure 1   Secondary electrons images made by a primary scanning electrons beam of quenched
samples;

   a) 20,000:1; b) 50,000:1.

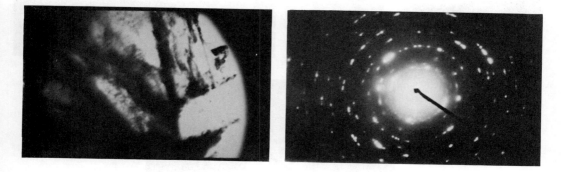

Figure 2 Images of electronic microscopy on thin foils
   a) image obtained by transmitted electrons,
   b) images obtained by diffraction of electrons.

Figure 3 Image of scanning
electron microscopy
magnified 10,000:1.

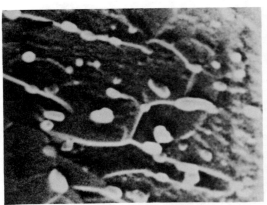

Figure 4 Images of electron microscopy obtained by:
a) scanning; 30,000:1
b)transmission; 30,000:1
c)diffraction of electrons

Materials Science Forum Vols. 163-165 (1994) pp. 87-92

THE ACTIVITY OF CARBON IN THE TWO-PHASE FIELDS OF THE Fe-Cr-C, Fe-Mn-C AND Fe-Si-C ALLOYS AT 1173 K

M. Przylecka, M. Kulka and W. Gestwa

Technical University of Poznan, Poland

ABSTRACT

The effects of various alloying elements on the activity of carbon in austenite have been studied by a number of investigators. The influence of chromium, manganese and silicon on the activity of carbon in alloyed austenite has been presented in the papers [1,2]. These observations included a number of compositions in the two-phase fields, γ + carbides. The effect of chromium on the activity of carbon in the two-phase fields at 1273 and 1373 K has been accurately studied [3,4].

Carburized layer formed on machine elements made of ŁH15 grade steel (52100 grade steel according to Standard ASTM-A295-70) considerably increases their functional properties. Its presence increases resistance of these elements to frictional wear and fatigue, hardness and it decreases friction coefficient [5]. The best functional properties of layer are connected with appearance in its surface zone of carbon within from 1.8 to 2.0 pct C.

In this paper the activity of carbon in the Fe-Cr-C, Fe-Mn-C and Fe-Si-C alloys has been studied by equilibration with controlled carburizing atmospheres at 1173 K and for composition up to about 1.55 pct Cr, 1.7 pct Mn and 1.53 pct Si. Research shows, that activity of carbon is diminished by chromium and manganese and is increased by silicon. The results in the austenite field are in agreement with the papers [1,2]. In two-phase fields of the Fe-Cr-C and Fe-Cr-Mn-Si-C alloys activity of carbon is given by new experimental equations according to carbon content in alloy and carbon potential of atmosphere. This relation makes possible controlled carburizing of ŁH15 grade bearing steel.

EXPERIMENTAL PROCEDURE

The carbon activity of the Fe-Cr-C, Fe-Mn-C and Fe-Si-C alloys has been studied by equilibration with controlled carburizing atmospheres at 1173 K and for composition up to about 1.55 pct Cr, 1.7 pct Mn and 1.53 pct Si. The atmosphere of fixed compositions

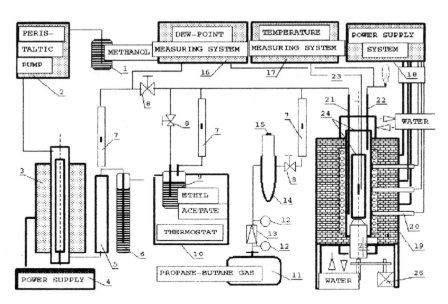

Fig.1 The arrangements used by experiments. 1-methanol; 2-peristal-
tic miniflow pump; 3-generator of atmosphare; 4-power supply system; 5-water cooler; 6-manostat; 7-rotameter; 8-valve; 9-ethyl acetate; 10-thermostat; 11-propane-butane gas; 12-manometer; 13-pres-
sure-reducing valve; 14-U-tube manometer; 15-ruff; 16-dew-point
measuring system; 17-temperature measuring system; 18-electronic
power supply system; 19-carburizing furnace; 20-regulating thermoelement; 21-gas input; 22-gas output; 23-measuring thermoelement;
24-internal and external quartz tubes; 25-ventilator; 26-motor of
ventilator

(cracked methanol with propane-butane gas or with ethyl acetate)
was used. This gas mixture doesn't protect against internal
oxidation, but carbon transfer coefficient and secondary carbon
availability of this atmosphere obtain high values. The
arrangements are shown in figure 1. Specimens were shaped in foils
of about 60 by 10 by 0.05 mm. Two specimens: pure iron and Fe-Cr,
Fe-Mn or Fe-Si alloy were put into the quartz tube and were
carburized up to obtain equilibrium with atmosphere. The final
carbon content of specimens was determined by chemical analysis.
Carbon content in pure Fe-C alloy corresponds to carbon potential
of atmosphere. The same gas mixture was used in the research of
carbon profile of carburized layer formed on LH15 steel. The
specimens in the form of roller were carburized at 1173 K. Three
fixed times of carburizing (2, 5 and 15 hours) were used. Carbon
contents were determined by chemical analysis of successive layers.

RESULTS AND DISCUSSION
      The influence of carbon potential (carbon content in pure Fe-C
alloy) on the carbon content in Fe-Cr-C, Fe-Mn-C or Fe-Si-C
specimens by equilibrium with carburizing atmosphere is shown in
figure 2. Experimental results are in good agreement with the
following equations:

- for Fe-Cr-C alloys:

$$C_{FeCr} = C_{Fe} \left[1 + \frac{Cr}{10} + \{ 1 - exp[-(C_{Fe} - A)^{10}] \} \right] \qquad (1)$$

- for Fe-Mn-C alloys:

$$C_{FeMn} = C_{Fe} \left(1 + \frac{Mn}{30.3} \right) \qquad (2)$$

- for Fe-Si-C alloys:

$$C_{FeSi} = C_{Fe} \left(1 - \frac{Si}{8.38} \right) \qquad (3)$$

where:

C_{Fe} - carbon potential (carbon content in pure Fe-C alloy);
C_{FeCr} - carbon content in Fe-Cr-C alloy [wt pct];
C_{FeMn} - carbon content in Fe-Mn-C alloy [wt pct];
C_{FeSi} - carbon content in Fe-Si-C alloy [wt pct];
Cr,Mn,Si - respectively: chromium, manganese and silicon
        content [wt pct];
A - experimental value depended on chromium content:
        A = 2.4942 exp(-1.6874 Cr)

These expressions are presented in comparison with results of papers [1,2,3]. In these papers alloying factors are given by equations (for austenite of the Fe-Cr-Mn-Si-C alloys):
- by Sauer, Lucas and Grabke [6]:

$$log \frac{C_{Fe}}{C_{FeCrMnSi}} = - 0.04\%Cr - 0.01\%Mn + 0.075\%Si \qquad (4)$$

- by Neumann and Person [7]:

$$log \frac{C_{Fe}}{C_{FeCrMnSi}} = - 0.057\%Cr - 0.016\%Mn + 0.062\%Si \qquad (5)$$

- by Gunnarson [8]:

$$log \frac{C_{Fe}}{C_{FeCrMnSi}} = - 0.04\%Cr - 0.013\%Mn + 0.055\%Si \qquad (6)$$

The activity of carbon was calculated from the carbon content of the pure iron specimen using the following equation [9]:

$$log\ a_C = \frac{2300}{T} - 0.92 + \frac{3860}{T} Y_C + log\ Z_C \qquad (7)$$

where:

$$Y_C = \frac{n_C}{n_{Fe}} \ ; \ Z_C = \frac{n_C}{n_{Fe} - n_C}$$

n_C, n_{Fe} - the number of atoms of C and Fe respectively in an
        alloy;
T - temperature [K].

Experimental results of carbon activity in function of carbon content in the alloys used are presented in figure 3 and are in good agreement (for Fe-Cr-C and Fe-Mn-C alloys) with data of papers [1,2] for austenite. In Fe-Si-C alloys the biggest influence of internal oxidation was obserwed.

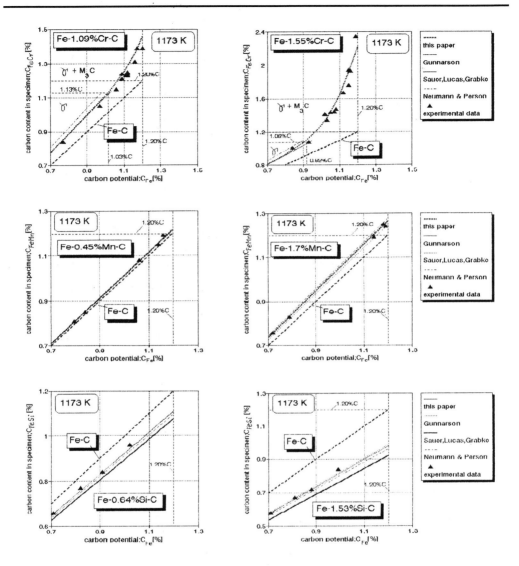

Fig.2 The influence of carbon potential on the carbon content in Fe-Cr-C, Fe-Mn-C or Fe-Si-C specimens by equilibrium with carburizing atmosphere at 1173 K.

Basing on the equations 1, 2 and 3 carbon content in Fe-Cr-Mn-Si-C alloy by equilibrium with carburizing atmosphere may be expressed as:

$$C_{FeCrMnSi} = C_{Fe}\left[1 + \frac{Cr}{10} + \frac{Mn}{30.3} + \frac{Si}{8.38} + \{1 - \exp[-(C_{Fe} - A)^{10}]\}\right]$$

(8)

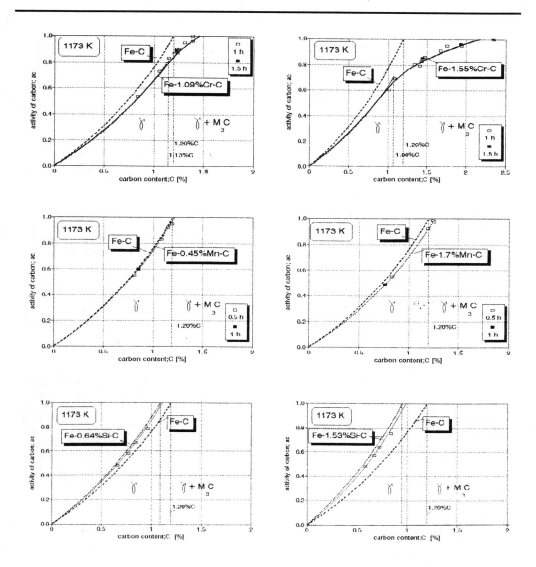

Fig.3 Activity of carbon in the Fe-Cr-C, Fe-Mn-C and Fe-Si-C alloys
    in relation to the carbon content at 1173 K.

Equation 8 may be used for austenite and for two-phase fields
(austenit + cementite). This expression was taken advantage of to
calculate carbon activity of the Fe-Cr-Mn-Si-C alloys and to the
prognosis of the maximal carbon content at the surface in
carburized layer formed on ĹH15 steel (figure 4). The increase of
carburizing time increases carbon content at the surface until
equilibrium with carburizing atmosphere is obtained. The example of
carburizing time influence on the carbon profile in ĹH15 steel is
presented in figure 5 [10]. Carbon potential 1.2 pct C and

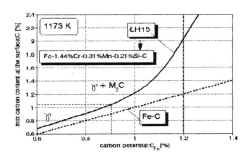

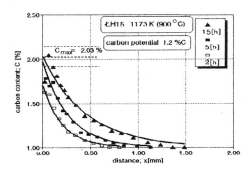

Fig.4 The influence of carbon potential on the maximal carbon content at the surface of carburized ĿH15 steel.

Fig.5 Carbon profiles of carburized layer formed on ĿH15 steel at 1173 K (fixed carburizing times). Carbon potential 1.2 pct C.

temperature 1173 K were used. The maximal carbon content at the surface (2.03 pct C) was calculated by equation 8. In figure 5 the influence of carbon potential changes (from 1.18 to 1.22 pct C) on the maximal carbon content at the surface is shown (interrupted lines).

CONCLUSION

In this paper the equation 8 to the calculations of carbon activity of the Fe-Cr-Mn-Si-C alloys in austenite and in two-phase fields at 1173 K, has been determined. This expression was taken advantage of to the prognosis of the maximal carbon content at the surface in carburized layer formed on ĿH15 steel after controlled carburizing. The results of carbon profile research in carburized layer formed on ĿH15 steel are in good agreement with maximal carbon content at the surface calculated by this equation. Carbon concentration at the surface obtained by carbon potential 1.2 pct C corresponds to the best functional properties of carburized ĿH15 steel.

REFERENCES

1) Wada T., Wada H., Elliott J.F., Chipman J., Met.Trans.,1972,3, 1657-1662
2) Wada T., Wada H., Elliott J.F., Chipman J., Met.Trans.,1972,3, 2865-2872
3) Bungardt K., Preisendanz H., Lehnert G., Archiv für das Eisenhüttenwesen, 1964, 35, 999-1007
4) Brandis H., Preisendanz H., Schüler P., Thyssen Edelst. Techn. Ber., 1980, 6, 155-167
5) Przyłęcka M., Technical University of Poznan, Dissertation, 1988, Nr 202 (ISBN 0551-6528)
6) Sauer K.-H., Lucas M., Grabke H.J., Härterei-Techn. Mitt., 1988, 43, 45-53
7) Neumann F., Person B., Härterei-Techn. Mitt., 1968, 23, 296-308
8) Gunnarson S., Härterei-Techn. Mitt., 1967, 22, 293-295
9) Ban-ya S., Elliott J.F., Chipman J., Met.Trans., 1970, 1, 1313
10) Kulka M., doctor's thesis, Technical University of Poznań, 1993

Materials Science Forum Vols. 163-165 (1994) pp. 93-98
© *1994 Trans Tech Publications, Switzerland*

THERMAL TREATMENT OF DEFORMABLE WHITE IRONS HAVING PLASTICITY INDUCED BY CARBIDE TRANSFORMATION (PICT-IRONS)

P.F. Nizhnikovskaja

DMK Tek, Inc., 717 E. Huron, Ann Arbor, MI 48104, USA
and
Dnepropetrovsk Metallurgical Institute, 4 Prospekt Gagarina,
Dnepropetrovsk 320635, Ukraine

ABSTRACT

Thermal treatment of deformable white irons is discussed. Based on investigations into structural and phase changes in ledeburitic white irons alloyed with carbide forming elements, the significance of preliminary annealing for the preparation of the material to forming is shown and conditions of such annealing are suggested. A non-conventional approach to improving the ductility of the brittle cementite phase is put forward that consists in carrying out during metal forming a phase transformation according to the equation $(Fe,M)_3C \Rightarrow MC + Fe_3C + Fe_\gamma$, where M is a carbide forming element.

On forming, the white iron possessing plasticity induced by carbide transformation, or PICT-iron, has a structure consisting of isolated carbide particles uniformly distributed in the metallic matrix. Owing to this structure, on soft annealing the cast iron acquires a hardness low enough for efficient machining, while hardening and tempering bring about a combination of high values of hardness, wear resistance and strength with satisfactory toughness.

INTRODUCTION

Deformable white irons possessing plasticity induced by carbide transformation, or PICT-irons, make a new family of structural/tool materials pioneered in the Ukraine.

Iron containing about 3% C and about 2% alloying elements is melted in an electric furnace and poured to ingots about 1 t in weight; these are annealed and subjected to hammer forging or rolled in a blooming and subsequently in a bar mill. On further annealing, final products are fabricated by machining and then hardened and tempered.

The outstanding fabricability and excellent service properties of PICT-iron such as formability, machinability and a combination of hardness, toughness and wear resistance to a considerable extent depend on thermal treating.

PRELIMINARY THERMAL TREATMENT

Conventional near-eutectic white irons are not suited for plastic forming because their eutectic cementite whose content is in excess of 30% is very brittle.

PICT-iron features phase transformations that take place in the cementite during forming and render it sufficiently plastic.

The eutectic cementite formed in the solidification of PICT-iron is supersaturated with carbide forming elements and is capable of decomposition accompanied by the formation of special carbides of alloying elements [1].

However, for these transformations to occur in forging or rolling on a blooming mill, the ingots should be subjected to a special preliminary thermal treatment. Direct observations of phase transformations and structural changes occurring in the eutectic during heating and high-temperature holding were carried out using heating chambers of JSM-35 scanning electron microscope and JEM-1000-9 transmission electron microscope (both of JEOL). Structural characterization of specimens subjected to annealing over a wide temperature range with holding times from 1 min to 200 h was performed with the use of light microscopy.

Unmixing of cementite is observed in heating of the experimental iron. At relatively low temperatures, down to 500 °C, a structure form which coarsen as the temperature is increased (figure 1).

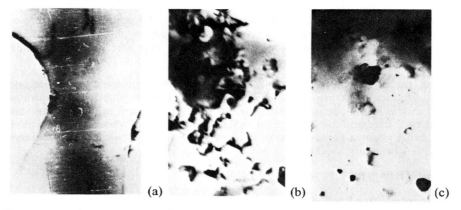

                (a)   (b)   (c)

Figure 1.   Cementite unmixing in heating in JEM-1000-9 microscope:
             (a) cementite before heating, 1200x
             (b) at 550 °C, 8000x
             (c) at 900 °C, 1500x

The electron probe analysis revealed that at temperatures within the range of 850 to 1000 °C V-poor cementite microregions and V and C-rich areas close in composition to $(Fe,V)_2C$ formed in the cementite containing vanadium and chromium. The boundaries between the above areas and the cementite are fuzzy which is an indication of coherent structure of the boundaries.

Any attempt of forming as-cast iron heated to 900-1050 °C failed because even at small amounts of reduction, cracks appeared at the cementite/austenite interface. However, the situation is radically changed when the high-temperature and forming is preceded by annealing with double recrystallization $\gamma \Leftrightarrow \alpha$. An example of such annealing is shown in figure 2.

The initial eutectic austenite has practically no boundaries or subboundaries before

$\gamma \Leftrightarrow \alpha$ transformation has occurred. This, together with the small number of growth defects in the cementite, is a result of a special nature of the crystallization of eutectic grains which are bicrystals [2]. While in single-phase solidification a loss of stability of a planar liquid-solid interface and formation of a cellular interface bring about an increase in defect density, in the coupled growth of the eutectic, as soon as the freezing front of one phase develops a nonplanar topography, the other phase immediately grows into the spaces between the hills and the microtopography becomes fixed.

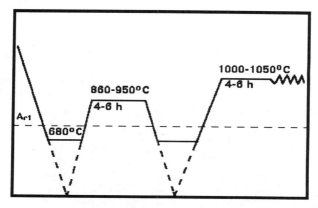

Figure 2.   PICT-iron annealing before forming

The phase recrystallization $\gamma \Leftrightarrow \alpha$ results in the formation of a network of grain boundaries and subboundaries in the austenite. During subsequent high-temperature holding, projections appear at points where a boundary or subboundary contacts austenite-carbide interface, figure 3.

Figure 3.   Formation of carbide projections at a junction of an austenite-carbide interface and a subboundary of austenite. SEM, 6600x

The reason is that a planar interface becomes stable at a junction with a boundary or subboundary. The angle of projection may be estimated if the vector sum of surface tensions is taken equal to zero, whence

$$2 \text{ arc cos } \tau/\tau_{AC} = \theta \quad (1)$$

τ being the tension on the austenite boundary (subboundary) at the projection, τ_{AC} the austenite-carbide interfacial tension.

Eventually, the eutectic cementite-austenite interface which was smooth on the solidification, becomes rough. This surface roughness enhances the effect of stresses external with respect to the cementite that develop in cooling, during phase transformation and in further heating. Local stresses σ_{loc} attain values high enough for dislocations to be generated in the surface areas having projections, that is

$$\sigma_{loc} \sim \sigma\, D/R \qquad (2)$$

σ being the external stress, D the projection height, R the projection tip radius.

Particles of a new carbide phase, e.g. VC, precipitate on the dislocations; the particles are observed near the interface associated with cementite projections, figure 4.

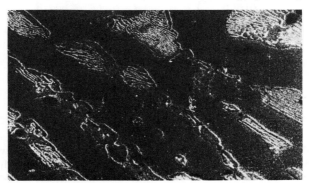

Figure 4.   VC precipitation in a surface layer of eutectic cementite $(Fe,V)_3C$.
             Phase-contrast SEM, 1200x

FORMING

During forming which follows the preliminary annealing, VC grains present in the cementite enhance the effects of stresses external with respect to the cementite and facilitate the formation of new dislocations in it. Further decomposition of cementite superasaturated with carbide forming elements proceeds as an autocatalytic reaction with new VC particles precipitating on the dislocations formed.

Concurrently, VC grains grow and part of the cementite transforms into austenite. These processes are accompanied by the generation of a large number of vacancies and the ductilization of the cementite which forms fibers, figure 5.

As the amount of reduction is built up, the density of dislocations free from VC precipitates increases, and polygonization occurs in the cementite. Cementite deformation is effected by shear along subboundaries. The divided cementite is distributed in the metal matrix.

SUBSEQUENT ANNEALING

On forming and air cooling PICT-iron hardness is about 43 Rc. In order to reduce hardness to 23-27 Rc and thus provide good machinability, bars, rods, tubes, plates or sheets whose diameter (thickness) may vary from 90 to 2 mm are annealed in conditions shown in figure 6. The as-annealed structure contains fine carbides whose distribution in ferrite is fairly uniform.

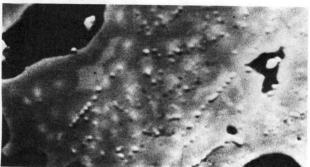

Figure 5.   Formation of VC in cementite during forming.
               Thermal etching, 2000x .

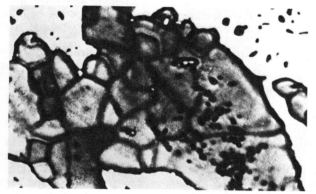

Figure 6.   Structural changes in PICT-iron during forming.
               Electrolytic etching, 1000x

FINAL THERMAL TREATMENT OF PARTS

Having a structure with fine carbides distributed in the metallic matrix, the white iron allows hardening. It has been ascertained that both surface (induction) and bulk hardening is possible. A maximum hardness of 68 Rc is ensured by iron hardening from 860 °C. Fine acicular martensite and about 8% retained austenite are present in the cast iron structure. If hardened from 860 °C and tempered at 150-200 °C, the deformed cast iron has the following properties: Rockwell C hardness 65-67 (figure 8), tensile strength 1300 to 1500 MPa, and impact strength 210-250 kJ/m^2 (unnotched specimens).

APPLICATIONS

The above combination of properties is ideal for cold rolls. PICT-iron rolls were fabricated and tested in rolling of stainless steel band. Work rolls made of PICT-iron had a life 4 times as long as those of ESR-steel with 1% Cr and 1% V.

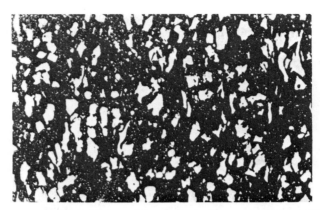

Figure 7.   As-formed PICT-iron (40 mm bar). Thermal etching, 600x

PICT-iron rolls feature high and uniform surface hardness, ensure a high quality of rolled product, have an excellent subsurface strength and offer low sensitivity to surface defects such as cracks, spalling etc. Rolls made of the deformed white iron retain a good hardness during annealing and local heating (62 Rc at 300-350 °C).

Comparison between the structures and properties of Sendzimir mill rolls made of PICT-iron (2.7% C, 0.5% Cr and 1.5% V, carbide content 33%) and D1 and D4 steels (1.2-2.25% C, 12% Cr, 0.7-1.2% Mo, maximum carbide content 22%) suggests that PICT-iron rolls will have longer life in spite of their lower cost.

CONCLUSIONS

The unique capacity to permit forming such as rolling or forging in conventional production equipment is conveyed to white iron containing carbide forming elements dissolved in the cementite by the structural and phase changes taking place in the preliminary annealing and in the forming process per se. A single high-temperature heating before forming brings about the decomposition of the eutectic cementite after a spinodal pattern and does not improve ductility. What facilitates the ductilization of the cementite is the annealing in which double $\gamma \Leftrightarrow \alpha$ transformations and high-temperature holdings ensure that dislocations are generated in the cementite surface layer, and on these dislocations special carbides precipitate which in further deformation intitiate autocatalytic decomposition of the cementite. The structure of the deformed cast iron consists of hard carbide grains uniformly distributed in a ductile metallic matrix. Owing to such a structure, the cast iron acquires a unique combination of mechanical properties and fabricability and so may be used as a comparatively inexpensive tool/structural material.

REFERENCES

1) Nizhnikovskaja, P., Taran, J.N., and Snagovski, L.M.: ISATA 25 Proc., Dedicated Conf. New and Alternative Materials for the Automotive Industries, Florence, Italy 1-5 June, 1992, 185.

2) Nizhnikovskaja, P.F.: Microstructural Design by Solidification Processing, Proc. Symp. Materials Week'92, Chicago, Illinois, USA 1-5 November, 1992, 217.

3) Taran, Ju.N.: Ibid., 193.

Materials Science Forum Vols. 163-165 (1994) pp. 99-106
© 1994 Trans Tech Publications, Switzerland

BORON HARDENABILITY EFFECT IN CASE-HARDENING ZF-STEELS

M. Kocsis Baán

University of Miskolc, Department of Mechanical Technology,
H-3515 Miskolc, Hungary

ABSTRACT

In Hungary research works have been focused on case-hardening boron steels for more than ten years. Within a dissertation - as a part of a research project supported by the Hungarian National Scientific Research Foundation between 1986-1990 - experiments were performed to investigate the influence of thermal history and austenitizing circumstances on the boron hardenability effect. It is well-known that this effect is determined by the amount of soluble boron. Based on thermodynamical equilibrium of Al-B-N system of steels, a computer program was elaborated on IBM PC for calculation of it. In addition, technical literature of boron steels - including more than 200 papers - was overviewed from the investigated respects.

Analysing the reasons of insufficient reproducibility of boron steel production, reliability of the qualification system - especially that of the Jominy testing method - was investigated. The high rate of microstructural inhomogenity causing insufficient control of boron hardenability effect proved to be the a consequence of differences in thermal history, in spite of normalizing applied according to the prescriptions.

Introduction

Boron was first considered as a potential alloying element in steels at the beginning of the century. At the end of 1940s boron was recognized to be by far the most effective agent for hardenability when compared with conventional alloying elements. However, its wide spreading application was delayed for some decades on account of the seemingly unreproducible influence of boron on hardenability, difficulties in detection, as well as a few problems of brittleness and hot shortness attributed to boron, in some cases. By the end of 1970s intensive research activity and progress in methods of materials research led to a far better understanding of the hardenability improving mechanism, providing the ability of producing consistent, high-quality products [1,2].

In Hungary boron steel production is dated from the same time, applying the licence of the German Zahnradfabrik (Friedrichshafen) for producing components of gears. Hungarian steelmakers had to face to reproducibility problems when developing the steelmaking technology in compliance with the specifications of ZF-standard. Partial research projects were started involving the enterprises on behalf of steelmakers and processing machine industry, as well as experts from some departments of Hungarian technical universities. However, these investigations had not been able to result in final and sophisticated solution to the reproducibility problem up to the middle of 1980s. Only the coordinated research project supported by the Hungarian National Scientific Research Foundation between 1986-1990 provided the chance to clarify those effects which influence the reproducibility of boron steel production. Working at the Department of Mechanical Technology at the University of Miskolc, which acted as the coordinator in this research project, my investigations were performed within this framework.

Reproducibility of Jominy-test

For analysing the reasons of experienced insufficient reliability of steelmaking process the reliability of the qualification system itself had to be controlled. Timeliness of such an issue could be proved by similar investigations on the reproducibility of Jominy test for the cases of diminished Jominy band in prescriptions [3]. Performing parallel qualifications on 19 heats of ZF steels at the two enterprises, - measuring 60-80 Jominy curves for each - a great deal of data were obtained and evaluated by mathematical statistical methods. It was concluded that
- for some heats (4 from the 19) the standard deviation was experienced to be low enough, and the probability of meeting the specifications of hardenability test (4 HRC Jominy-band at four distances) proved to be as high as 90% or more.
- In case of 7 heats these values are lower than 70%, while the average probability was found ~75%.
- Considering the allowance of ±1.5 HRC involved in prescriptions, the average probability increases to ~97%, with the minimum value of ~83%.

Based on these results it can be stated that the reproducibility of the hardenability testing method - by itself - does not explain the low reproducibility of production. Obviously, the reproducibility of production could be equal to that of the testing method, if the mean value of the measured feature is equal to the mean value of the prescribed interval. Analysing the differences of these mean values at the prescribed distances from quenched end, tendential deviations were experienced, shown in Fig.1.

The typically high hardness values obtained at the distances of 5 and 10 mm and low hardness at 25 and 50 mm are the consequences of insufficient boron hardenability increasing effect. Theoretical curves show the aimed shape of Jominy curve: increased hardenability effectiveness (B) , combined with lower C-content (C) would result in a profile of better matching. Consequently boron hardenability mechanism in ZF steels should be investigated in order to increase and control its effectiveness. On the other hand, the large variety in the probability values experienced for the different heats, drew attention to the microstructural inhomogenity of the given heats.

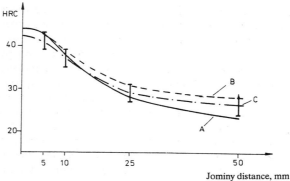

Fig.1: Prescribed Jominy values for ZF7B steel grade
A - Experimental Jominy curve (average for the tested heats)
B - Theoretical Jominy curve for better boron effectiveness
C - as A, but for lower C-content - aimed shape of Jominy curve

Effectiveness of Boron on Hardenability

Numerous theories has been published to explain the exceptionally strong effect of boron on hardenability [4,5]. Nearly all of them focus on the action of boron at high temperature austenite grain boundaries, retarding the nucleation of ferrite. It was found, that appropriate amount of soluble boron (3-5 ppm) segregated to the grain boundary at a certain degree can provide the maximum effect. As boron combined with oxygen and nitrogen is ineffective for hardenability, good deoxidisation and adequate protection against nitrogen by adding nitride-forming elements (Ti, Zr, Al) can be applied in order to maximise and control the boron effect. Though boron combines readily with carbon as well, borocarbides does not cause the loss of hardenability effect, as they can be dissolved at an adequate austenitizing temperature.

In production of ZF steels, addition of strong nitride-forming elements is not allowed, as the primary aim of boron alloying is stated to improve toughness properties due to insoluble boron, combined in nitride. Mentioned as a particular German practice, this effect was not confirmed by other authors [6,7], and the observed higher energy absorption was estimated to be related to the scavenging effect of boron, reducing the oxygen and nitrogen content of the steels. Anyway, hardenability-prescription is a requirement of capital importance, and the diminished Jominy band is reasonable for providing the predictable and controllable distortion of parts during the case hardening process.

In lack of adding Ti or Zr, and provided that the steel is well deoxidized, the ratio of soluble and insoluble boron is determined by the nitride forming processes of Al-B-N system of steels. Theoretical calculation based on the equations of solubility products of AlN and BN, and the mass conservation equations has been published in the literature, illustrating the alloying practice of Japanese steelmakers [8]. Applying increased amount (0.06-0.08%) of aluminium proved to be an appropriate method to provide the required 3...5 ppm soluble boron at the given temperature even if the nitrogen and boron content of the steel vary within the usual concentration interval. Searching for a similar, theoretically founded alloying strategy, a computer program for IBM PC was elaborated for calculations. This program represents the results in 3D figures, offering excellent possibilities for overviewing the nitride forming processes for different combinations of the chemical composition and austenitizing temperatures. As an example, amounts of soluble boron are shown (Fig.2), illustrating the wide range of optimum soluble boron for the Japanese practice and the very narrow interval of that for the ZF prescriptions.

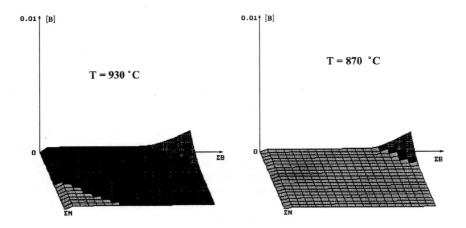

Fig. 2: Ranges of optimum soluble boron
(3...5 ppm, dark field at different austenitizing temperatures)
(Al_a =0.06%, N= 60-120 ppm, B=0-60 ppm)

Based on these thermodynamical calculations, it can be drawn, that the hardenability effect of boron in ZF steels, under these circumstances is very limited. There is no possibility to elaborate a theoretically founded alloying strategy for ZF steels with keeping these prescriptions.

On the other hand these calculations can be used to clarify the order of AlN and BN precipitation in equilibrium cooling conditions, as well as the temperature of different nitride forming processes. For the usual interval of nitrogen contents of steels, Fig.3 shows the threshold temperatures of AlN and BN precipitation.

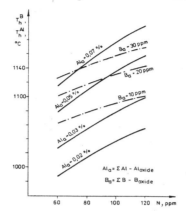

Fig.3: Threshold temperatures of AlN and BN precipitation
under equilibrium cooling conditions

It can be concluded, that these processes are expected to perform within the temperature interval of hot working, so they should be strongly influenced by its parameters. At the same time, it became clear as well, that the prescribed temperature of normalizing is not high enough to dissolve these nitrides, and so to eliminate the effect of thermal history on properties.

Effect of hot working parameters on the hardenability of ZF steels

The significance of the above mentioned problem can be explained by the fact, that the specimens used for the qualification procedure were forged independently from the large mass of the heat. Moreover, very large scattering was experienced for the parameters (e.g. temperature and time of preheating) of these operations. Our suspicion became stronger and stronger, that the low level of reproducibility may characterise only the preparatory operations of specimens and not the heats themselves. Preheating temperature and time have effect on the rate of dissolution of previously formed boron compounds, while the hot working conditions influence the simultaneous processes of recrystallization and nitride precipitation. Considering that the forging operation may result in a far lower consistency of the influencing factors during hot working, it may account for the significant degree of microstructural inhomogenity - experienced in some heats causing lower reproducibility of Jominy-test - as well.

Experiments on the same heat were carried out in order to examine the effect of hot working parameters on boron hardenability. Jominy tests were performed on rolled and forged specimens produced with different parameters and normalised according to the prescriptions (940°C, 1 hour). Effect of increased normalising temperature and/or time was examined, as well, in order to find a reliable previous heat treatment which can eliminate the effect of different thermal history on hardenability, providing consistent results of Jominy tests.

Comparing both the mean values and the standard deviations of the parallel probes, it was found, that
-random variation in hot working conditions may influence the final result of qualification, causing differences of 1.5-2.3 HRC in the mean value,
-standard deviation of Jominy results was the higher, as the preheating time was shorter, the starting temperature of forging was lower,
-prescribed normalising process was not found to be adequate method for eliminating the effect of differences in thermal history on boron hardenability effect, increasing of temperature and/or time of normalising have not led to better reliability of results.

Conclusions

Based on theoretical considerations and experimental results, it was concluded, that the reliability of ZF steel production is significantly influenced by the low reproducibility of the qualification system. Within this, the reproducibility of the testing method was found less critical factor, while the insufficient control of the boron hardenability effect - influenced by the random variations in hot working conditions - seems to be more responsible for the experienced inconsistencies. Considering the difficult interactions of hot working, recrystallization and nitride precipitation processes, elaborating of a more reliable, sophisticated qualification system requires further investigations on the boron effect in ZF steels.

Acknowledgements

This paper is based on studies performed at the Department of Mechanical Technology of the University of Miskolc and the Diósgyőr Metallurgical Stock Corporation, with the financial support of Hungarian National Scientific Research Foundation. The author also wishes to thank her colleagues at both research laboratories for their helpful discussions and practical assistance with this paper.

References

[1] Ph. Maitrepierre, J. Rofes-Vernis, D. Thivellier in "Boron in Steel" (ed. S. K. Banerji, J. E. Morral) TMS/AIME, Warrendale, 1980. 1. p.

[2] P.D. Deely, K.J.A. Kundig: Review of Metallurgical Applications of Boron in Steel, Ferrolegingar A.G., Shieldalloy Corporation, Newfield, 1982.

[3] P.G. Dressel, et. al. : Stahl und Eisen, v. 106,1986. Nr.24, 1354 p.

[4] J.E. Morral, T.B. Cameron in "Boron in Steel" (ed. S. K. Banerji, J. E. Marral) TMS/AIME, Warrendale, 1980. 19.p.

[5] G.M. Pressouyre et al. Proc. Conf. HSLA Steels'85, Beijing, China,1985. 335.p.

[6] D.T. Llewellyn, W.T. Cook : Metals Technology, Dec. 1974, 517 p.

[7] B. M. Kapadia : J. Heat Treating, v.5. No.1, 1987, 41 p.

[8] R. Habu et al. : Trans ISIJ, v.18. 1978. 492 p.

Materials Science Forum Vols. 163-165 (1994) pp. 107-114
© *1994 Trans Tech Publications, Switzerland*

EFFECT OF TEMPERING LEVEL ON MECHANICAL PROPERTIES AND TOUGHNESS OF SPRING STEELS

E. Gariboldi, W. Nicodemi, G. Silva and M. Vedani

Politecnico di Milano, Dipartimento di Meccanica,
P.zza L. Da Vinci 32, I-20133 Milano, Italy

ABSTRACT

the influence of tempering level and of environmental temperature on the mechanical properties of four spring steels was studied. Despite the similar tensile behaviour of the steels tempered at the same hardness level, the toughness properties significantly differed. The importance of a proper material and heat treatment selection is therefore emphasized.

INTRODUCTION

Recent trends towards energy saving, reduction of costs and better performances are forcing components in several kinds of structures to be exploited up to the maximum level allowed by the material strength. A full knowledge of the material properties is therefore desired for proper design of springs. In particular, spring steels are the result of the development of specific chemical compositions and thermal treatments mainly optimized to emphasize the required high strength and elastic behavior. When choosing both the alloy and its thermal treatment, the designer should thus be aware of the complex implications on the finished product.

It is well known that the final tempering treatment, for a given carbon content and alloy composition, is the main factor affecting strength and toughness of the material. Lowering the tempering temperature enhances the tensile and fatigue behaviour but also affects other factors, particularly the material toughness, of importance above all in safety components. Other critical operating conditions for such property include dynamic loading or low temperature applications due to the possible transition to brittle fracture behaviour of ferritic and martensitic steels.

MATERIALS AND EXPERIMENTAL PROCEDURES

A series of alloy steels ranging from conventional spring steels, such as the Cr-V and the Si-Cr grades, to Ni-Cr-Mo structural steels was studied. The aim

of this work was to evaluate the general mechanical properties, as related to
the possible application for spring parts. Table I lists the chemical
compositions of the materials investigated.

Table I. Chemical compositions (wt.%) of the materials investigated.

Material	C	Mn	Si	Ni	Cr	V	Mo	P	S
50CrV4	0.49	0.86	0.31	—	0.97	0.14	—	0.014	0.007
60SiCr8	0.58	0.80	1.92	—	0.25	—	—	0.008	0.005
39NiCrMo3	0.40	0.73	0.25	0.95	0.84	—	0.21	0.021	0.005
40NiCrMo7	0.41	0.78	0.27	1.68	0.80	—	0.22	0.016	0.017

Tensile tests, fracture toughness and Charpy KV-notch impact tests were
performed at different tempering levels, featuring nominal hardness values of
40, 45 and 50 HRC. The latter hardness value was studied only for the
40NiCrMo7 grade to investigate the concurrent effects of high Ni alloying and
of high hardness levels on toughness.

The thermal treatments were carried out according to the parameters summarized
in table II. The materials, in the form of commercial bars of diameter ranging
from 27 mm to 35 mm, were first rough machined to a diameter of 15 mm for the
tensile and KV-notch Charpy specimens, and of 24 mm for the fracture toughness
samples, heat treated in bunches homogeneous in specimen type and material,
and finally finished to the specimen size.

Table II. Summary of the heat treatment parameters.

Material	Austenitizing ($^\circ$C)	Quenching	Tempering ($^\circ$C)		
			40 HRC	45 HRC	50 HRC
50CrV4	890	oil (80°C)	540	460	—
60SiCr8	890	oil (80°C)	575	500	—
39NiCrMo3	890	oil (80°C)	530	440	—
40NiCrMo7	890	oil (80°C)	485	390	300

The second investigation line was the influence of the environmental
temperature on the mechanical behaviour of the steels. Since low temperatures
are commonly encountered in transportation industry applications, these
service conditions were simulated by performing the tensile and fracture
toughness tests both at room temperature and at 0°C, -20°C and -40°C. Further,
Charpy KV-notch impact tests were chosen for the determination of the
transition curves of the different materials in a temperature interval ranging
from -60°C to +120°C.

RESULTS

Tensile Strength

Table III lists the main conventional tensile parameters obtained from the
tests. For each condition, at least two specimens (gauge length: 45 mm
diameter: 8 mm) were pulled to fracture.

Table III. Comparison of the tensile data as a function of hardness and testing temperature. UTS: ultimate tensile strength; YS: yield strength; A: elongation to fracture. For each tempering level the actual specimen hardness is indicated in parenthesis.

Material	T(°C)	UTS (MPa) Experimental	UTS (MPa) Calculated	YS (MPa) Experimental	YS (MPa) Calculated	YS/UTS Experimental	YS/UTS Calculated	A (%) Experimental	A (%) Calculated
50CrV4		(40.7 45.6)	(40) (45) (50)	(40.7 45.6)	(40) (45) (50)	(40.7 45.6)	(40) (45) (50)	(40.7 45.6)	(40) (45) (50)
	+20	1369 1546	1344 1524	1293 1448	1271 1429	0.944 0.937	0.946 0.937	9.2 8.2	9.3 8.3
	0	1372 1582	1342 1556	1312 1493	1286 1471	0.956 0.944	0.958 0.945	9.5 5.8	9.7 5.9
	-20	1406 1580	1381 1559	1334 1481	1313 1463	0.949 0.937	0.951 0.939	9.9 7.0	10.1 7.1
	-40	1422 1633	1392 1607	1344 1526	1318 1504	0.945 0.934	0.947 0.936	9.5 7.1	9.7 7.2
60SiCr8		(41.3 45.0)	(40) (45) (50)	(41.3 45.0)	(40) (45) (50)	(41.3 45.0)			
	+20	1354 1504	1301 1504	1202 1372	1142 1372	0.888 0.912	0.878 0.912	13.1 8.8	13.6 8.8
	0	1375 1556	1311 1556	1214 1409	1145 1409	0.883 0.906	0.873 0.906	12.5 11.0	13.1 11.0
	-20	1370 1515	1319 1515	1219 1401	1155 1401	0.890 0.925	0.876 0.925	13.9 9.0	14.4 9.0
	-40	1408 1633	1329 1633	1254 1478	1175 1478	0.891 0.905	0.884 0.905	10.3 9.1	10.9 9.1
39NiCrMo3		(38.4 44.4)	(40) (45) (50)	(38.4 44.4)	(40) (45) (50)	(38.4 44.4)			
	+20	1257 1524	1328 1551	1180 1396	1238 1418	0.939 0.916	0.932 0.914	14.3 9.9	13.6 9.7
	0	1301 1551	1368 1576	1219 1421	1273 1441	0.937 0.916	0.931 0.914	14.5 9.9	13.8 9.8
	-20	1312 1565	1379 1590	1221 1442	1280 1464	0.931 0.921	0.928 0.921	14.5 9.1	13.8 9.0
	-40	1320 1608	1397 1637	1229 1485	1297 1511	0.931 0.924	0.929 0.923	14.8 9.5	14.0 9.3
40NiCrMo7		(41.7 46.8 49.7)	(40) (45) (50)	(41.7 46.8 49.7)	(40) (45) (50)	(41.7 46.8 49.7)			
	+20	1366 1674 1797	1263 1565 1810	1303 1518 1554	1231 1442 1558	0.954 0.907 0.865	0.975 0.921 0.861	12.0 10.3 10.7	12.9 10.0 10.6
	0	1400 1672 1816	1309 1576 1831	1321 1506 1585	1259 1441 1593	0.944 0.901 0.973	0.962 0.914 0.870	11.5 8.5 10.1	12.5 9.0 10.0
	-20	1404 1702 1820	1305 1597 1832	1325 1548 1601	1251 1469 1606	0.944 0.910 0.880	0.959 0.920 0.877	12.6 10.0 10.7	13.5 10.6 10.6
	-40	1439 1715 1857	1347 1618 1872	1369 1556 1636	1307 1490 1644	0.951 0.907 0.881	0.970 0.921 0.878	12.0 10.3 11.1	12.8 10.9 11.0

As expected [1-3], there was little difference in the ultimate tensile strength (UTS) values of the various steels tempered at the same nominal hardness level. The scatter of some groups of experimental data can be accounted for by the actual hardness of the specimens which had been heat treated in a semi-industrial plant. In an attempt to compare the steels at exactly the same hardness, each tensile parameter was also numerically corrected to offset the slight hardness differences noticed. Corrections of the UTS and YS data were obtained, for each material and testing temperature, by linear interpolation while the elongation to fracture, A, was calculated by supposing the following factor to be a constant.

$$I_A = \left(\frac{UTS}{10} + 11 \right) A$$

Where the above index is a commonly used parameter, empirically determined on the basis of the inverse proportionality existing between UTS and A.

A comparison of the materials on the basis of the yield strength shows that the 50CrV4, the 39NiCrMo3 and the 40NiCrMo7 grades were very similar whereas the 60SiCr8 steel experienced a lower 0.2% yield strength, above all at the 40 HRC level. This observation was also confirmed by the YS/UTS trends. Further, it is worthwhile noting that the latter steel was the only one showing an increasing trend of YS/UTS with increasing hardness.

The major differences between the alloy steels occurred in the elongation to fracture, referred to a gauge length of 5 times the diameter. Table III shows that the lowest values were experienced by the 50CrV4 alloy while the 39NiCrMo3 and the 40NiCrMo7 were the most ductile steels, as expected.

As far as the influence of testing temperature is concerned, all the materials under investigation showed comparable increases in strength when the environmental temperature was lowered. Notwithstanding the uncertainty in evaluating the strength sensitivity on temperature due to the narrow interval examined and the experimental data scatter, it was possible to state that, for a decrease of 60°C the rise in both the UTS and the YS was of the order of 50-100 MPa. It was also apparent that the more the hardness increased the wider became the gap between the strength at -40°C and that at +20°C.

Finally, it is worth reporting that no marked influence on testing temperature on fracture elongation was noticed.

Charpy Impact Toughness

Figure 1 depicts the Charpy KV-notch impact resistance curves of the materials
examined. The average values of the specimen hardness are also reported near
the curves.

The ductile-to-brittle transition of the specimens tested at the nominal
hardness level of 40 HRC lay within, or close to, the examined temperature
range. A marked transition to ductile behaviour could be recognized for the
50CrV4 (70-80°C), the 39NiCrMo3 (10-20°C) and the 40NiCrMo7 (10-20°C) steel
grades whereas the increasing slope of the 60SiCr8 curve suggested that the
transition to a completely ductile fracture behaviour was close to the upper
temperature bound.

Tempering at higher hardness levels (45 and 50 HRC) resulted in the shift
toward higher temperatures of the transition regions. In the examined
temperature range, the impact energy profiles represented therefore the
brittle fracture regions. The impact toughness values, measured for different
materials at these hardness levels, were fairly similar, even though a
slightly improved behaviour could be recognized for the two Ni-containing
steels.

Fracture Toughness

The fracture toughness tests were carried out in the same temperature range of
the tensile tests, namely from -40°C to +20°C. Owing to the limited size of
the starting bars, three point bending specimens with dimension: B=11mm,
W=22mm and S=85mm were used for all the materials, independently from the
expected toughness properties. This allowed the determination of the K_{IC}
values according to the ASTM E399 standard only for the most brittle materials
while, for the other conditions, the J_{IC} integral had to be calculated
according to ASTM E813 standard. The results thus obtained are shown in figure
2. Again, near the trend lines the actual specimen hardness values are given.

The toughness properties allowed to distinguish more easily the influence of
alloying elements and testing temperature at constant hardness levels. The
50CrV4 and the 60SiCr8 grades showed a marked sensitivity to testing
temperature in the examined interval, above all at lower hardness levels. For
these steels a regular determination of the K_{IC} parameter was carried out in
almost all the conditions and temperatures tested. The only exception was the
60SiCr8 steel in the softer condition (which incidentally experienced a low
hardness of 36.5 HRC) at 0°C and at +20°C.

As expected, the Ni-containing steels showed improved toughness properties. A
direct comparison amongst the two grades was rather complex due to the
differences in the actual hardness values. Furthermore, it is to consider that
the choice of fixed hardness levels for the material comparison, forced the
40NiCrMo7 grade to be tempered within the critical embrittling temperature
interval which partially impaired the expected toughness properties. This
phenomenon, well described in literature [4-5] and already noticed in spring
steel applications [1], is usually identified as one-step embrittlement or
350°C embrittlement. It appears as a trough, ranging approximately from 270°C
to 400°C, on a toughness vs. tempering temperature plot.

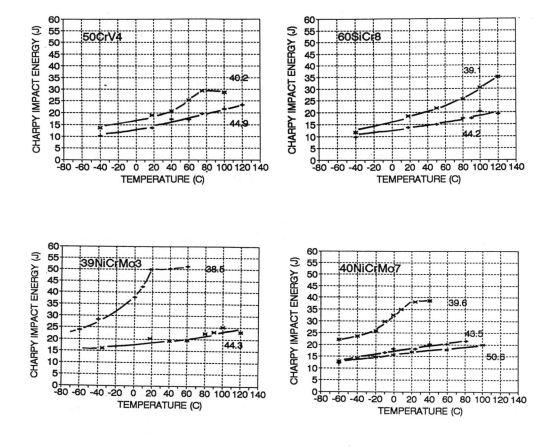

Figure 1. Charpy KV-notch energy vs. temperature curves.

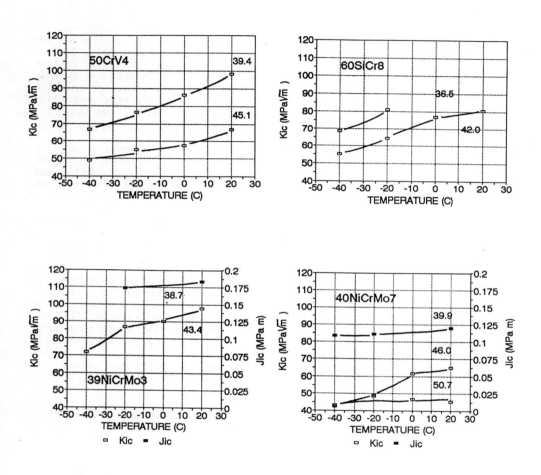

Figure 2. Effect of testing temperature on fracture toughness.

DISCUSSION AND CONCLUDING REMARKS

Literature reports only a limited number of papers dealing with the basic mechanical properties of spring steels [1-3, 6]. These data are indeed of great importance for the design of spring parts, above all at the stage of alloy selection, when a well defined chemistry and thermal treatment must be chosen. In the present paper, it was shown that the tensile properties of the steels, studied at similar hardness levels, were roughly comparable. With a certain approximation, this is also likely to be true for the fatigue strength, due to the strict dependence of fatigue on the tensile resistance, provided the surface finish is good, in order to avoid the influence of notch sensitivity of the different materials. Investigations on fatigue properties of the materials here described are currently in progress and will be presented in a future publication.

It was determined from the data that the resulting low-toughness properties are inevitably associated to a number of material use conditions which must be satisfied to reach high strength levels required by cost-reduction and performance-improvement necessities. Amongst these, the following points can be listed.
- Low tempering temperature, chosen to achieve high strength levels.
- Possibility of falling within the embrittling tempering temperature intervals (270-400 $^\circ$C and 500-580 $^\circ$C).
- Service of springs in low temperature environments which significantly deplete toughness.

The picture of strength and toughness properties is exposed in an attempt to emphasize the great influence of tempering level and of environmental temperature on toughness properties, a feature of primary importance for the design of spring parts according to safety and reliability criteria. The data wish to highlight the subtle effects of increasing strength in spring steels, not always correctly considered. Some non conventional steel grades are also proposed for spring applications on the basis of a criteria devoted to improve toughness properties, still maintaining the strength at acceptable levels.

ACKNOWLEDGEMENTS

The authors would like to thank Dr. R. Galli (ILMA, I-20032 Cormano, Milan) for having encouraged the research and supplied the materials.

REFERENCES

1. Kenneford A.S., Ellis G.C.:J. Iron and Steel Inst., 1950, 3, 265-277.
2. Assepour-Dezfuly M., Brownrigg A.: Metall. Trans., 1989, 20A, 1951-1959.
3. Metha K.K., Kemmer H., Rademacher L.: Thyssen Edelst. Techn. Ber., 1979, 5, 175-180.
4. Olefjord I.: Int. Metal Rev., 1978, 4, 149-163.
5. Briant C.L., Banerji S.K.: Int. Metal Rev., 1978, 4, 164-199.
6. Wanke K.: Draht, 1987, 6, 38, 482-488.

Materials Science Forum Vols. 163-165 (1994) pp. 115-132
© *1994 Trans Tech Publications, Switzerland*

THE RELATIONSHIP BETWEEN PROCESSING, MICROSTRUCTURE AND PROPERTIES IN TLP-50-AZ STEEL FOR OFF-SHORE PLATFORMS

J. Asensio [1], B. Fernández [2], J.I. Verdeja [1] and J.A. Pero-Sanz [3]

[1] ETS Ingenieros de Minas, Universidad de Oviedo, Spain

[2] Departamento de Investigación y Desarrollo,
Laboratorios Centrales, ENSIDESA, Spain

[3] ETS Ingenieros de Minas, Universidad Politécnica de Madrid, Spain

ABSTRACT

The microstructures of a 0.03%Nb controlled rolled steel subjected to different processing conditions have been characterised by optical microscopy and quantitative metallographic techniques to correlate changes in processing with microstructure and mechanical properties.

It is found that heterogeneities in the ferrite grain size distributions increased with the plate thickness and with high finishing temperatures during the thermomechanical treatment. Those factors appear to induce statistical dispersions on the mean linear intercept $\bar{L}_3(\alpha)$.

The discrepancies between the theoretical and experimental mechanical properties of the steel in relation to the thickness and position in the plate are discussed and related to the hot plate mill conditions.

INTRODUCTION

The development of oil resources, with the oil crisis as a stimulus has promoted a number of projects looking for oil resources which due to the present scarcity, have been shifted to cold deep waters. Structures for use in icy sea waters require sufficient safety at operating temperatures of -40°C to -60°C. It is important for the steel plates in ice-service structures to be easy to transport and to repair. The general requirements for these plates could be summarized as follows:

(1) Sufficient strength for withstanding icy sea conditions: *high strength steels and heavy gage thickness.*
(2) *Low temperature toughness* for withstanding loads and impacts at low temperatures specially in the welds.
(3) Easy fabrication and/or repair: *high weldability.*
(4) Possibility of giving complicated shapes to the plate without altering the mechanical properties of the steel: *uniform properties in the thickness direction.*

Steel plates designed for artic applications could be standarized according to its yield point class and its process route. The international standards more commonly accepted are:

· U.K. guidance: Guide for the design and construction of marine structures (off-shore), 1977 (Department of Energy in the United Kingdom).
· DnV (Det Norske Veritas): Regulations for the design, construction and instalation of marine structures, 1977. Norway.
· DnV proposal (Proposals made to the Committee number IIW-IX): consisting

of a set of proposals for structural steels in fixed sea structures, 1981.

The typical requirements and grades according to API 2W standards are summarized in Table I.

GRADE	42	50	50T	60
Yield Strength Ksi. [MPa]				
t ≤ 1 in.[25mm]	42-67 [290-462]	50-75 [345-517]	50-80 [345-552]	60-90 [414-621]
t > 1 in.[25mm]	42-62 [290-427]	50-70 [345-483]	50-75 [345-517]	60-85 [414-586]
Tensile Strength min. Ksi [MPa]	62 [427]	65 [448]	70 [483]	75 [517]
Elongation in 2 in. or 50 mm.- min. (%)	24	23	23	22
Elongation in 8 in. or 200 mm.-min. (%)	20	18	18	16

Table 1.

Niobium controlled rolling steels have a desirable combination of strength and toughness for marine applications. Niobium precipitates in the austenite restricting recrystallization during controlled rolling favouring a pancaked structure which remains unaltered until the transformation and thus ensuring a fine ferrite grain size. It also provides some strengthening by precipitation in the ferrite. The combined hardenability effect of manganese and niobium result in a lower temperature of transformation .

EXPERIMENTAL PROCEDURE

Base Chemistry

It should be noted that is not the purpose of the purchaser to tell the steelmaker how to make the plate for an oil sea platform in particularly aggressive enviroments. Specifications should primary concern to the properties and dimensions in regard to construction and operational factors, and ideally how to reach those should be the steelmaker task.

The composition of the steel used in the present work is listed in table 2 and corresponds to a TLP-50-AZ steel (Tension Leg Platform Steel) cast and controlled rolled in ENSIDESA - Spain, to three final thicknesses of 20 mm, 24 mm and 30 mm according to the processing characteristics listed in Tables 3, 4(a), (b) and (c).

Element	C	Mn	Si	P	S	Al	Cr	Ni	Nb	N_{total}
%	0.096	1.46	0.31	0.007	0.003	0.027	0.021	0.026	0.033	0.006

Table 2. Chemical composition of the TLP-50-AZ steel subjected to study.

In accordance to the API 2W standard the steel must have a tensile strength of 448 MPa or bigger and optimum resistance to lamellar tearing in the z - direction (across the thickness).

Two specimens were cut from the two ends of each plate which in all the text that follows will be referred as FRONT END and TAIL END.

The critical transformation temperatures were calculated as follows indicating that the finishing rolling ended in the austenite region in all the cases.

Transition Temperatures :

$$A_e \ (^\circ C) \ = \ 727-10.7 \cdot Mn-16.9 \cdot Ni+29.1 \cdot Si+16.9 \cdot Cr+290 \cdot As \ = \ \underline{720.34 \ ^\circ C}$$

$$A_{C3} \ = \ 912-203\sqrt{C}-30 \cdot Mn+15.2 \cdot Ni-11 \cdot Cr+44.7 \cdot Si+400 \cdot Al+700 \cdot P \ = \ 1107 \ K \ = \ \underline{834 \ ^\circ C}$$

In order to quantify the weldability it has been calculated the following parameters:

Equivalent Carbon contents :

$$P_{CM} \ = \ C+\frac{Mn+Cu+Cr}{20}+\frac{Si}{30}+\frac{V}{10}+\frac{Mo}{15}+\frac{Ni}{60}+5B \ = \ \underline{0.1808}$$

$$C_{eq}^A \ = \ C+\frac{Mn}{6}+\frac{Si}{24}+\frac{Ni+Cu}{15}+\frac{Cr+Mo}{10} \ = \ \underline{0.386}$$

$$C_{eq}^B \ = \ C+\frac{Mn}{6}+\frac{Cr+Mo+V}{5}+\frac{Cu+Ni}{15} \ = \ \underline{0.3453}$$

The steel composition has low C_{eq} (or P_{CM}) values indicating that the weldability and thus the hardenability of the steel is acceptable.

It should be noted that welding a platform by the construction contractor must be done on a material free from defects, aiming to operate in a safe manner to its predicted life. Numerous factors contribute towards to facilitate this operation in which it should be considered other parameters like the wall thickness and the type of welding process applied.

Figure 1 corresponds to indication of weldability of different HSLA steels and it can be observed that for the Carbon content and the calculated P_{CM} values, it has similar values than the API X-70.

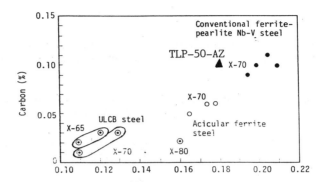

Figure 1.   Indication of weldability for different microalloyed steels.

Rolling Parameters of the Three plates.

(a) Soaking of the continuously cast slabs.

PLATE IDENTIFICATION	SOAKING PROCESS	
	TEMPERATURE (ºC)	TIME (minutes)
t = 20 mm	1213	258
t = 24 mm	1197	260
t = 30 mm	1203	273

Table 3 .- Soaking parameters for the plate.

(b) Rolling parameters.

ROUGHING ROLLING				
PLATE IDENTIFICATION	TEMPERATURE (°C)		INITIAL THICKNESS (mm)	FINAL THICKNESS (mm)
	INITIAL	FINAL		
t = 20 mm	1051	1025	230 .	40
t = 24 mm	1066	1060	230	48
t = 30 mm	1083	1083	230	60

Table 4 (a). Roughing rolling characteristics.

FINISHING ROLLING				
PLATE IDENTIFICATION	TEMPERATURE (°C)		INITIAL THICKNESS (mm)	FINAL THICKNESS (mm)
	INITIAL	FINAL		
t = 20 mm	840	794	40	20
t = 24 mm	835	809	48	24
t = 30 mm	839	825	60	30

Table 4 (b). Finishing rolling characteristics.

ROLLING STAGE	PLATE ID	TEMPERATURE (°C)		TRUE VALUE OF THE DEFORMATION	
		START	FINISH	$\varepsilon'_{PSC} = \ln\dfrac{h_1}{h_0}$	$\varepsilon = \dfrac{2}{\sqrt{3}}\,\varepsilon'_{PSC}$
ROUGHING	t=20mm	1051	1025	(-)1.75	(-)2.02
	t=24mm	1063	1060	(-)1.57	(-)1.81
	t=30mm	1083	1075	(-)1.34	(-)1.55
FINISHING	t=20mm		794		
	t=24mm	840	809	(-)0.69	(-)0.80
	t=30mm		825		

Table 4 (c). Summary of the rolling parameters.

2. **RESULTS**

2.1 Mechanical Testing Characterization:

(a) Tensile and Z-test

The data pertaining to the tensile test of the steel on samples taken from the three plates in both ends are listed in table 5.

SPECIMEN		Yield Strength (MPa)	Maximun Tensile Strength (Mpa)	Yield/Tensile	Elongation in 50 mm (%)	Reduction in area	Z - test σ_M (MPa)
20 mm	FRONT	456	564	0.8085	27	69.9	588
	TAIL	454	562	0.8078	23	63.3	588
24 mm	FRONT	425	552	0.7699	26	67.3	590
	TAIL	421	549	0.7668	26	67.3	587
30 mm	FRONT	431	544	0.7923	27	69.9	585
	TAIL	427	545	0.7835	27	73.5	595

Table 5. Mechanical Characterists of the steel.

(b) Absorbed Energy in a Charpy Test.
The results of experiments carried out at 20ºC, 0ºC, -20ºC, -40ºC, -60ºC and -80ºC in the three samples (both ends) are presented in Figure 2.

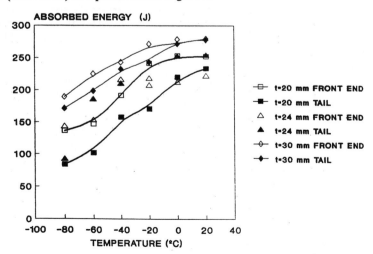

Figure 2. Absorbed Energy in Joules for different temperatures.

2.2. Metallography

2.2.1. Optical Metallography.

Metallographic observations were carried out on the through thickness parallel to the rolling direction on samples taken from the front and the back end in the three plates. They were mechanically polished and chemically etched with nital-2 following the ASTM standards: ASTM E-3-80 and ASTM E-2-62 (1964).

Microscopic observations were done using a NIKON EPIPHOT metallographic bench, and micrographs were taken at 400X and 600X. Figure 3 corresponds to the front and tail ends of the three plates at 400X.

2.2.2. Quantitative Metallographic Determinations.

(a) Volume Fraction of Pearlite.

It was carried out on photographs at 400X following a manual point counting technique in which a grid is overlapped in randomly selected fields until counting a sufficient

number of points of the phase of interest to reach a given relative error $\dfrac{\sigma_x}{\overline{X}}$. All the

results are summarized in Table 6.

SPECIMEN IDENTIFICATION		NUMBER OF FIELDS	TOTAL NUMBER OF POINTS	$V_v \pm CL_{95\%}$	RELATIVE ERROR
t = 20 mm	FRONT	10	138	0.1617 ± 0.0078	0.025
	TAIL	9	151	0.1973 ± 0.0094	0.024
t = 24 mm	FRONT	7	99	0.1655 ± 0.0112	0.035
	TAIL	8	116	0.1698 ± 0.0099	0.030
t = 30 mm	FRONT	12	155	0.1515 ± 0.0762	0.021
	TAIL	10	138	0.1623 ± 0.0783	0.025

Table 6

A semiautomatic image analyzer AMS model 40-10 was also used in combination with a NIKON OPTIPHOT microscope to determine a first estimate of the pearlite volume fraction. Table 7 shows the observed differences between the manual and the semiautomatic determinations. The discrepancies could be attributed to the subjective

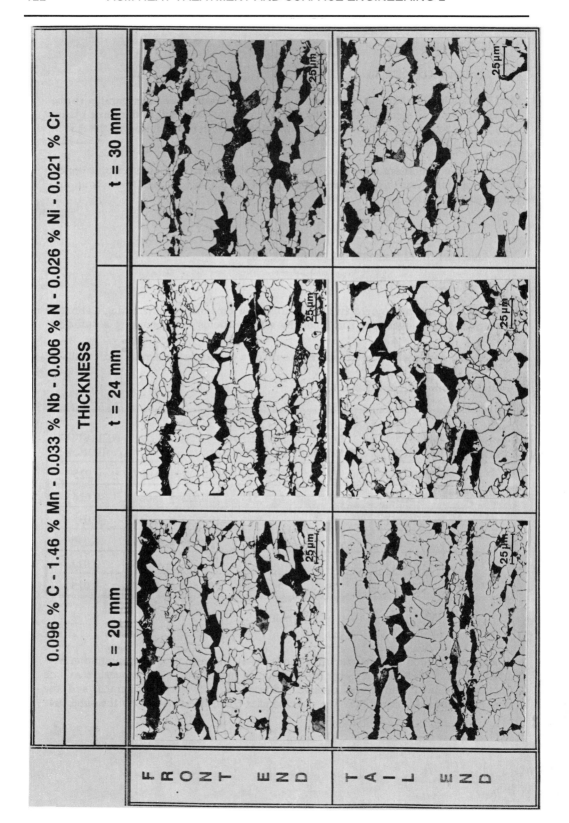

error of the operator as the grey level is controlled manually.

SPECIMEN IDENTIFICATION (thickness and position)		VOLUME FRACTION OF PEARLITE	
		MANUAL (Point Counting)	SEMIAUTOMATIC (Areal Analysis)
t = 20 mm	FRONT END	0.1617 ± 0.0078	0.1381
	TAIL END	0.1973 ± 0.0094	0.1748
t = 24 mm	FRONT END	0.1655 ± 0.0112	0.1630
	TAIL END	0.1698 ± 0.0099	0.1678
t = 30 mm	FRONT END	0.1515 ± 0.0762	0.1640
	TAIL END	0.1623 ± 0.0783	0.1626

Table 7

(b) Grain Size Measurement.

The measurement of the ferrite grain size was carried out by linear analysis counting the total umber of intercepts of test lines on micrographs at 400X with ferrite grain boundaries.

PLATE THICKNESS	POSITION	$N_L \pm CL_{95\%}$ (mm^{-1})	$\bar{L}_3 \pm CL_{95\%}$ (mm)	$G \pm CL_{95\%}$ (ASTM number)
t = 20 mm	FRONT END	123.5±13.52	$(6.78\pm0.74) \cdot 10^{-3}$	$11.05\pm 8.6\cdot10^{-2}$
	TAIL END	102.0± 5.87	$(7.87\pm0.45) \cdot 10^{-3}$	$10.62\pm 9.9\cdot10^{-2}$
t = 24 mm	FRONT END	130.7±17.70	$(6.39\pm0.86) \cdot 10^{-3}$	$11.22\pm 10.0\cdot10^{-2}$
	TAIL END	93.5± 8.84	$(8.88\pm0.84) \cdot 10^{-3}$	$10.27\pm 11.0\cdot10^{-2}$
t = 30 mm	FRONT END	98.9± 9.29	$(8.53\pm0.80) \cdot 10^{-3}$	$10.39\pm 9.6\cdot10^{-2}$
	TAIL END	96.7± 6.81	$(8.66\pm0.61) \cdot 10^{-3}$	$10.34\pm 9.7\cdot10^{-2}$

Table 8

Once it has been determined the total number of intercepts per unit length of test line P_L, it was estimated the number of grains per unit length N_L. Thereafter the mean linear intercept of ferrite L_3 (α) is calculated from the following relation:

$$\overline{L_3}\ (\alpha) = \frac{V_v\ (\alpha)}{N_L\ (\alpha)}$$

The results for the mean linear intercept and for the ASTM grain size with its repective 95% confidence limit are presented in Table 8.

SPECIMEN IDENTIFICATION (thickness and position)		ASTM GRAIN SIZE		
		MANUAL (Linear Analysis)	SEMIAUTOMATIC (Areal Analysis)	
			40 X	60 X
t = 20 mm	FRONT END	11.05	10.33	10.7
	TAIL END	10.622	9.91	10.41
t = 24 mm	FRONT END	11.226	9.90	10.41
	TAIL END	10.275	9.90	9.80
t = 30 mm	FRONT END	10.449	10.110	10.58
	TAIL END	10.347	9.75	10.08

Table 9.   Comparison between ASTM number from manual and semi-automatic determinations.

Semiautomatic image analysis was also carried out for the determination of the ASTM ferritic grain size. The data processor works using an algorithm based on HEYN method for which the grain size is given by:

$$G = \frac{N_A}{\log 2} - 2.954$$

where N_A is the number of grains per mm^2. When the structure is mainly formed of equiaxed grains, then:

$$N_A = \left(\frac{I}{F} \right)^2$$

where I is the total number of intercepts and F the test area subjected to observation. The results of both semiautomatic and manual (linear analysis) are shown in Table 8.

(c) <u>Grain Size Distribution of the Ferrite</u>.

From the microstructures observed in **Figure 3** it could be noticed the existence of populations of grains of different size. Mixed structures of grains - oftenly called duplex grain structures when there are two populations of grains easily distinguishable - are typical of partially recrystallized materials and of structures developing abnormal grain growth. They present as a common feature a bimodal distribution for the mean linear intercept $\bar{L}_3\ (\alpha)$, one for the coarser and one for the smaller grain population.

In the present work it has been observed the presence of two overlapped distributions for α grains: one containing clusters of small grains and other presenting a relatively random association of big and medium grains (Figure 4).

The analysis of the grain size distribution was based on a technique proposed by Van der Voort. This technique consists of measuring the mean linear intercept of a set of parallel lines spaced a magnitude equal to the smallest grain and measuring the length between grain boundaries of α grains. The same operation has to be repeated at 45º, 90º and 135º to the first position to get the spatial grain distribution. For commodity reasons photographs enlarged to 1050X and sets of parallels spaced 5 mm in between were used.

3.- <u>COMMENTS AND CONCLUSIONS</u>.

(1) The microstructure present in the three plates were ferrite+pearlite, showing a banded structure in the section parallel to the rolling direction. It could be observed that both the volume fraction of pearlite and the ASTM grain size decrease with an increase in the plate thickness (bigger values of the temperature for the end of finishing rolling).

The biggest difference for the volume fraction between front and tail ends corresponds to the 20 mm plate, which cools at the highest rate.

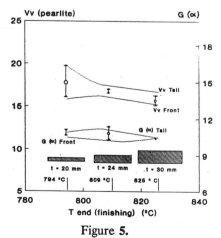

Figure 5.

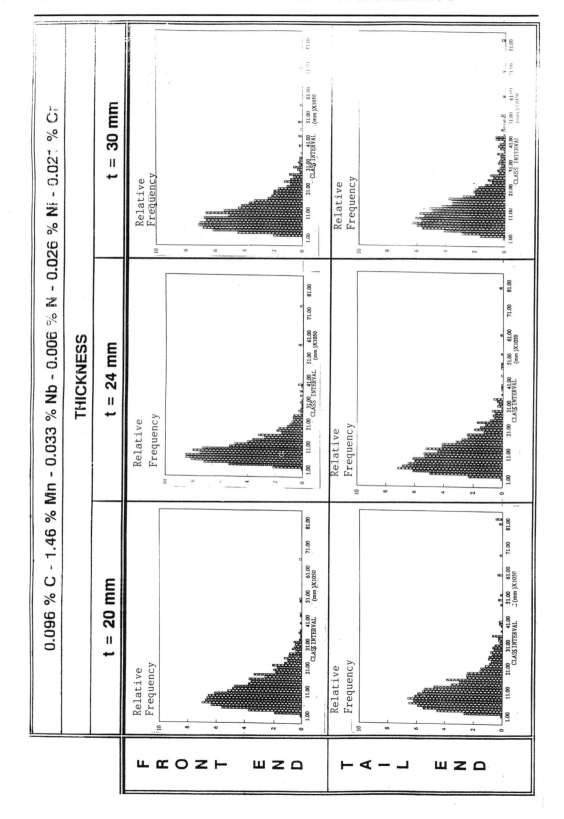

(2) The mean size of the α grains was comprised between 10 - 7 μm and a detailed observation of the microstructure suggests the existence of more than one population of sizes, which appears to be related to the plate thickness, the cooling rate and the finishing temperature during controlled rolling. From the present work it can be observed that the ferritic grains are systematically bigger by a 4.5% or more in the front end, compared to the tail end.

Subjective errors in the determination of the ASTM grain size with the image analyzer appear to be related not only to the bias introduced by the operator but to the magnification used. Those values were closer to the manually calculated ones when higher magnification were used and bigger number of fields counted to reach a given relative error.

(3) It has been found in all the plates that the α grains distribute heterogeneusly in size.

GROUP CLASS	t = 20 mm		t = 24 mm		t = 30 mm	
	FRONT END	TAIL END	FRONT END	TAIL END	FRONT END	TAIL END
SUBTOTAL 1	11.1	11.5	12.4	11.8	12.0	11.5
SUBTOTAL 2	7.6	8.1	9.8	8.9	9.1	8.0
SUBTOTAL 3	6.5	6.6	8.1	7.6	7.5	6.6
SUBTOTAL 4	5.1	5.2	7.4	6.3	6.9	6.2
SUBTOTAL 5					5.6	5.6
AVERAGE	10.77	10.84	11.16	10.90	10.90	10.50

Table 10 (a). ASTM grain size populations.

GROUP CLASS	t = 20 mm		t = 24 mm		t = 30 mm	
	FRONT END	TAIL END	FRONT END	TAIL END	FRONT END	TAIL END
SUBTOTAL 1	84.4	72.6	48.1	61.2	52.5	71.7
SUBTOTAL 2	14.2	23.3	41.3	25.5	40.2	21.3
SUBTOTAL 3	0.5	1.5	8.9	9.8	4.0	2.2
SUBTOTAL 4	0.9	2.7	1.7	3.5	2.0	1.3
SUBTOTAL 5					1.3	3.4
TOTAL	100	100	100	100	100	100

Table 10 (b). Volume fractions corresponding to the populations in table 9 (a).

Table **10** (a) and (b) show the group classes in which it has been divided the total. The ASTM number renged from 12 to 6 ASTM. It is interesting to note that the bigger volume fractions corresponded to the finer grains. Besides it has been distinguished four groups of grain populations in the 20 and 24 mm thickness plates, whereas in the 30 mm plate an extra set of 5 ASTM grain size accounting for a 3% in volume has been included.

It is also relevant that the heterogeneity of the structure increases with an increase in the plate thickness and that it seems to be connected to the temperature for the end of the finishing. The higher is this temperature the more heterogeneous is the resulting structure.

(4) Table **11** show the experimental and predicted values for the yield strength.

FACTOR	t = 20 mm		t = 24 mm		t = 30 mm	
	FRONT	TAIL	FRONT	TAIL	FRONT	TAIL
Constant	54	54	54	54	54	54
Mn effect	47	47	47	47	47	47
Si effect	26	26	26	26	26	26
N_{free} effect	--	--	--	--	--	--
Grain Size	211	196	218	185	188	187
TOTAL (Predicted)	338	323	345	312	315	314
EXPERIM'L	456	454	425	421	431	427
Δ	118	131	80	109	116	113

Table **11**. Experimental and predicted values for the yield stress. Predicted values were calculated according to Pickering's formulation:

$$\sigma_y = 15.4 \cdot (3.5 + 2.1 (\%Mn) + 5.4 (\%Si) + 23 (\%N_{free})^{1/2} + 1.13 (\overline{L_3})^{-1/2})$$

Figure **6** shows the variations of the yield strength with the heterogenity in the ferrite grain size. It seems that despite of the differences in the grain size distribution in the front and in the back ends, the essential factor conditioning the yield strength is the volume fraction of the finer α grains.

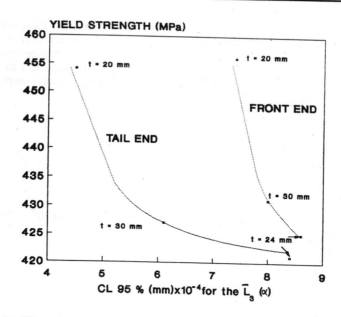

Figure 6. Variation of the experimental yield strength with the 95% Confidence Limit of the ferrite mean linear intercept $\bar{L}_3$ (α).

The contribution of the ferrite grain size accounts almost in the same proportion than the volume fraction of pearlite to the tensile strength value. The differences between experimental and theoretical values can be seen in Table 12.

FACTOR	t = 20 mm		t = 24 mm		t = 30 mm	
	FRONT	TAIL	FRONT	TAIL	FRONT	TAIL
Constant	294	294	294	294	294	294
Mn effect	40	40	40	40	40	40
Si effect	26	26	26	26	26	26
Pearlite %	62	76	64	65	58	63
Grain Size	94	87	96	82	83	83
TOTAL (Predicted)	516	524	521	508	502	506
EXPERIM'L	564	562	552	549	544	545
Δ	48	38	31	41	42	40

Table 12. Experimental and Predicted values for the Tensile Strength . Predicted values were calculated according to Pickering's formulation:

$$\sigma_m = 15.4 \cdot (9.1 + 1.8 \cdot (\%Mn) + 5.4 \cdot (\%Si) + 0.25 \cdot (\%V_v(p)) + 0.5 \cdot (\bar{L}_3 (\alpha))^{-1/2} .$$

(5) However when plotting the theoretical Impact Transition Temperature *vs.* the experimental Tensile Strength it can be seen that the tendency in the front and in the tail ends differ notably. That is, an increase in the tensile strength in the front end corresponds to better ITT values. On the other hand ITT values remain basically unaltered in the tail, being detected the biggest differences in the ITT value between both ends in the thinnest plate.

$$ITT(^{\circ}C)= -19+44\cdot\%Si+700\cdot\%(N_F)^{1/2}+2.2\cdot\%V_V (p)-11.5[\bar{L}_3 (\propto)]^{-1/2} \quad (\text{Pickering})$$

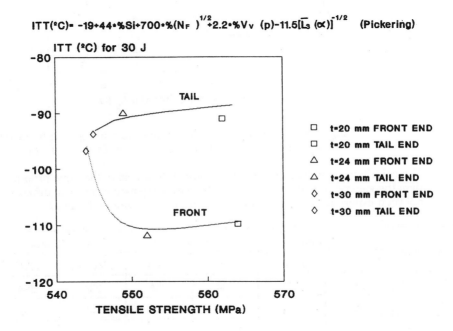

Figure 7. Variation of the theoretical Impact Transition Temperature (°C) with the experimental tensile strength.

Acknowledgments

The authors want to thank to the FICYT for the funding of the project and to ENSIDESA for the experimental material. Special thanks must be given to Ing. Katya Arredondo and to Mr. J.O. García for their valuable contribution and help.

References

1.- "Steel Plates used for Offshore Structures in the North Sea". Kenro I. et al., Nippon Steel Technical Report no. 24, Dec.1984, pp. 25-33.

2.- "New HT50 Steels with superior weldability for Marine Structures", Yasuo S. et al., Nippon Steel Technical Report no. 24, Dec.1984, pp. 1-16.

3.- "The relationship between the Microstructure and Fracture Toughness", Knott J.F., 150th Anniversary of the Discovery of Vanadium, The Metals Society, 1981, pp 8/1 - 8/19.

4.- "Grain Size Measurement" Vander Voort G.F.. Practical Applications of Quantitative Metallography. ASTM STP 839, pp 85-131, 1984.

5.- "Strengthening Mechanisms in High Strength Microalloyed Steels", Gawne D.T. and Lewis G.M.H., Materials Science and Technology, Vol. I, No. 3, 1985, pp. 183-191.

Materials Science Forum Vols. 163-165 (1994) pp. 133-138
© 1994 Trans Tech Publications, Switzerland

STEEL HEAT TREATMENT WITH FRESNEL LENSES

G.P. Rodríguez Donoso, A.J. Vázquez Vaamonde and J.J. de Damborenea Gonzalez

Centro Nacional de Investigaciones Metalúrgicas (CENIM/CSIC),
Av. Gregorio del Amo, 8, E-28040 Madrid, Spain

ABSTRACT

The concentration of solar energy with a Fresnel lens let the possibility to get power densities enough high to produce structural surface transformation of materials. The present paper informs about a installation consisting in a Fresnel lens 895 mm in diameter and the results obtained on surface transformation of steels.

1.- INTRODUCTION

Recently the use of concentrated solar energy is a research topic applied in many laboratories to produce surface modification of materials, mainly steel, such as hardening by steel (1-3) and cast iron (4) phase transformation, CVD (5), cladding (5-7) and after treatment of coatings obtained by laser, plasma, PVD (8), etc.

Many types of solar facilities may be applied to treat materials (9). In this paper the possibility to work with the solar energy concentrated with a Fresnel lens is described. This type of installation concentrates the sun to get enough high power density to produce the melting of thin steel sheets. That means it is possible to work with enough power density to produce the phase transformation of steels.

The Fresnel lens refracts the sun's rays through the facets that vary slightly their angle with respect to the optical axis. The result of this sun beams' deviation is their concentration on a small spot on the focal point of the lens.

The main advantages of this type of installation in comparison with the more traditional as flat mirrors and central receiver or the different types of parabolic mirrors are the low cost of the lens, the low cost of the installation and the easy access to the installation, i.e., it is possible to work with the hands just near the focal point without any risk of hand burning. Moreover, if the equipment is a lens and not a mirror, the surface of the target is upright and this is a more favourable position for surface alloying or cladding, etc.

Also brazing and soldering is possible with this equipment as report in 1979 Kaddou and Abdul-Latif (10). Nevertheless more recent papers on this subject were not found.

Figure 1.- Fresnel lens

2.- EQUIPMENT

The size of the lens used in this work was 895 mm in diameter with a focal length of 855 mm (Fig. 1).

The lens is on an automatic equipment that follows the sun track. This equipment let to maintain the plane of the lens always normal to the sun beam direction.

The target samples lay on a one axis transversal table.

3.- EXPERIMENTAL

3.1. Size and position of the focal spot. Power of the lens

To identify experimentally the focal distance of the lens thin plain carbon steel sheets were used. The sheets, 0.5 mm thickness and 100 mm side length, were put at different distances during enough time to get the beginning of the melting.

Different prints of these tests can be seen in Figure 2. At the focal spot, 855 mm from the lens, the diameter of the hole is about 8 mm.

When a 1.0 mm thickness sheet, 50 mm x 50 mm in size, was used it was not possible to get a melting hole due to the strong cooling effect of the sheet and

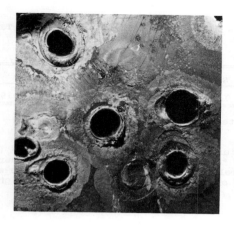

Figure 2.- Melting holes in a 0.5 mm thickness sheet

to the small size of the focus. To reduce this cooling effect small square samples, 10 mm x 10 mm in size with 1 mm and 1.5 mm in height were prepared and put inside a ceramic block. When the samples were heated, the piece of steel was entirely melted in 15 and 25 seconds respectively.

Two small pieces 3 mm thickness and 10 mm x 10 mm square in size were put together and the welding was performed.

3.2. Tests on rebar steel samples

At the beginning some tests were performed on 12 mm diameter and 20 mm height steel rebar samples. This steel is not the best to be treated by hardening but the tests were devoted to identify the

thermal aspects of depth of hardening in different situations.

The samples were positioned at the focal spot and were heated to a temperature above Ac_3 and after that cooled in cold water. The results are presented in Table 1.

Temperature	842 °C	842 °C	970 °C
Time	100 s	125 s	145 s

Table 1.- Heating conditions of the samples

The samples were cut in two halves normal to the heating plane and some measurements of the Vicker hardness were performed. The load was 1 kg and the variation of the hardness with reference to the distance to the treated surface are presented in Fig. 3.

The thickness of the quenched zone in the sample treated at the lower time (100 s) and temperature (1115 K <> 842 °C) is very low , 2 mm, and the depth of the total heat affected zone amounts to 4 mm. With a larger time (145 s) and temperature (1243 K <> 970 °C) the whole sample is heat affected.

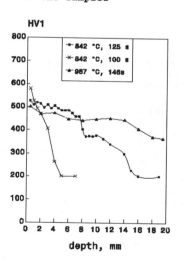

Figure 3.- Vickers hardness vs. depth

3.4.- Tests with martensitic stainless steel

The composition of the martensitic stainless steel used is presented in Table 2. The samples were cylindrical with 5 mm thickness and 35 mm diameter.

C	Cr	Mn	Si	Ni	S	P
0.32	12	0.38	0.5	0.34	0.27	0.015

Table 2.- Composition of the martensitic stainless steel (% weigth)

The tests were performed at three different temperatures above Ac_3: 1293 K (920 °C), 1263 K (990 °C) and 1283 K (1010 °C). The variation of temperature during the tests were registered with a Chromel Alumel thermocouple welded at the side of the sample. In Fig. 4 the curve of one of the samples heated at the highest temperature is shown.

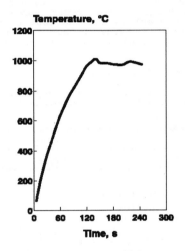

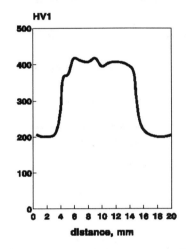

Figure 4. Variation of temperature vs. time in a sample heated with a Fresnel lens.

Figure 5. Distribution of the Vickers hardness in the sample treated with Fresnel lens

The initial structure of the steel corresponded to a globular annealing. The microstructure was thus formed by globular carbide precipitates of the $(Cr, Fe)_{23}C_6$ type, distributed homogeneously in a ferritic matrix. The hardness of the material in this state was 200 HV. After the treatment with the Fresnel lens we a martensite metallographic structure was obtained. The Vickers Hardness values are according to this·structure in the whole thickness of the sample.

The relatively large austeniting times at a temperature above Ac_3 produces the whole transformation of the relatively thin sample so in the back of the sample the temperature obtained is also above Ac_3. As a consequence of the large diameter of the sample in comparison with the spot size at the focal spot only the central part of the sample could be transformed.

The profile of hardness in a track just under the exposed surface, about 0.5 mm, is presented in Figure 5. This profile shows the size of the spot of the concentrated beam, about 8 mm in diameter, similar to that obtained in the melt tests of the thin sheets above.

Nevertheless the whole thickness of the samples quenched, the temperature was not enough high during enough time to produce the total dissolution of the carbides in the as received material. Martensite is, in consequence, lower in carbon content to that corresponding to the steel, 0.32 %, and this fact explains the relatively low values of hardness obtained. It will be necessary to reach higher temperatures to produce the total dissolution of carbides and to obtain the right hardness values.

3.5.- Thermal cycles

It is very simple to produce thermal cycles on the surface of the sample. The first procedure is to defocalize the sample, the second and easiest is to cover and discover the lens. This second procedure is easily made putting the cover between the lens and the target where the density is enough low.

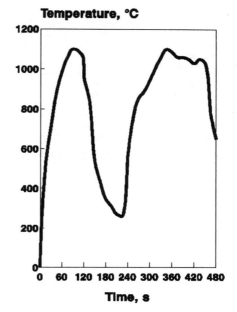

Figure 6. Temperature cycles with Fresnel lens

In Fig. 6 an example of the shape of the cycles shows the variation of the temperature in a sample of 7 mm thickness.

4. CONCLUSIONS

With a Fresnel lens of enough diameter it is possible to make heat treatments of steels.

The maximum power obtained with a Fresnel lens 895 mm in diameter and 855 mm of focal length is enough high to produce a hole of 8 mm in diameter by melting of a steel sample 0.5 mm in thickness

In consequence, it is possible to treat samples with a transformation temperature from 1273 K and above.

It is also weld steel samples with small size if the conduction loss are enough low according to the power input of the concentrated beam

5. ACKNOWLEDGEMENTS

Authors wish to thanks to the IER - CIEMAT, Madrid, for their facilities in the experimental work and the European Community for funding this project as part of the program on Large Scale European Research Facilities.

5.- REFERENCES

1) Yu Z.-K. et al.: J. Heat Treating. 1982, $\underline{2}$,(4), 344.
2) Mayboroda V.P. et al.: Metalloved Term. Obrad. Met., $\underline{1}$, 59.
3) Vázquez A.J. et al.: Solar Energy Materials, 1991, $\underline{24}$, 751.
4) Yu Z.-K. et al.: J. Heat Treating, 1983, $\underline{3}$ (2), 120.
5) Korol A.A. et al.: Poroshkovaya Metallurgiya, 1983, $\underline{4}$, 39.
6) Yu Z.-K. et. al.: Surface Engineering, 1987, $\underline{3}$, 41.
7) Pitts J.R. et al. in P.A. Nelson, W.W. Schertz, and R.H. Till, Eds.:
Proceedings of the 25th Intersociety Energy Conversion Engineering Conference,
1990, American Institute of Chemical Engineers, NY, 1990, $\underline{6}$, 262.
8) Stanley J.T. et al.: Advanced Materials & Processes, 1990, $\underline{12}$, 16.
9) Rodriguez G.P. et al.: Proc. 25th Int. Symp. on Automotive Technology and
Automation (ISATA) - Laser Applications in the Automotive Industries, Florence,
Italy, 1992.
10) Kaddou et al.: Solar Energy, 1969, $\underline{12}$, 377.

Materials Science Forum Vols. 163-165 (1994) pp. 139-150
© 1994 Trans Tech Publications, Switzerland

THE EFFFECTS OF QUENCHANT MEDIA SELECTION AND CONTROL ON THE DISTORTION OF ENGINEERED STEEL PARTS

R.T. von Bergen

Houghton Europe BV, Birmingham, UK

ABSTRACT

The paper commences by defining the types of distortion that can occur during heat treatment, and is followed by a comprehensive review of the factors which influence the distortion of engineered steel parts during quenching. The paper provides an in depth examination of the effects of quenchant type (including mineral oils and polymer quenchants), quenching characteristics, operating temperature, method and degree of agitation and quench system design upon the nature of distortion. Methods of controlling distortion through effective quenchant media selection and control are discussed with numerous actual case histories.

INTRODUCTION

Distortion is perhaps one of the biggest problems in heat treatment.(1) Comparatively little information has been published on the subject and this paper is intended to provide a useful guide for the practical heat treater to help identify reasons for distortion and suggest possible corrective action which can be taken.

Distortion can be defined as a permanent and generally unpredictable change in the shape or size of a component during processing. It invariably costs money in terms of rectification or straightening processes, or in the provision of increased grinding allowances and may also result in difficulties during assembly or the scrapping of costly machined components. All too often the heat treatment department gets the blame for distortion, mainly because it is only during subsequent grinding or assembly operations that the distortion becomes apparent. However before considering distortion during heat treatment it is perhaps worthwhile briefly mentioning some possible causes for distortion before heat treatment.

DISTORTION BEFORE HEAT TREATMENT

Dimensional changes can occur in the soft condition for a number of reasons.

a) **Relief of internal stresses.** Relief of internal stresses during machining operations can lead to springing out of shape, particularly if components are made from cold worked or drawn steel.

b) **Excessive collet pressures.** This can lead to distortion due to work being deformed elastically during machining and springing back upon release from the collet. The machining of thin section bearing rings from cold reduced tube can result in this type of distortion.

c) **Machine slackness and failure to comply with drawing limits.** This may result in components being out of tolerance before they enter the heat treatment furnace.

d) **Fabrication techniques.** Processes such as welding or those involving cold plastic deformation such as stamping or extrusion can introduce stresses into the component which may cause distortion.

DISTORTION DURING HEAT TREATMENT
This can be classified into two types and it is important to distinguish between them.

I) Size distortion or movement.

2) Shape distortion or warpage.

Size distortion
It is well known that the volume of a component changes during heat treatment due to the structural changes occurring in the steel. These are influenced by steel composition, austenitising temperature, soaking time, quenching rate and tempering treatment.

Changes in heat treatment conditions which influence the microstructure will affect the finished size of the part. For example, if the quenching speed is slower than the critical cooling rate non-martensitic products will be formed. Alternatively, during hot oil quenching of bearing rings, changes in oil temperature may affect the proportion of retained austenite. Both will influence the overall size changes occurring in the component.

However for a particular steel and heat treatment conditions, these size changes are generally predictable and can therefore be allowed for in machining and grinding allowances.

Shape distortion
This is the most troublesome form of distortion since it is often unpredictable. There are many different forms of shape distortion.

Out of round	Taper
Out of flat	Dishing
Bending	Bowing
Buckle	Closing in of bores

Shape distortion can occur either during austenitisation or during quenching.

Shape distortion during austenitisation
a) **Relief of internal stresses.** Stresses introduced during prior metal forming or machining operations will be relieved during austenitisation. Annealing and straightening operations may be needed before heat treatment.

b) **Sagging or creep.** This can be due to inadequate support of the work in the furnace. Careful attention must be paid to the condition of jigs and fixtures, furnace hearths and work trays and to the way in which components are placed in the furnace.

> Long shafts should be suspended vertically, and rings should be laid flat. Overhangs on large complex components such as dies should be supported.

c) **Mechanical damage in furnace hot zone.** Therrnal expansion of components can cause jamming across furnace hearths. Automatic vibratory feeding systems on continuous furnaces can force work into the hot zone, and hence distort components that are already at temperature.

d) **Temperature gradients in furnace.** Temperature variations due to blocked burners, faulty heating elements or poor atmosphere circulation can result in uneven heating rates and hence cause internal stresses in components. This is particularly important with large distortion prone components where it is important to ensure that these are in the the central part of the hot zone where temperature uniformity is best.

e) **Component size variation**. Variations in component section size will lead to uneven heating rates, resulting in the possible formation of internal stresses. Wherever possible, low heating rates should be employed on critical components having large changes in section thickness in order to avoid uneven heating.

Shape distortion during quenching
Shape distortion during quenching generally occurs as a result of an imbalance of internal residual stresses, which, taken to its extreme limit can lead to cracking, either microcracking or bulk failure of the component.

A number of factors, either individually or in combination, can influence the nature and extent of shape distortion during quenching.

Steel composition and hardenability
Component geometry
Mechanical handling
Type of quenching fluid
Temperature of quenchant
Condition of quenchant
Circulation of quenchant

1. Steel composition and hardenability

This dictates the critical cooling rate required to achieve specific structure and mechanical properties and governs the volume changes (and hence residual stresses) formed during quenching. (Fig. 1)

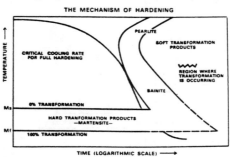

Fig. 1. Critical cooling rate for complete martensitic transformation.

In order to minimise distortion a quenchant should be selected which just exceeds the critical cooling rate of the steel and which provides a low cooling rate in the Ms to Mf transformation temperature range. However, it is often necessary to select a compromise in cooling rate to enable the processing of a wide range of steels of varying hardenability.

2. Component geometry

The section thickness of a component will influence the critical cooling rate required for full hardening and hence the selection of quenchant. Variations in section thickness on a particular component will lead to differential cooling rates and may result in variations in the level of residual stress with consequent distortion. Consideration should be given to quenching at the slowest possible speed dictated by the largest section or the possible use of hot oil quenching techniques.

3. Mechanical handling

In the austenitic condition steel is only about one-tenth as strong as it is at room temperature, as shown in Fig. 2 for a high carbon chromium ball bearing steel. [2]

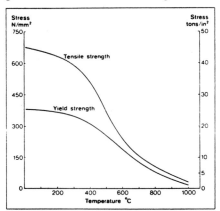

Fig. 2. The strength of high carbon chromium steel as a function of temperature.

Components are therefore very susceptible to mechanical handling damage when at elevated temperature. It is therefore important to handle components carefully, and to avoid dropping critical components into the bottom of the quench tank. Particular attention should be paid to quench chute design on continuous furnaces. Even inclined chutes can cause damage if thin section components strike pick-up slats on conveyor systems.

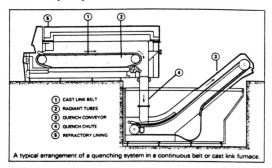

A typical arrangement of a quenching system in a continuous belt or cast link furnace.

Fig. 3. Schematic continuous heat treatment installation.

An interesting example of this kind of distortion occurred at a manufacturer of automotive water pump spindles. Stepped shafts were quenched from a continuous furnace and dropped vertically approximately 3m down a quench chute onto the conveyor. Many of the shafts were bent and this was initially attributed to residual stresses due to the change in section size. However, closer investigation revealed the presence of flats on the end of certain shafts diametrically opposite the direction of bend, confirming that the distortion had been caused by mechanical impact. A re-design of the quench chute to prevent this impact damage eliminated the problem.

4. Type of quenching fluid

The three stages of cooling which occur during quenching i.e. vapour phase, boiling phase and liquid cooling phase are well documented and understood. However, the relative duration of the three stages and the cooling rates during each stage can vary widely for different quenchants and can consequently have a significant effect on distortion.

The ideal situation is for each stage of the quenching process to occur uniformly on the component to provide uniform transformation with minimum residual stress. Unfortunately this is not usually the case due to such factors as

- changes in component section thickness
- surface finish - the vapour phase tends to be more stable on smooth machined or polished surfaces, and break-up more readily with the onset of boiling at sharp corners, holes or surface irregularities.

Vapour Phase Characteristics

Although the quench rate in the convection phase generally considered to be the most critical parameter because this is within the Ms - Mf temperature range where transformation occurs, the characteristics of the vapour stage can also be very important. This is perhaps best illustrated with an example of gear hardening.

Precision automotive transmission gears are frequently assembled with no further finishing operations to gear teeth. Requirements for weight saving have led to progressive reductions in gear mass and tooth section with consequent potential for increased distortion. During oil quenching, vapour retention in tooth roots, combined with the onset of boiling on the flanks can cause undesirable "unwinding" of thin section gears (Fig. 4) This can be minimised by the use of accelerated oils where special additives are used to reduce the stability of the vapour phase and promote boiling - thereby giving greater uniformity of cooling.

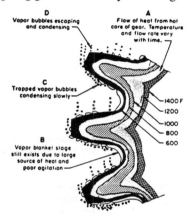

Fig. 4. Vapour retention in gear tooth roots during oil quenching.[3]

Convection Phase Characteristics

The quenching characteristics in the convection phase are best illustrated by the differential cooling rate curve. The cooling rate at 300°C has been fairly universally accepted as being critical for distortion since this is within the Ms - Mf temperature range of a wide range of engineering steels. Typical values for some commonly used types of quenchant are shown below.

Quenchant		Cooling rate at 300°C (°C/sec)
Normal speed oil		5 - 15
Accelerated oil		10 - 15
Polymers	PAG	30 - 80
	ACR	10 - 25
	PVP	10 - 25
	PEO	10 - 30

It can be seen that PAG quenchants have higher cooling rates in the convection phase than quenching oils. Components are therefore more susceptible to the possibility of distortion and careful consideration has to be given to steel hardenability, component section size and surface finish before adopting a PAG quenchant. In general, PAG quenchants are suitable, and widely used for plain carbon, low alloy or carburised steels or higher alloy steel components of large section size.

More recently developed polymer quenchants such as ACR, PVP or PEO have much lower cooling rates at 300°C, similar to those of quenching oils. This has enabled the extension of the use of water-based technology to critical alloy steel components which would not otherwise be suitable for PAGs.

5. Temperature of quenchant

With modern quenching oil formulations, the temperature of operation does not have a significant effect on quenching speed. (Fig. 5)

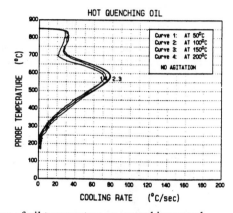

Fig. 5. The influence of oil temperature on quenching speed.

However, as is well known, the temperature of operation can have a dramatic influence on distortion and oils have been developed specially for use at elevated temperatures for mar-quenching or mar-tempering applications.

This process is generally applied to high precision engineered components, such as thin section bearing rings or automotive transmission gears or shafts where close dimensional control during quenching is essential. The process is designed to minimise the residual stresses generated during quenching by promoting uniform transformation.

During conventional quenching the surface, or the thinner sections of a component cool more rapidly than the centre or thicker sections. This can mean that some areas in the component have cooled to Ms and are hence beginning to transform whereas other areas are still in the soft, austenitic condition.(Fig. 6) When these subsequently transform volume changes may be restricted by the previously formed hard, brittle martensite, creating surface tension or imbalance of stress which can lead to distortion or quench cracking.

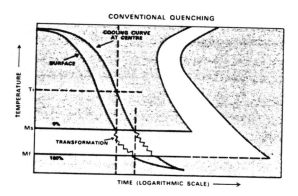

Fig. 6. The formation of stresses and distortion during conventional quenching.

During hot oil quenching, components are quenched into specially formulated oils generally at temperatures within the range 120°C - 200°C depending upon component complexity and distortion tendency and held at this temperature for sufficient time to allow for equalisation of temperature gradients across the section. The components are then withdrawn from the oil and during subsequent slow cooling in the furnace atmosphere transformation occurs uniformly throughout the section. (Fig. 7) This minimises the generation of internal stresses and reduces distortion.

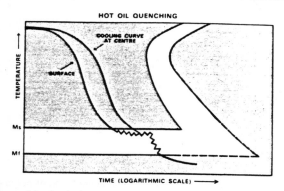

Fig. 7. Hot oil quenching techniques to reduce distortion.

Whilst the Ms temperature of typical engineering steels lies within the range 250°C to 350°C experience has shown that mar-quenching into hot oil at 150°C to 200°C can lead to significant reduction in distortion on complex precision engineered parts.

6. Condition of quenchant

The quenching characteristics of fluids can, and do, change during use for a variety of reasons. Most lead to an increase in quenching speed, particularly in the convection phase, and consequently to an increased risk of distortion and cracking.

Possible causes for changes in quenching speed include:

- contamination of quenching oil with water. As little as 0.05% of water in quenching oil has a very dramatic effect on quenching properties, by increasing maximum cooling rate and cooling rate in the critical convection phase. [4] (Fig. 8)

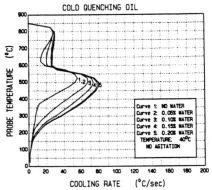

Fig. 8. The effect of water contamination in quenching oil.

- oxidation of quenching oil. Oxidation of mineral oils reduces the stability of the vapour phase and increases maximum cooling rate. [4] (Fig. 9)

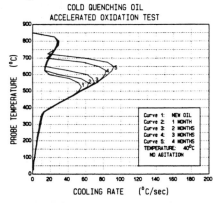

Fig. 9. The effect of oil oxidation on quenching characteristics.

- contamination or degradation of polymer quenchant. Thermal degradation of polymer quenchants generally leads to an increase in cooling rate in the convection phase which can, to some extent, be compensated for by an increase in concentration. (Fig. 10)

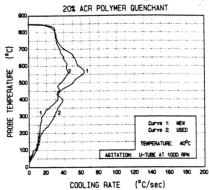

Fig. 10. The effect of polymer degradation on quenching characteristics.

It is therefore important, particularly in critical applications where close control of distortion is essential, to implement regular monitoring of the quenching fluid. Physical tests for acidity and water content in quenching oils, and the concentration of polymer quenchant solutions should be considered the minimum, and in practice, periodic evaluation of quenching characteristics is highly desirable.

7. Circulation of quenchant

Circulation of quenchant, either by pump or propeller is very important, to maintain uniform bath temperature and assist in the breakdown of the vapour phase of the quenching process.

The degree of agitation has a significant influence on the cooling late of both quenching oils and polymer quenchants, but a different influence on distortion.

Quenching oil

As agitation increases there is a decrease in the duration of the vapour phase, an increase in maximum cooling rate and perhaps most important of all, an increase in the cooling rate in the convection phase.(Fig. 11) This latter effect may increase the risk of cracking and therefore excessive agitation should not be used.

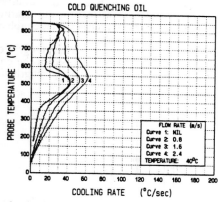

Fig. 11. The effect of agitation on quenching oil.

Polymer quenchant

With polymer quenchants, while there is a pronounced effect on the vapour phase and maximum cooling rate, increasing agitation has very little effect on the cooling rate in the convection phase. (Fig. 12) Vigorous agitation is normally recommended to ensure uniform quenching characteristics and this can be applied without increasing the risk of cracking.

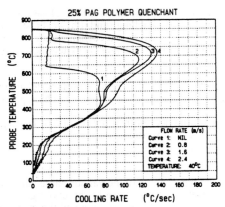

Fig. 12. The effect of agitation on polymer quenchants.

The direction of flow of the quenchant over the workpiece surface can also influence distortion. This can perhaps best be illustrated by a recent case history in the bearing industry which demonstrates that uni-directional flow of quenchant over a surface can cause problems.

During the oil quenching of cylindrical bearing outer rings in an integral quench furnace, a predominant upward flow of quenchant through the charge resulted in the collapse of the bore towards the top of the ring and failure to clean up the OD during subsequent grinding.

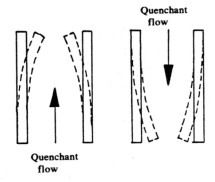

Fig. 13. The effect of quenchant flow on distortion.

Reversing the flow of quenchant produced the opposite effect. The answer to the problem was found by reducing the quenchant flow to the minimum required for circulation through the coolers and providing agitation by alternate up and down movement on the quench lowerator. Result -uniform contraction of bores as would be expected from theory.

SUMMARY
To summarise, distortion during quenching is a complex subject which can be influenced by many factors including polymer steel composition and hardenability, component geometry and surface finish, mechanical handling and quench system design, type of quenching fluid, temperature of quenchant, condition of quenchant and amount of circulation.

The significance of distortion depends very much upon the type of component and the need for subsequent processing operations. For example, a few rnm bend on a large bar or forging may be of no consequence or capable of simple rectification, whereas movement at micron level on precision gears or bearings receiving minimal subsequent processing may render the components completely unfit for service.

If distortion should be a problem much can be done to minimise it by careful and systematic examination of possible causes. I hope that my paper today will be of help to the practical heat-treater in identifying some of the causes and suggesting possible solutions.

REFERENCES

1. Quenching Principles and Practice, Houghton Vaughan plc, UK.

2. G.E. Hollox and R. T. von Bergen, Heat Treatment of Metals, 1978.2.

3. MEI Course 6, Heat Processing Technology, ASM, 1977.

4. R. T. von Bergen, Heat Treatment of Metals, 1991.2.

Materials Science Forum Vols. 163-165 (1994) pp. 151-156
© *1994 Trans Tech Publications, Switzerland*

APPLICATIONS OF "STANDARD" QUENCHANT COOLING CURVE ANALYSIS

D.L. Moore [1] and S. Crawley [2]

[1] Instruments and Technology INC., Naperville IL, USA

[2] Drayton Probe Systems, Trentham, Stoke-on-Trent, UK

ABSTRACT

There has been much activity in recent years toward a "standard" cooling curve analysis (CCA) test for determining quenchant cooling characteristics. This paper reviews typical applications of the "standard" test.

INTRODUCTION

The interest in establishing a "standard" cooling curve test resulted from recognition of CCA as the most meaningful quenchant cooling test [1], the necessity to establish minimum technical requirements and the need for uniformity in the equipment and procedure to assure comparability of results.

THE STANDARD TEST

The standard CCA test for oil quenchants as currently specified by the International Federation for Heat Treatment and the International Organisation for Standardisation [2] is based on a specification [3] developed under the auspices of the Wolfson Heat Treatment Centre. A Similar standard has been recommended by the American Society for Metals (Quenching and Cooling Committee) to the American Society for Testing and Materials. As indicated it is a laboratory test, conducted with an Inconel 600 12.5mm dia probe, heated to 850 C, quenched into a 2000ml volume of sample at 40 C without agitation. These and other details define the "standard" test. Equipment suitable for conducting the test is available in two versions, laboratory (stationary), and portable.

Tests may be conducted at other sample temperatures, agitation conditions, etc. but these constitute extended uses of CCA as does testing in the quench tank. This discussion relates only to applications of the standard laboratory test.

The data for a one minute test, 480 data sets, can be plotted simply as probe temperature versus time (cooling curve). However, by plotting the derivative (cooling rate) against temperature, the three stages of oil quenching (vapour, boiling, convection) are clearly portrayed as are transition temperatures between phases, the maximum cooling rate and the temperature at which it occurs. The standard test also calls for reporting six key characteristic values :

Maximum Cooling Rate	Time to reach 600 C
Temperature at Maximum Cooling Rate	Time to reach 400 C
Cooling Rate at 300 C	Time to reach 200 C

Figure 1. shows a test report with a cooling curve, cooling rate curve and six key characteristic values.

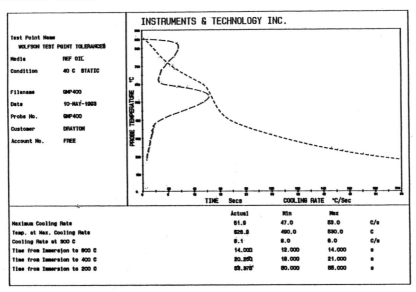

Figure 1. Standard CCA Report.

APPLICATIONS OF STANDARD CCA TEST
Some application areas of the standard CCA test are :
    Quenchant Development
    Quenchant Quality Control
    Quenchant Comparison Studies
    Heat Treat Process Control
    Trouble Shooting

CCA can provide valuable assistance in quenchant development. The basic shape of an oil quenchant cooling curve is determined by the base oil as shown in figure 2. Additions such as accelerators are made to alter the performance as shown where A1 and A2 could be different accelerants or different concentrations of the same accelerant. Other additives such as anti-oxidents and rust retardents may also alter the curve. With the use of cooling curve testing it is possible to establish families of curves for potential base oils and potential additives from which to select formulations for desired behaviour and/or minimum cost.

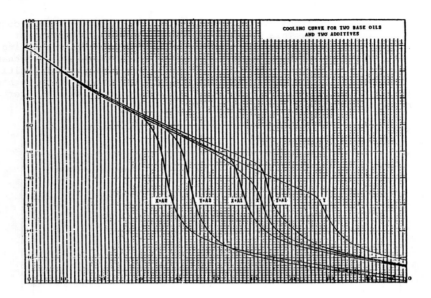

Figure 2. Cooling Curves for two base oils and two additives.

The total performance of a quenchant must also take into account its ageing
ability. In this sense ageing may refer not just to time in use but the number of
cycles or tonnage throughput. A prediction of ageing can be achieved using the
stationary type test system. As the system is configured to drop a heated probe
into the sample it can be programmed to repeatedly heat and quench a (dummy) probe
into the sample. After a given number of cycles the test probe is reinstalled and
the sample retested. This procedure can be repeated and the key characteristic
values charted to establish the effect of cumulative cycles on the behaviour of
the quenchant.

For a quenchant in production the same cooling/cooling rate curves and six key
characteristic values plus appropriate control limits can be used for product
quality control testing. Sample control limits are included in the report shown in
figure 1.

The standard test provides the ability to compare commercially available
quenchants on a common basis. Suppliers' data secured on various in-house systems,
while quite acceptable for internal use, presents a dilemma for a prospective user
attempting to relate competitive data and for a user, attempting to relate shop
test data to the suppliers' data. From the purchasing aspect it can be quite
valuable if competitive quenchants can be shown to be equivalent. From the
standpoint of process engineering it also enables specifying (or selecting) a
quenchant via such performance criteria as Stage A time, Stage B start
temperature, etc..

One very important and sometimes overlooked element of heat treat process control is quenchant process control. This consists of several aspects : testing of incoming quenchants whether new or reprocessed, periodic testing and charting of quenchant/s in use and retesting quenchant/s after replenishment. Oil quenchants can degrade due to contamination, oxidation or preferential dragout of additives [1]. Figure 3 shows an as-used quench oil having a shorter vapour stage, a higher maximum cooling rate and a lower boiling temperature than when new. This certainly is cause for investigation and corrective action.

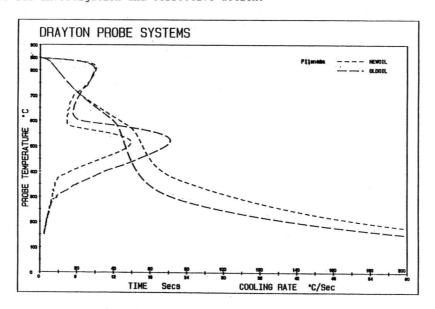

Figure 3. Cooling/Cooling Rate Curves for a quenching oil, new and used.

A classic contaminant in oil quenchants is water. Its effect can be very deleterious to hardening. As little as 0.1% water can also cause a bath to foam excessively and greatly increase the danger of fire. Figure 4. shows the sensitivity of cooling curve testing to the presence of water at least as low as 0.1%.

In every quench tank there is potential for change in quenchant behaviour. A change in key characteristic cooling curve values can be the alarm! Implementing a process control program requires setting control limits for six characteristic values and determining the desired frequency of testing. Should trouble develop in part hardness, cracking or distortion the quenchant is often suspected. If a quenchant process control program is in place and the date of the processing is known the quenching media behaviour is also known.

A significant cost savings is available if quenchant replacement/reprocessing is done when process control data shows it is warranted. In any arbitrary schedule there is the risk of unnecessary cost from early replacement versus rejection cost if not done when needed.

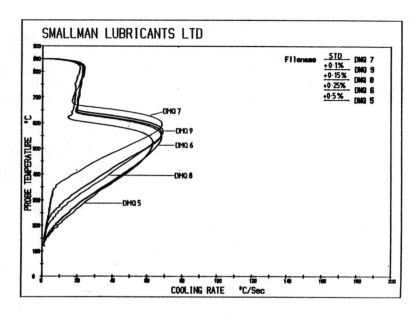

Figure 4. Cooling Rate Curves for Water in a Quenching Oil.

Should trouble develop and a quenchant process control program not be in place immediate cooling curve analysis is called for. Shifting of the cooling/rate curves for a sample of the used oil compared to those of a retained sample of the new oil indicates possible contamination, oxidation, or dragout.

A suspect situation may also warrant immediate CCA. Figure 5. shows the significant effect from known failure in a hydraulic furnace door. An assumption that some additional oil in the quench tank didn't matter would have been quite wrong.

AQUEOUS POLYMER QUENCHANTS
The standard test as now published is directed to oil quenchants. A similar standard is in preparation for aqueous polymer quenchants. It currently is based on the same equipment, probe and procedure as the standard for oil, except for the addition of agitation, the nature and degree of which has not been determined.

EXTENDED USES OF CCA
There are many uses of CCA beyond those discussed here. Within the limits of the equipment and safety considerations variations may be made in the temperature of the sample and various types and degrees of agitation may be used. It is thus possible to build a model of quenchant behaviour versus conditions and to approach process conditions.

Another extended use is testing in the quench tank with a portable system. Such test parameters as sample volume and probe positioning are sacrificed for the ability to secure tests of the total quench system, i.e. including the effects of tank shape, type and degree of agitation and operating temperature.

TEST EQUIPMENT

CCA test equipment is produced in both laboratory and portable models by Drayton Probe Systems of Stoke-on-Trent, England.

The Laboratory QUENCHALYZER system includes a drop through furnace and motor driven probe handling mechanism which provides maximum repeatability in test execution and minimum operator involvement. It is totally enclosed to provide maximum safety.

The Portable QUENCHALYZER allows for testing in the quench tank, if accessible, and may also be used for laboratory tests. Both systems include a PC with hard disk storage and a colour plotter. System software provides for data handling, screen displays and plotting. It also permits user input of control limits for six key characteristic values. Systems are in use by quenchant suppliers and users world wide.

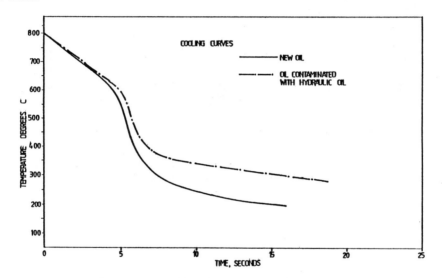

Figure 5. Effect of Hydraulic Oil contamination in a Quenching Oil.

References :

[1] G.E. Totten, C.E. Bates, and N.A. Clinton. Handbook of Quenchants and Quenching Technology, American Society for Metals, 1993.

[2] ISO/DIS 9950, Industrial Quenching Oils - Determination of Cooling Curve Characteristics - Laboratory Test Method, to be released in 1993.

[3] "Laboratory Test for Assessing the Cooling Characteristics of Industrial Quenching Media", Wolfson Heat Treatment Centre (Engineering Group Specification), 1982.

[4] H.E. Boyer and P.E. Cary, Ed., Quenching and Control of Distortion, ASM International, 1988.

III. HEAT TREATMENT OF NON-FERROUS ALLOYS

Materials Science Forum Vols. 163-165 (1994) pp. 159-172

THEMOMECHANICAL SURFACE TREATMENT OF TITANIUM ALLOYS

L. Wagner [1] and J.K. Gregory [2]

[1] TU Hamburg-Harburg, D-21073 Hamburg, Germany

[2] GKSS Research Center, D-21502 Geesthacht, Germany

ABSTRACT

Thermomechanical treatments are widely used to optimize the properties of high-strength titanium alloys for a given application. Mechanical surface treatments such as shot peening generally - but not always - result in improved fatigue behavior of titanium parts. Shot peening changes the surface properties in three respects: surface roughness, residual stresses, and degree of work-hardening are altered. Because these three parameters can independently influence crack nucleation and crack propagation, the overall fatigue life is a complex function of the surface condition. By separating the individual contributions of these parameters to the total fatigue life, it can be determined that an increase in surface roughness leads to early crack nucleation, while an increase in surface strength due to work-hardening delays the crack initiation phase. Furthermore, once surface cracks are present, these propagate faster in work-hardened material, but their growth is retarded by compressive residual stresses. Based on this analysis, recommendations for surface treatments for service at high temperatures (where the beneficial compressive stresses may anneal out) will be presented. Novel methods for improving fatigue performance by combining mechanical surface treatments and thermal treatments will be described.

1. MECHANICAL SURFACE TREATMENT

Mechanical surface treatments such as shot peening, polishing, or surface rolling can be utilized to improve the endurance limit in titanium-base materials [1]. In most cases, three surface properties are altered (figure 1):

• Surface roughness

• Degree of cold work (dislocation density)

• Residual stresses

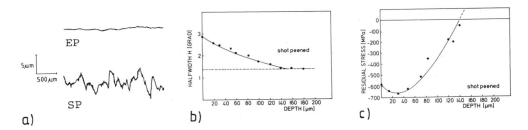

Figure 1: Surface properties altered by shot peening in Ti-6Al-4V (equiaxed) a) surface roughness b) degree of cold work c) residual stresses [Ref. 2]

As fatigue failure represents the sum of both crack nucleation and crack propagation life, the changes induced by such treatments can have contradictory influences on the fatigue strength. The surface roughness determines whether fatigue strength is primarily crack nucleation controlled (smooth) or crack propagation controlled (rough). For smooth surfaces, a work-hardened surface layer can delay crack nucleation owing to the increase in strength. In rough surfaces, the crack initiation phase can be absent, and a work-hardened surface layer is detrimental to crack propagation owing to the reduced ductility. Near-surface residual compressive stresses are clearly beneficial, as they can significantly retard crack growth once cracks are present. This is summarized in the table below.

	Crack Nucleation	Crack Propagation
Surface Roughness	accelerates	no effect
Cold Work	retards	accelerates
Residual Compressive Stresses	minor or no effect	retards

It must be kept in mind that the changes induced by a mechanical surface treatment are not necessarily stable. In particular,

• Residual stresses — can be reduced or eliminated by a stress-relief treatment (figure 2a).

• Degree of cold work (dislocation density) — can be removed by recrystallizing.

• Surface roughness — can be reduced by an additional surface treatment (e.g., polishing)

Furthermore, the beneficial residual compressive stresses can be reduced by cyclic plastic deformation, i.e., during fatigue loading in service (figure 2b).

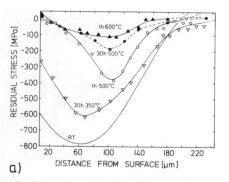
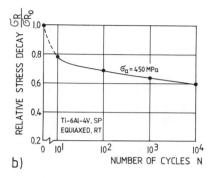

a)   b)

Figure 2: Reduction in shot-peening induced residual stresses caused by a) heat treatments and b) cyclic loading [Ref. 3]

These factors can be taken into account to design surface treatments which are appropriate to the application. For example, figure 3 shows a series of S-N curves for Ti-6Al-4V at room temperature (figure 3a) and at elevated temperature (figure 3b). As a reference condition, an electropolished surface (EP) which is free of residual stresses and cold work and with a mirror finish is considered. Compared with the EP condition, shot-peening (SP) significantly improves the endurance limit at room temperature. At elevated temperature, however, shot peening lowers the fatigue limit to values below that of the reference condition. This can be explained by separating out the individual contributions of the surface properties to fatigue life. For example, stress relieving 1 h at 600 °C after shot peening (SP + SR) decreases the endurance limit of the shot peened condition at room temperature, but does not alter the endurance limit at 500 °C. In effect, the stress relief treatment is redundant when cyclic loading occurs at high temperature. If the compressive residual stresses were the only mechanism operating, one would conclude that shot peening cannot be used to improve fatigue performance at elevated temperatures. However, an additional surface treatment which reduces the surface roughness, in the present case electropolishing (SP + EP), demonstrates that work hardening in the surface layer can also be exploited to improve the endurance limit, irrespective of whether a stress relief treatment is applied (SP + SR + EP) or not. The degree of cold work in the surface layer is not significantly altered by the stress relief treatment (figure 4) and thus raises the fatigue limit.

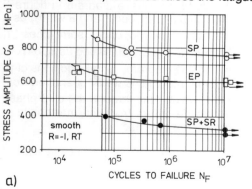
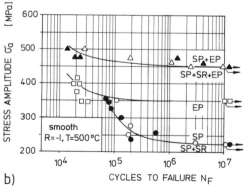

a)   b)

Figure 3: S-N curves for Ti-6Al-4V with a fine lamellar microstructure [Ref. 3]
a) room temperature   b) 500 °C

Figure 4: Transmission electron micrographs showing the near surface dislocation density in Ti-6Al-4V (fine lamellar) a) before surface treatment b) after shot peening and heat treating 1 h 600 °C [Ref. 3]

2. PROCESSING AND PROPERTIES IN TITANIUM ALLOYS

2.1 Background   Pure titanium exists in two stable crystal structures: the low temperature, hexagonal α phase and the high temperature, body-centered-cubic β phase. The β phase is readily workable owing to its high ductility, while the α phase, though not exactly brittle, is considerably less ductile and exhibits anisotropic properties [4-8]. (These texture effects, which can be very pronounced in titanium alloys, will not be discussed in this paper.) The widely quoted transition temperature of 882 °C is only valid for laboratory-pure titanium; the inevitable presence of oxygen in commercially available titanium raises this transition temperature [9]. The single most important consideration for the heat treat-ment or thermomechanical processing of titanium alloys is this transition, or β-transus temperature.

The most important alloying element in high-strength titanium alloys is aluminum, which stabilizes the α phase and, in amounts of more than roughly 4 wt. %, forms a streng-thening Ti_3Al (α_2) precipitate [10]. This α_2 precipitate is coherent with the α matrix and forms very rapidly. Oxygen promotes this phase reaction, thus commercial usually contain up to 0.2 % oxygen. The schematic pseudo-binary phase diagram in figure 5 depicts the phase stability in titanium alloyed with aluminum (between 3 and 8 wt. %) as a function of temperature and amount of β-stabilizers (usually V, Mo, or Nb, sometimes Cr, Fe) and provides guidelines for heat treatment. Titanium alloys are classified as α, $\alpha+\beta$, or (meta-stable) β-alloys depending upon their location in this diagram. Small additions of β-stabilizers make possible a martensitic transformation if material is water-quenched from high temperature. The martensite start temperature, which runs through the $\alpha+\beta$ phase field, is shown as the broken line. Titanium martensite differs from the martensite in carbon steels in that no appreciable gain in strength can be obtained.

The table on the next page lists selected commercial titanium alloys of current interest according to type.

2.2 α - Alloys The α-alloys are considered to be non-heat-treatable, since they either contain little or no alloying elements which influence phase stability, or else they contain so much aluminum that the precipitate reaction $\alpha \rightarrow \alpha + \alpha_2$ is accelerated to the point

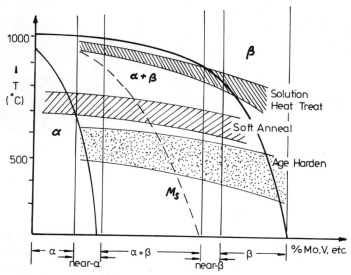

Figure 5: Schematic pseudo-binary phase diagram for Ti-xAl alloyed with β-stabilizing elements

α-Alloys
C.P. Titanium Grades 1-4 (commercially pure)
Ti-0.2Pd (Grade 7)
Ti-0.2Pd (Grade 11) (reduced oxygen and iron content compared to Grade 7)
Ti-0.8Ni-0.3Mo (Grade 12)
Ti-5Al-2.5Sn (Grade 6)

near-α-Alloys
Ti-6Al-2Sn-4Zr-2Mo (Ti-6242)
Ti-6Al-2.7Sn-4Zr-0.4Mo-0.45Si (Ti-1100)
Ti-5.8Al-4Sn-3.5Zr-0.7Nb-0.5Mo-0.35Si-0.06C (IMI 834)

α+β-Alloys
Ti-6Al-4V (Grade 5)
Ti-6Al-4V ELI (Grade 5 ELI) (reduced oxygen content compared to Grade 5)
Ti-6Al-6V-2Sn
Ti-6Al-2Sn-2Zr-2Cr-2Mo (Ti-62222)
Ti-3Al-2.5V

near-β-Alloys ("solute-lean")
Ti-10V-2Fe-3Al (10-2-3)3
Ti-5Al-2Sn-4Zr-4Mo-2Cr-1Fe (Beta-CEZ)

β-Alloys ("solute-rich")
Ti-15V-3Al-3Cr-3Sn
Ti-3Al-8V-6Cr-4Mo-4Zr (Beta-C)
Ti-15Mo-2.7Nb-3Al-0.2Si (Beta 21S)
Ti-6V-6.2Mo-5.7Fe-3Al (Timetal 125)

Figure 6: Microstructures of Ti-8Al a) fine grained b) coarse grained [Ref. 11]

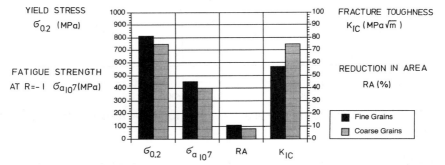

Figure 7: Representative mechanical properties for Ti-8Al aged 10 h 500 °C [Refs. 11,12]

where it cannot be suppressed by even a water quench.   Aside from cold working, the only possibility for microstructural control is via grain size (figure 6). This can be achieved by the traditional methods of cold-working and recrystallizing. Fine-grained material generally exhibits higher strengths and fatigue limits than coarse-grained material, while coarse-grained material has better resistance to fatigue crack growth and higher fracture toughness (figure 7) [11-13].

2.3 (α+β)- and near-α-Alloys The (α+β)-and near-α alloys offer wide possibilities for microstructural control [1]. Both the amount and distribution of the α and β phases can be varied, and the α phase can be hardened by α_2 precipitates. The near-α alloys are usually those selected for service temperatures up to about 600 °C. The dominant microstructural feature for these alloys is the phase morphology.

If heated above the β-transus temperature, the high-temperature microstructure consists exclusively of β grains which *transform* upon cooling to a structure which reveals the prior β grain size. The α phase tends to form as lathes or plates whose dimensions depend on cooling rate (figure 8a). A slow (furnace) cool produces a coarse lamellar microstructure, while a fast (air) cool produces a fine lamellar microstructure (figure 9). At sufficiently high cooling rates, as in the case of a water quench for thicknesses of less than about 30 mm, martensite (hexagonal α' or orthorhombic α") is generated [14].

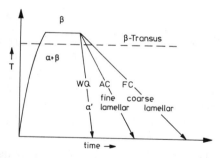

Figure 8: Heat treatment cycles (schematic) which generate transformed β-microstructures in (α+β) alloys

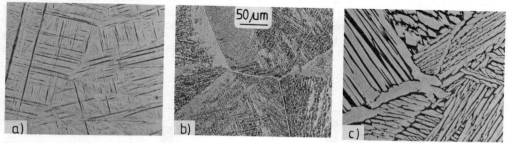

Figure 9: Transformed β microstructures in Ti-6Al-4V a) martensite b) fine lamellar c) coarse lamellar

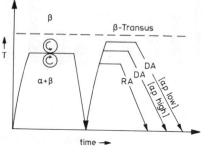

Figure 10: Thermomechanical processing cycles which generate fine microstructures in (α+β) alloys

Figure 11: Representative fine microstructures in Ti-6Al-4V
a) equiaxed b) duplex (40 % α_p) c) duplex (20 % α_p)

Completely different microstructures are obtained by hot working in the α+β phase field and then recrystallizing at sub-transus temperatures (figure 10). The presence of both phases during hot-working prevents the coarsening which occurs when only β grains are present. If the subsequent recrystallization anneal (RA) is performed below the martensite start temperature, a fine microstructure consisting of α and β grains (figure 11) forms. Above the martensite start temperature, the fine β grains transform in the same way as described earlier, and the microstructure consists of α grains and fine lamellar regions ("duplex" microstructure [7,15,16]). The duplex annealing (DA) temperature determines the relative amounts of α and fine lamellae (transformed β) according to the lever rule. A final aging treatment of 2-24 h at 450-600 °C brings out the α_2 phase.

The transformed β, or lamellar microstructures generally exhibit good fracture mechanics behavior (fracture toughness, long through-crack da/dN) and creep resistance, but poor low cycle fatigue (LCF) strength and ductility. Fine microstructures possess higher ductilities and LCF strength, but lower fracture toughness, higher crack growth rates and poorer creep resistance than the lamellar microstructures [15-19]. This is summarized in figure 12.

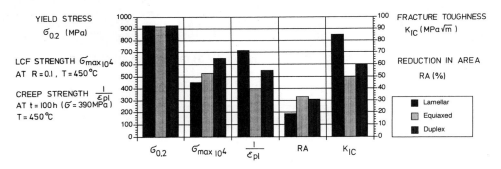

Figure 12: Typical mechanical properties for selected microstructures in Ti-6242 [Ref. 17]

2.4 β -Alloys The (metastable) β-alloys are defined as those alloys whose compositions are within the α+ β phase field, but whose β-stabilizer content is so high that the martensite start temperature is below room temperature [20, 21]. In principle, micro-structural development is the same as for transformed β microstructures in (α+β) alloys. However, nucleation and growth of the α phase are extremely slow, owing to the high amount and molecular weight of the β-stabilizing elements. The β-alloys can be heat treated by solutionizing and aging. The solute-lean sub-group of the metastable β-alloys, (also known as near-β alloys) can exhibit volume fractions of primary (coarse) α comparable to those found in (α+β)-alloys. As in the (α+β) alloys, primary α tends to be lathlike unless material is (α+β) worked and recrystallized. Solute-lean β-alloys should be solution heat treated below the β-transus to avoid even the remote possibility of a martensitic transformation upon cooling. The transformation kinetics in the more heavily stabilized solute rich β-alloys are so sluggish that primary α does not form. Microstructures in these alloys thus consist either of β grains only or β grains with secondary (precipitate) α. Solute-rich β-alloys are solution heat treated above the β-transus, where the choice of temperature is dictated by whether material should recrystallize or not. The distinction between primary and secondary α is arbitrary, since the only difference is the size of the α phase. Exactly this pronounced dependence of the α precipitate size on temperature makes the β alloys heat treatable. The critical parameter is the <u>age hardening</u> temperature. Low aging temperatures (below roughly 500 °C) produce higher strengths, but require longer aging times than higher temperatures. Cold working is effective in accelerating the β→β + α transformation, and is recommended when very high strengths (above 1500 MPa) are desired [21]. Heat treatment cycles are shown schematically in figure 13. Typical microstructures showing β-grains, β-grains with coarse primary α and β-grains with fine secondary α are shown in figure 14.

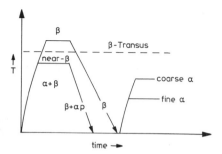

Figure 13: Heat treatment cycles for metastable β-alloys (schematic)

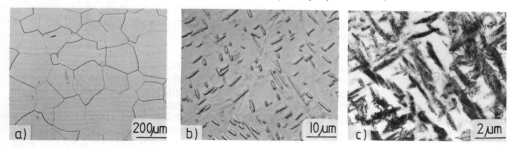

Figure 14: Microstructures in β-alloys a) Ti-3Al-8V-6Cr-4Mo-4Zr as-SHT b) Ti-10-2-3 with coarse primary α c) Ti-3Al-8V-6Cr-4Zr-4Mo aged to produce fine secondary α

If no aging treatment is performed, a solutionized solute-rich β-alloy is characterized by a very high ductility and good fracture toughness, but low strength and fatigue limit. Aging increases strength and (up to a point) fatigue limit at the expense of ductility and fracture toughness. Figure 15 illustrates two distinctly different possibilities for Ti-3Al-8V-6Cr-4Mo-4Zr. By varying the aging temperature, it is possible to achieve property combinations between these two extremes.

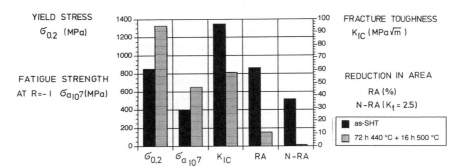

Figure 15: Mechanical property profiles for Ti-3Al-8V-6Cr-4Mo-4Zr in two extreme conditions a) as solution heat treated b) aged to achieve a high strength condition [Ref. 22,23]

3. THERMOMECHANICAL SURFACE TREATMENT

As shown in the previous section, the mechanical properties associated with a given microstructure can be markedly superior in some respects while inferior in others, or may represent a compromise. Since the surface of a mechanically loaded part often experiences different service conditions than the bulk, it can make sense to "tailor" microstructural variations from the surface to the interior to meet the differing requirements, e.g., as in the carburizing of steels. As demonstrated in the following examples, cold-working induced by mechanical surface treatments can be utilized to develop a surface microstructure which is different from that in the bulk, thus combining the optimum features of both, even in cases where conventional thermomechanical processing may not be practical, as in thick sections. A distinct advantage to be gained by altering the surface microstructure is that such alterations are more stable than those induced by mechanical surface treatments alone.

3.1 α -Alloys   A mechanical surface treatment in combination with a subsequent recrystallization offers the possibility to combine the high strengths and endurance limits associated with fine grains with the superior long through-crack fatigue crack growth behavior and fracture toughness of the coarse grains. To maximize the total fatigue life in thicker sections, fine grains are needed on the surface, where good resistance to crack initiation is critical, and coarse grains in the interior, where they can reduce the driving force for long crack growth. Shot peening followed by a heat treatment of 1 h 820 °C was

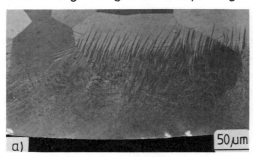

Figure 16: Near-surface microstructures in Ti-8Al a) as-shot peened b) shot peened and locally recrystallized [Ref. 24]

performed on coarse grained Ti-8Al to <u>cold-work and recrystallize</u> the surface (figure 16). The improvement in fatigue limit owing to the fine (20 μm) surface grains as opposed to the coarse grains in the bulk (100 μm) is significant, roughly 50 MPa at 350 °C.

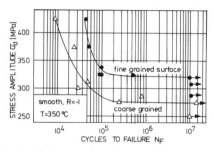

Figure 17: S-N curves for coarse grained Ti-8Al at 350 °C with and without thermomechanical surface treatment for local grain refinement [Ref. 24]

<u>3.2 (α+β)- and near-α-Alloys</u> Because these alloys are often intended for high temperature service, for example in gas turbines, creep resistance is an important consideration. On this basis, lamellar microstructures would be preferable. However, these microstructures have poor fatigue resistance, particularly in the LCF regime, where surface crack growth determines fatigue life. In such cases, a variation in phase morphology between the surface and the core can be desirable. Figure 18 shows examples for an (α+β) and a near-α alloy where fine surface microstructures were obtained by mechanically working the surface by shot peening and then heat treating. The improvement in S-N behavior (at high temperature) gained by this thermomechanical surface treatment is shown for Ti-6242 with a creep-resistant, fine lamellar core and a fatigue-resistant, fine equiaxed surface layer in figure 19.

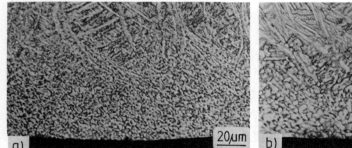
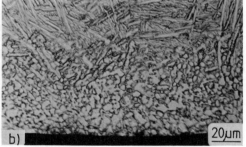

Figure 18: Near-surface microstructures after thermomechanical surface treatment
a) Ti-6Al-4V with a lamellar core and equiaxed surface microstructure
b) Ti-6242 with a lamellar core and duplex surface microstructure [Ref. 24]

<u>3.3 β -Alloys</u> Both shot peening and surface rolling in combination with specially developed aging treatments have been applied to Ti-3Al-8V-6Cr-4Mo-4Zr to <u>selectively age harden</u> only the surface [25,26]. This new thermomechanical surface treatment shows promise for improving properties of high strength springs and fasteners. Figure 20 shows the near-surface region for shot-peened material both without and with a selective surface aging (SSA) treatment. The high strength of the surface increases the fatigue limit

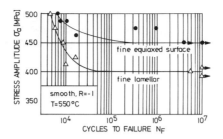

Figure 19: S-N curves for Ti-6242 at 550 °C with a fine lamellar microstructure with and without a thermomechanical surface treatment [Ref. 24].

to values above those of a conventional bulk-aging treatment, while the high ductility of the solution heat treated (SHT) condition in the interior provides good notched ductility [26] and toughness. Fully reversed notched fatigue behavior for this alloy after surface rolling with and without SSA is shown in figure 21. Rolling alone increases the notched fatigue limit of the SHT condition (expressed as nominal stress times stress concentration factor) from the low value of 400 MPa to 1100 MPa. Depending on the subsequent aging treatment, the S-N behavior can deteriorate slightly (SSA1) or be even further improved (SSA2). These results suggest that the residual compressive stresses present in the as-rolled condition are significantly relieved by the SSA1, but not the SSA2 treatment.

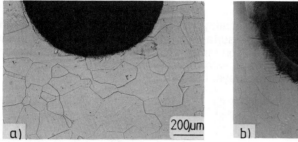

Figure 20: Near-surface microstructures in Ti-3Al-8V-6Cr-4Mo-4Zr a) after shot peening b) after selective surface aging following shot peening

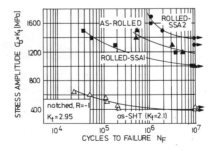

Figure 21: S-N curves for Ti-3Al-8V-6Cr-4Mo-4Zr in the as-SHT and after rolling both with and without subsequent aging treatments

CLOSURE

Mechanical surface treatments such as shot peening and surface rolling can be applied alone or in combination with heat treatments to obtain optimum properties in mechanically loaded titanium parts. The particular treatment applied should reflect the type of alloy (α, $\alpha+\beta$, or β), to make use of its characteristic response to heat treatment and/or thermo-mechanical processing.

ACKNOWLEDGMENTS

The authors would like to thank Prof. Kloos, TH Darmstadt, for providing the facilities for surface rolling and Mr. Schaffner for performing the rolling as well as Mr. Mann (GKSS) for technical assistance. Part of this work was financed by the Deutsche Forschungs-gemeinschaft.

REFERENCES

1) Properties of Titanium Alloys, ASM Metals Handbook, Ninth Edition, Vol. 3, 1980, 372.

2) Wagner, L. and Lütjering, G.: in Shot Peening, ed. A. Niku-Lari, Pergamon Press, Oxford, 1982, 453.

3) Gray, H., Wagner, L. and Lütjering, G.: les editions de physique, 1988, 1895.

4) Larson, F.R. and Zarkades, A.: Metals and Ceramics Information Center Report, Columbus, Ohio, June 1974.

5) Sommer, A.W. and Creager, M.: Air Force Materials Laboratory Technical Report 76-222, June 1974.

6) Bowen, A.W. Acta Metall., 1975, 23, 1401.

7) Peters, M.: Doctoral Thesis, Ruhr University Bochum, 1980.

8) Lütjering, G. and Wagner, L.: in Directional Properties of Materials, ed. H.J. Bunge, Deutsche Gesellsschaft für Material kunde, Oberursel, 1988, 177.

9) Okazaki, K. and Conrad, H.: in Titanium Science and Technology, eds. J.C. WIlliams and A.F. Below, Plenum Press, 1982, 429.

10) Blackburn, M.J. and Williams, J.C.: Trans. ASM, 1969, 62, 389.

11) Müller, C., Gysler, A. and Lütjering, G.: in Fracture Control of Engineering Structures, EMAS, Ltd., West Midlands, UK, 1986, 1879.

12) Wagner, L., Gregory, J.K., Gysler, A. and Lütjering, G.: in Small Fatigue Cracks, eds. R.O. Ritchie and J. Lankford, TMS-AIME, Warrendale, PA, 1986, 117.

13) Lütjering, G., Gysler, A. and Wagner, L.: les editions de physique, 1988, 71.

14) Bagariatskii, L.A., Nosova, G.I. and Tagunova, T.V.: Soviet Phys. Doklady English Transl., 1959, 3, 1014.

15) Jaffee, R.I., Wagner, L. and Lütjering, G.: les editions de physique, 1988, 1501.

16) Wagner, L., Lütjering, G. and Jaffee, R.I.: in Fatigue '90, Materials and Component Engineering Publications, Birmingham, UK, 1990, 261.

17) Saal, S., Wagner, L., Lütjering, G., Pillhöfer, H. and Däubler, M.A.: Zeitschrift Metallkunde, 1990, 90, 535.

18) Williams, J.C., Chesnutt, J.C. and Thompson, A.W.: in Microstructure Fracture Toughness and Fatigue Crack Growth Rate in Titanium Alloys, eds. Chakrabati and J.C. Chesnutt, TMS-AIME, Warrendale, PA, 1987, 255.

19) Chesnutt, J.C., Rhodes, C.G. and Williams, J.C., ASTM-STP 600, Philadelphia, PA, 1976, 99.

20) Duerig, T.W. and Williams, J.C.: in Beta Titanium Alloys in the 1980s, eds. R.R. Boyer and H.W. Rosenberg, TMS-AIME, Warrendale, PA, 1984, 19.

21) Ankem S., and Seagle, S.R.: in Beta Titanium Alloys in the 1980s, eds. R.R. Boyer and H.W. Rosenberg, TMS-AIME, Warrendale, PA, 1984, 107.

22) Gregory, J.K. and Wagner, L.: VII International Meeting on Titanium organized by Ginatta Torino Titanium, Turin, Italy, November 15, 1991) also GKSS Report GKSS 92/E/7, Geesthacht, Germany.

23) Wagner, L. and Gregory, J.K.: in Beta Titanium Alloys for the 1990s, ed. D. Eylon, TMS-AIME, Warrendale, PA, in press.

24) Gray, H., Wagner, L. and Lütjering, G.: in Shot Peening, eds. H. Wohlfahrt, R. Kopp and O. Vöhringer, Deutsche Gesellschaft für Materialkunde, Oberursel, 1987, 467.

25) Gregory, J.K., Wagner, L. and Müller, C.: in Surface Engineering, ed. P. Mayr, 1993, Deutsche Gesellschaft für Materialkunde, Oberursel, 435.

26) Gregory, J.K. and Wagner, L.: in Fatigue '93, Montreal, Canada (in press)

Materials Science Forum Vols. 163-165 (1994) pp. 173-180

THERMOMECHANICAL TREATMENT OF AA6061

J. Mironi

RAFAEL, Armament Development Authority,
P.O. Box 2250, Haifa, Israel

ABSTRACT

By definition, a thermomechanical treatment (TMT) is considered to be a combination of thermal and mechanical (plastic) deformation aiming to change the structures and properties of an alloy. During TMT the modifications obtained are greater than the results obtained by adding those induced by thermal and mechanical treatment taken separately.

AA6061 is one of the most important alloys employed in engineering due to its natural corrosion resistance. Its use is limited by the low mechanical properties, compared to the 2xxx and 7xxx families.

The aim of this presentation is to describe a TMT process aiming to improve the mechanical properties well over the natural properties of the alloy, while retaining an acceptable ductility.

The process involves a preliminary low temperature ageing at 80°C, plastic deformation of 40–90% and final artificial ageing at 165°C. As a result of this TMT, the ultimate and yielding strengths are enhanced by 40% over the minimum specification requirements, while losing only 2% in elongation.

INTRODUCTION

As stated by McQueen [1] a thermomechanical treatment (TMT) is a combination of thermal treatment and plastic deformation whose result is a change in microstructure and properties. The metallurgical mechanisms involved in the TMT of aluminum alloys are:

- Phase precipitation by the Guinier Preston (GP) three stages mechanism.
- Dislocation creation, multiplication, rearrangement and annihilation.
- Grain boundaries changes.
- Interaction between dislocations grain boundaries and precipitates.

According to [2], there are three types of TMT, considering the temperature of the plastic deformation stage relative to the recrystallization temperature:

- Low temperature TMT – performed under 0.5 Tm (150–160°C for aluminum alloys)
- High temperature TMT – performed above 0.5 Tm
- Preliminary TMT – the plastic deformation is performed together with the thermal treatment and prior to the hardening thermal treatment.

Experimental work on TMT of aluminum alloys was performed for the following: 7075 [3–8]; 2024 [9–10]; 5083 [11–14]. The improvements obtained were in mechanical properties, thermal stability and stress corrosion.

Data concerning the TMT of AA6061 is limited [15–19] and deals with artificial–clean compositions [15–16] or with controlled Si/Mg ratio (Mg_2Si precipitates) [17–19], but not with commercial alloys. Using various TMT cycles (preageing, plastic deformation and final ageing) [15] succeeded to improve the yield and ultimate strength of a pure 6061, the strength being similar to that of an AA2024–T351 alloy but with ductility half the minimum specification requirements.

The aim of this research was to find a TMT able to improve the mechanical properties of a commercial grade AA6061–T651 alloy without losing ductility under the minimum specified by QQ–A–250/11F.

EXPERIMENTAL

Plates 7/8" thick of AA6061–T651 commercially manufactured by Kaiser–Aluminum–USA were used. The mechanical properties were tested (six longitudinal specimens) and compared with the minimum requirements of QQ–A–250/11F.

Generally, the TMT proposed consists of several steps, as follows:

AR + W + Pre Ageing + Plastic Deformation + Final Ageing
                T°C/t hrs   T°C %   T°C/t hrs.

W – Solution treating: 540°C for 1½ hours, quenching in 20% Aqua Quench solution
PA – Preageing at 80; 100; 120°C and 1÷24 hrs
PD – Plastic deformations at room temperature, 80°C; 120°C and 25%; 40%; 90% deformation by rolling
FA – Final aging at 165; 180; 200°C and 1÷24 hrs.

The program was based on three phases according to the PD.

Phase I – 25% Plastic Deformation

The TMT cycle was as follows:

W	+	PA	+	PD	+	FA
		80;100;120°C		RT;80;100;120°C		165;180;200°C
		1 to 24 hrs.		25%		1 to 24 hrs.

Due to the multiplicity of combinations (cca 3000) a first evaluation was performed using hardness DPH_{5kg} as an indicating value, as performed by Kaneko [18]. The T651 hardness was 90–100 DPH. After having tested thermomechanically treated cupons, the hardness changed in the range 82–118 DPH covering values between 10% lower and 20% higher than the initial T651 condition. It was observed that 120°C PA is inefficient as well as 200°C FA, the hardness being in the lower 10% range. Tensile specimens representing some extreme TMT cycles were tested to find a possible correlation between hardness and tensile properties. The TMT conditions were as follows:

PA	+	PD	+	FA
80°C, 4 and 24 hrs		RT;80;100°C;25%		165; 180°C; 4 and 24 hrs.

Also, a set of specimens without PA was checked in order to evaluate its influence and whether the treatment proposed is a TMT by definition or not. The tensile testing results (six specimens for each set of conditions) are summarized in Table 1.

Table 1: Mechanical Properties After Phase I (25% PD)

W	PA		PD		FA		σ_y [kg/mm²]	σ_{uts} [kg/mm²]	$\in$ [%]	
	T(°C)	t(h)	T(°C)	t(h)	T(°C)	t(h)				
540°	80	4	25	RT	165	4	34	36	14	
		4				24	34	36	14	
		24				4	34	37	16	
		24				24	35	38	14	
540°	80	4	25	RT	180	4	35	36	13	
		4				24	35	37	14	
		24				4	33	34	14	
		24				24	32	33	14	
540°	–	–	25	RT	165	8	31	33.5	15	Without PA
						18	29	30.5	13	
						24	32	33.5	12	
540°	–	–	25	RT	180	8	30.5	32	13	Without PA
						18	32	34	14	
						24	32.5	34	13	
540°	–	–	–	–	–	–	29.7– 31.1	31.2–32.6	14–17	6061– T651
540°	–	–	–	–	–	–	24.5	25.9	9	Std. min. required

It is evident that without the PA step there is no improvement compared with the T651 material. The tensile strength obtained for the tested TMT is 10% higher than that of the T651 material, meaning that the proposed treatment is really a thermomechanical one.

Phase II – 40% Plastic Deformation

A reduced schedule of TMT was used, as follows:

W   +   PA   +   PD   +   FA
        80; 100°C; 4 and 24 hrs.   RT;80;100°C;40%   165;180°C; 4 and
                                                                        24 hrs

After tensile testing of the specimens, it was found that:

– Mechanical properties improved by 25%.
– 24 hours of PA were more effective than 4 hours of PA.
– 4 hours of FA were more effective than 24 hours of FA.
– PD temperature does not influence the final results and room temperature is economically advantageous.
– No difference was observed while preageing at 80°C or 100°C.

Phase III – 90% Plastic Deformation

Finally, a limited schedule was selected, as follows:

W   +   PA   +   PD   +   FA
        80;100°C; 24 hrs.   RT; 90%   165°C; 4 hrs.

The mechanical testing results are summarized in Table 2, together with those obtained at 40%, 25% PD and the T651 condition.

Table 2: Mechanical Properties after TMT at Various % PD

TMT				Mechanical Properties		
W	PA	PD	FA	σ_y [kg/mm^2]	σ_{uts} [kg/mm^2]	$\in$ [%]
540°C–1½ h	80°C–24 h 100°C–24 h	90%–RT 90%–RT	165°C–4 h 165°C–4 h	39.1–39.5 38.0–38.2	40.7–41.5 39.0	6–9 4–5
	80°C–24 h 100°C–24 h	40%–RT 40%–RT	165°C–4 h 165°C–4 h	37.0 37.0	39.0 39.0	12 12
	80°C–24 h 100°C–24 h	25%–RT 25%–RT	165°C–4 h 165°C–4 h	34.0 36.0	37.0 38.0	16 14
6061–T651	–	–	165°C–18 h	30.0	31.7	15
QQ–250/11F	–	–	–	24.5	29.5	9

It is evident that the TMT at 80°C PA leads to higher mechanical properties. The strength values are 30% higher than the as received T651 condition and 40% higher than the minimum specified requirements, while the elongation obtained is slightly lower than the QQ requirements.

DISCUSSION

Looking at the results obtained during the experimental part of the research, it can be inferred that the mechanical properties present a complex dependency on the main stages of the TMT (PA, PD, FA) and their characteristic parameters (time, temperature, plastic deformation). The results obtained represent an improvement in tensile strength and a decrease in elongation. Looking at Table 2 and plotting the results as a function of the plastic deformation (Figure 1), it can be seen that in order to maintain the minimum elongation requirements for the T651 condition, a 65% plastic deformation is required thus lowering the yield strength to cca 38 kg/mm² and the ultimate strength to cca 40 kg/mm², both of which are well higher than those required by the specifications.

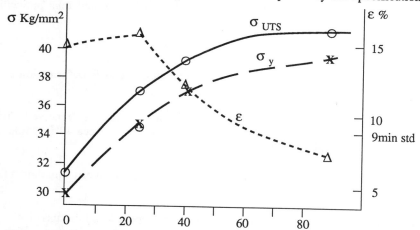

Figure 1. Mechanical Properties of TMT 6061 Alloy at Various PD Levels
[PA 80°C/24 hrs. + PD/RT + FA 165°C/4 hrs.)

During the PA stage the precipitation mechanism is active, creating a high density of GP zones. These GP zones do not allow free movement of dislocations, which are pinned [16]. The GP zones are stable up to 125°C [2] thus explaining the observation on the ineffectiveness of the 120°C preageing treatment.

During the PD stage, the dislocations induced are forced to cut the GP zones, leading to their multiplication, resolution or creation of new precipitation (new GP zones). Up to 25% plastic deformation a Tangled Dislocation Network (TDN) is created. Its distribution is determined by the existing GP zones [15]. Between 25–90% plastic deformation, the TDN is reordered. It coalesces and forms a grain substructure. During the final ageing (FA), the precipitation mechanism lasts until finer Mg_2Si precipitation occurs (due to previous higher density of GP zones) in grains having a marked subgrain structure. Consequently, the plastic deformation induced in the natural precipitation mechanism is responsible for the improved mechanical properties of the thermomechanically treated alloy.

The proposed thermomechanical treatment is:

Preageing (80°C/24 hrs) + Plastic Deformation (RT/90%) + Final Ageing (165°C/4 hrs)

representing the best cycle obtained during this research. We believe that it can be further improved or modified in order to shorten it and make it more attractive economically.

Further, the proposed TMT when employed as a fabrication cycle, can be used for rolling plates, cold extrusion of profiles, shear spinning of thin tubes or ring rolling of circular segments.

Certainly, the process is limited by the low temperature stability of the structure (cca 160–200°C) and material exposure to high temperatures has to be avoided thus preventing welding of the TMT alloy.

CONCLUSIONS

A practical TMT was defined. It is composed of preageing, a plastic deformation and a final ageing stage.

The mechanical properties obtained are higher than those of the as "received T651" material by cca 30% and higher than the minimum requirements by cca 40%, while still maintaining an acceptable elongation.

Further improvements on the proposed cycle are possible, aiming to shorten treatment time and improve ductibility.

A metallurgical mechanism for alloy strengthening was proposed and discussed.

REFERENCES

1)   McQueen, H.J.: Thermomechanical Processing of Aluminum Alloys, St. Louis, Missouri, October 18, 1978.
2)   Rabinovich, M.K.: Foreign Technology Division, Wright Patterson Air Force Base, Ohio, January, 1974.
3)   Conserva, M., Buratti, M., DiRusso, E. and Gatto, F.: Mat. Sci. Eng., 11, 1973.
4)   DiRusso, E., Conserva, M., Gatto, F. and Markus, H.: Met. Trans., 4, 1973.
5)   DiRusso, E., Conserva, M., Buratti, M. and Gatto, F.: Mat. Sci. Eng., 14, 1974.
6)   Waldman, J., Sulinski, H. and Markus, M.: Met. Trans., 5, 1974.
7)   Osterman, F.: Met. Trans., 2, 1971.
8)   Mehrpay, F., Kudsin, D.L. and Haworth, W.L.: Met. Trans., 7A, 1976.
9)   Buratti, M., Conserva, M. and DiRusso, E.: Aluminio, 13, 1974.
10)   Pattanaik, S., Srinivasan, V. and Bhatea, M.L.: Scr. Met., 6, 1972.

11) Titchener, A.L. and Ponniath, C.D.: Proc. of the Third Int. Conf. on Strength of Metals and Alloys, 1973, Inst. of Metals, London.

12) McEvily, A.J., Snyder, R.L. and Clark, J.B.: Trans. AIME, 227, 452, (1963).

13) Conserva, M. and Leoni, M.: Met. Trans., 6A, 189, (1975).

14) Edstrom, C.M. and Blakeslee, J.J.: Atomic International Division, Golden Co., Rocky Flats Plan, Contract AC04–76DP03533.

15) Rack, H.J. and Krenzer, R.W.: Met. Trans., 8A, February 1977.

16) Swearengen, J.C.: Mat. Sci. Eng., 10, (1972).

17) Benedyk, J.: Light Metal Age, April 1968.

18) Juniki Kaneko: J. Japan Inst. Light Metals, 27, (1977).

19) McQueen, H.J.: J. Met., February 1980.

Materials Science Forum Vols. 163-165 (1994) pp. 181-188
© 1994 Trans Tech Publications, Switzerland

INFLUENCE OF PRELIMINARY HEAT TREATMENTS ON MICROSTRUCTURE, MECHANICAL PROPERTIES AND CREEP BEHAVIOUR OF A 50Cr-50Ni NIOBIUM CONTAINING ALLOY

G. Caironi, E. Gariboldi, G. Silva and M. Vedani

Dipartimento di Meccanica, Politecnico di Milano,
P.za Leonardo da Vinci, 32, I-20133 Milano, Italy

ABSTRACT

Studies on mechanical behaviour and creep properties at $700°C$ were carried out on a 50Cr-50Ni-Nb casting alloy both in the as cast condition and after preliminary heat treatments at 1000, 1075, $1150°C$. The creep behaviour was related to the microstructure, which in the as cast condition undergoes drastic precipitation phenomena at this temperature. The preliminary annealing at the afore mentioned temperatures gave rise to a coarse precipitation that modified the mechanical properties of the alloy.

INTRODUCTION

A heat resisting 50Cr-50Ni niobium-containing casting alloy, particularly developed for petrochemical and power steam plants was studied. The material combines high temperature strength with excellent fuel ash corrosion resistance [1-3]. In industrial components, usually utilized in the as cast condition, the exposure to high temperatures during service gives rise to precipitation phenomena. Within a hundred hours of maintenance at $700°C$, for instance during slow pre-heatings of furnaces and boilers, the aging process causes a fine, massive, precipitation significantly altering the microstructure and embrittling the material [4-5].

Proper preliminary thermal treatments, leading to the occurrence of a first precipitation, change both the size and distribution of the precipitates, that develop during exposure to service temperature, thus modifying the mechanical characteristics of the material.

Studies on aging at $700°C$ were carried out either in the as cast condition or after annealing at 1000, 1075, $1150°C$. Microstructural modifications on a set of samples have been studied and compared, correlating them to the evolutions of mechanical behaviour brought about by aging.

MATERIAL

A nominal 50Cr-50Ni alloy corresponding to the ASTM A560 50Cr-50Ni-Nb alloy was studied. The material was supplied in cast ingots according to ASTM A370 standard. The main elements characterizing the chemical composition of the material are given in table I.

Table I - Chemical composition (wt.%) of the examined 50Cr-50Ni-Nb alloy.

Cr	Nb	C	N	Ni
48.35	1.50	0.06	0.07	bal.

The microstructure of the as cast alloy consisted in a γ-Ni phase matrix and in interdendritic regions rich in α-Cr phase, as appears from the micrographs in fig. 1a, representing the microstructural features of electrolytically etched (10 vol.% oxalic acid) samples. The cooling rate during the solidification of the casting did not allowed an equilibrium condition to be reached. Thus, both the γ matrix and α interdendritic regions are supersaturated in chromium and nickel, respectively [4,6].

EXPERIMENTAL AND RESULTS

Preliminary heat treatments allowed to investigate operative behaviour of components where precipitation had already taken place. Three annealing treatments, at 1000, 1075, 1150 °C for 1 hour, were accomplished. The development of precipitation phenomena was monitored at 700 °C, already known as the most severe operating condition [7] even though usual service temperatures lie in the range 800-900 °C. Thus, four samples were aged at this temperature for 100 h. Correspondingly, tensile tests were performed both on as cast or simply annealed material and on the aged conditions. A set of creep tests were carried out at 700 °C with 200MPa stress levels. This allowed to relate the tensile and creep behaviour to the microstructural and hardness evolutions already studied on a different heat of 50Cr-50Ni-Nb alloy and presented in another work [5].

The annealing treatments caused an oriented and elongated α-Cr precipitation to occur into the γ-Ni matrix, as illustrated in figure 1. A corresponding γ-Ni finer precipitation was noticed in the α-Cr phase of the above mentioned interdendritic regions. The size of α precipitates increased and they coarsened as the annealing temperature ranged from 1000 to 1075 and 1150 °C. Meanwhile, the amount of these precipitates, similar in the lower and medium temperature annealing conditions, is considerably smaller after heat treatment at 1150 °C. Furthermore, precipitate free zones, wider in this latter material condition, surrounded interdendritic Cr-rich zones.

Aging at 700 °C for 1h caused a fine precipitation to occur in all the examined conditions, as illustrated in figure 2. The precipitate size was smaller when a preceding depletion of chromium in the matrix had been brought about during annealing by the first α precipitation or by diffusion processes.

Data on the mechanical behaviour of the material in the examined conditions are listed in table II. In considering the results of tensile tests for the unaged conditions the effects of the harder α phase volume content, brought about by annealing treatments, and of the morphology of precipitates must be taken into account. The fairly good strength of the as cast condition was improved by the massive occurrence of α precipitation in the 1000 °C annealed material. At the same time, the finest size and the most elongated shape of the particles in this condition caused a drastic loss in ductility. Intermediate strength levels and elongations to fracture corresponded to intermediate microstructural features in other treatment conditions.

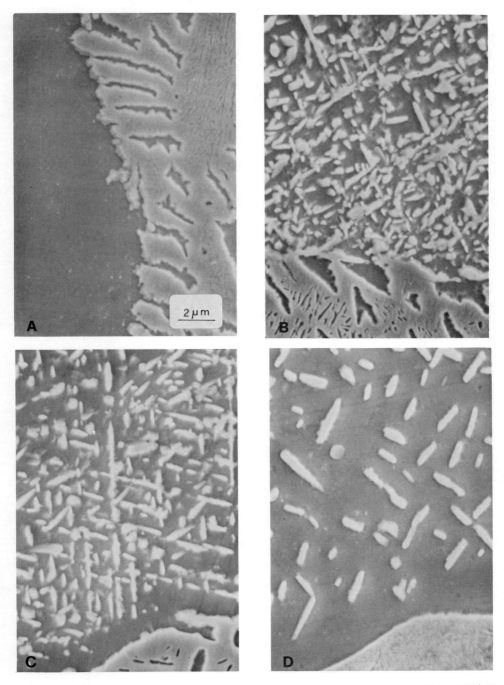

Figure 1 - Microstructure (SEM micrographs) of the examined 50Cr-50Ni-Nb alloy. a) as cast condition b) material annealed at 1000°C for 1h c) material annealed at 1075°C for 1h, d) material annealed at 1150°C for 1h

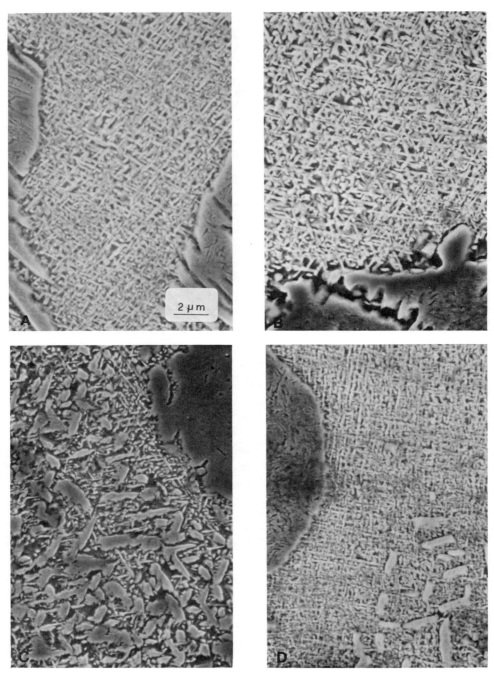

Figure 2 - Microstructural features (SEM micrographs) of the examined 50Cr-50Ni-Nb alloy. after aging for 100h a) as cast condition b) material annealed at 1000°C for 1h c) material annealed at 1075°C for 1h, d) material annealed at 1150°C for 1h.

Table II - Tensile data and Rockwell hardness of the examined material unaged or simply annealed at 1000, 1075, 1150°C and after aging for 100h. UTS: Ultimate Tensile Strength; YS: 0.2% offset Yield Strength; El: elongation; HRC: Rockewll Hardness C (in the as cast condition * indicates that Rockwell Hardness B was measured).

	unaged				aged for 100 h			
	UTS MPa	YS MPa	El %	HRC	UTS MPa	YS MPa	El %	HRC
as cast	705	448	22.3	97 *	1024	855	0.9	44
1000°C 1h	891	640	2.6	35	901	735	1.2	34
1075°C 1h	785	539	8.2	28.5	1023	772	2.2	39
1150°C 1h	743	467	15.7	20	1268	905	3.3	45

The secondary fine precipitation occurred during aging altered the outline of tensile behaviour of the four examined conditions. As expected [7], the as cast condition was strongly strengthened and embrittled by aging. The presence of both first and secondary precipitations of different morphology determined the differences between the characteristics of the annealed conditions. The lower the amount of secondary α precipitation, the lower the increment in strength. Thus, the massive secondary precipitation in the 1150°C annealed conditions led to the highest strengthening while almost constant ultimate tensile strength were noticed in 1000°C annealed condition. The secondary fine precipitation phenomena were also responsible for the depletion of ductility of the material with respect to the unaged conditions.

The HRC hardness values (HRB for the as cast condition) confirmed the correlation between this mechanical characteristic, the tensile strength and the amount of precipitates, this latter already observed by the authors in a study on the microstructural evolutions of the alloy in the same thermal treated conditions [5].

The creep tests, performed at elevated stress values, were aimed at developing a complete primary stage during the first hundred hours of testing, roughly corresponding to the complete occurrence of α precipitation phenomenon in matrix [5]. Thus, when the secondary stage was reached, an almost constant structure was supposed to be present. The evaluation of data during the first stage of the creep tests should take into account the structural phenomena occurring during pre-heating or associated to loading and creep deformation. The secondary creep rates, observed to be lower the greater the amount of the fine α particles, lay in the range 0.0007-0.01 %/h [9].

Examinations of the creep rupture surfaces evidenced that the fracture followed the interfaces between matrix and interdendritic regions, as illustrated in figure 3a. In the proposed example, the lower amount and coarsened morphology of the precipitates in the 1000°C condition (fig. 3b) evidenced the ductility of the matrix and allowed a greater strain to fracture to be reached.

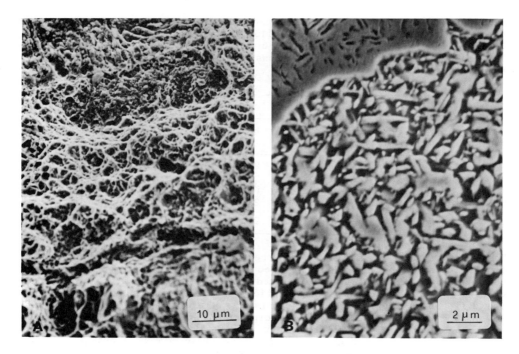

Figure 3 - Creep sample of material annealed at 1000°C (time to rupture 401h, 9.8% strain to fracture); a) fractography, b) microstructure (SEM micrograph).

CONCLUSIONS

Modifications of microstructure and mechanical behaviour brought about in a 50Cr-50Ni-Nb alloy by annealing at 1000, 1075, 1150°C for 1 hour were studied. They primarily consisted of the occurrence of an oriented, elongated precipitation of the α harder phase in the matrix. The particles were greater and coarser, even if in lower amount, as the annealing temperature passed from 1000 to 1075 and 1150°C. Correspondingly, the hardness and tensile strength increased with annealing temperature while elongation level, highest in the as cast condition, decreased when the annealing temperature was lowered.

The exposure at 700°C for 100h was observed to cause a secondary fine precipitation which strongly modified the mechanical behaviour of the as cast condition. The lower the amount of primary precipitation at the beginning of aging, the higher the amount of the secondary precipitation that increased the tensile strength and reduced the material ductility.

The creep strength at 700°C with a stress level of 200 MPa reflected the microstructural aspects of the material. Creep fracture surfaces preferentially included the boundary of the γ and α phases, this latter located in interdendritic regions.

ACKNOWLEDGMENTS

The authors would like to thank FIAS srl (I-21055 Gorla Minore) for having
supplied the material and technical information.

REFERENCES

1) Lewis, H., J. Inst. Fuels, 1966, Jan.
2) Penrice, P.J., Stepley, A.J., Towers, J.A., J. Institute of Fuels, 1966
Jan.
3) Swales, G.L., Ward, D.M., NACE Corrosion '79, Atlanta, Georgia, USA, 1979.
4) Balbi, M., Caironi, G., Silva, G., Tosi, G., Int. Conf. "Monitoring,
Surveillance and Predictive Maintenance of Plants and Structures", Taormina,
1989.
5) Caironi, G, Gariboldi, E., Silva, G., Vedani, M., EUROMAT '93, Paris, F,
1993.
6) Otero, E., Merino, M.C., Pardo, A., Biezma, M.V., Metallographia, 1988, 21,
217.
7) Ennis, P.J., INCO Europe Limited, 1974.
8) Balbi, M., Caironi, G., Silva, G., Tosi, G., Fonderia Italiana, 1979, 9,
225.
9) FIAS s.r.l.(Fonderie Italiane Acciai Speciali) technical data.

Materials Science Forum Vols. 163-165 (1994) pp. 189-194
© 1994 Trans Tech Publications, Switzerland

INTERDIFFUSION IN HIGH TEMPERATURE TWO PHASE, Ni-Cr-Al COATING ALLOYS

C.W. Yeung [1], W.D. Hopfe [2], J.E. Morral [2] and A.D. Romig, Jr. [3]

[1] Fanamation, Compton, CA 90220, USA

[2] Department of Metallurgy and Institute of Materials Science, University of Connecticut, Storrs, CT 06269, USA

[3] Sandia National Laboratories, Materials and Process Sciences, 1800 Albuquerque, NM 87185-5800, USA

ABSTRACT

Diffusion paths and concentration profiles of two phase Ni-Cr-Al diffusion couples were measured experimentally and predicted with an analytical model. Twelve $\gamma+\beta$ alloys were prepared and formed into six, two phase diffusion couples. After annealing at 1200°C for 100 hours and quenching in ice-water, image analysis, electron microprobe analysis and X-ray diffraction analysis were performed. The data collected were then converted to concentration profiles and diffusion paths. The diffusion path of each couple was found to follow a zigzag path as predicted by theory and the phase boundaries for the two phase $\gamma+\beta$ region were established. It was also found that diffusion in a two phase region can be modeled like diffusion in a single phase region.

INTRODUCTION

Two phase, $\gamma+\beta$, Ni-Cr-Al alloys are used commercially as overlay coatings for high temperature oxidation protection of Ni-base, $\gamma+\gamma'$ superalloys. Above 1000°C, interdiffusion between the coating, and substrate degrades the protective ability of the coating and the mechanical properties of the superalloy. Therefore, it is desirable to predict the extent of this interdiffusion with appropriate models. Understanding and ultimately predicting diffusion assisted degradation in even simple coating-substrate systems require first a comprehension of the interdiffusional behavior within the coating and the substrate themselves.

At present, diffusion in $\gamma/(\gamma+\beta)$ diffusion couples and γ_1/γ_2 diffusion couples have been studied extensively and modeled numerically by Nesbitt and Heckel [1,2,3]. But there have been no studies of $(\gamma+\beta)_1/(\gamma+\beta)_2$ diffusion couples. Until now, only diffusion between single phase alloys has been studied [1,2,3,4] analytically, while diffusion between two phase alloys has received little attention. The purpose of this work was to investigate diffusion paths and concentration profiles in two phase diffusion couples of the type $(\gamma+\beta)_1/(\gamma+\beta)_2$.

A theoretical study of diffusion in two phase regions by Hopfe and Morral [5] preceded the experimental work. In this theoretical study, the diffusion path and interdiffusion kinetics were modeled for the case when both the initial alloys and the diffusion path lie in a two phase region of a ternary phase diagram and the alloys exhibit only small difference in average composition. Under these conditions, the diffusion path was predicted to follow a zigzag path. In this work six diffusion couples with small concentration differences ($\leq$ 5 at.%) were made from twelve alloys that lie in the two phase $\gamma+\beta$ region of the Ni-Cr-Al phase diagram to test the model.

BACKGROUND

One dimensional diffusion in n-component phases can be described by a generalized form of Fick's Second Law of Diffusion:

$$\frac{\partial [C]}{\partial t} = [D]\frac{\partial^2 [C]}{\partial x^2} \tag{1}$$

in which [C] is a (n-1) column vector matrix describing the local composition of an alloy, t is time, x is distance, and [D] is the (n-1)x(n-1) diffusivity matrix. The elements of [D], i.e., the D_{ij}'s, are "diffusion coefficients" and they must remain constant in the reaction zone in order for Equation (1) to apply. These conditions may be approached by making the concentration differences across the initial interface sufficiently small.

The error function solutions for the two phase ternary diffusion couples have the same form as that for single phase couples except that the coefficients and the eigenvalues are now dependent on an effective diffusivity matrix [D^{eff}][5], which depends on the phase diagram. As a result, one of the eigenvalues of the effective diffusivity matrix [D^{eff}] is zero. The error function containing the zero eigenvalue becomes a Heaviside function H(x) as defined below:

$$C_1(x,t) = A_{10} + A_{11} \, erf\left(\frac{x}{2\sqrt{E_1 t}}\right) + A_{12} H(x) \tag{2}$$

$$C_2(x,t) = A_{20} + A_{21} \, erf\left(\frac{x}{2\sqrt{E_1 t}}\right) + A_{22} H(x) \tag{3}$$

where H(x)=1 for x>1, H(x)=0 for x<1. In equations (2) and (3), A_{10} and A_{20} can be determined from the initial conditions, A_{11}, A_{12}, A_{21}, and A_{22} can be determined from the boundary conditions and the effective diffusivity matrix [D^{eff}], and E_1 and E_2 are the eigenvalues of the [D^{eff}].

EXPERIMENTAL PROCEDURES

Twelve $\gamma+\beta$ alloys made of Cr, Al and Ni were prepared to make six diffusion couples. The atomic compositions of these samples prior to the arc melting procedure are listed in Table 1, and the average composition of each of the six diffusion couples was 17.5 at % Cr, 21 at. % Al and 61.5 at. % Ni. The microstructures of each alloy can be seen in the end portion of the micrograhph in Figure 1 and the configuration of these diffusion couples is shown in Figure 2.

Alloys buttons were made by arc melting the corresponding weighed elements on a water-cooled copper hearth in a high purity argon atmosphere. The samples were melted four times to encourage homogeneity. Each button was then drop cast into a cylindrical tapered copper mold with a diameter of about 6mm. Thin alloy disks (2.5mm thick with a diameter of 5mm) were then cut from the cylindrical rods by a diamond-impregnated saw. The samples were ultrasonically cleaned in ethanol, rinsed in distilled water, and then dried in compressed air.

The samples were brought together to form diffusion couples in a Type 304 stainless steel clamp which was then encapsulated in a quartz tube under a near vacuum to minimize oxidation. The couples were first annealed at 1200 $\pm$ 2 °C for 100 hours in a tube furnace and then immediately quenched and broken in ice water. The diffusion couples were then sectioned, mounted and polished using standard metallographic techniques. The reaction zones were studied first by using image analysis to determine the volume fraction of each phase present. X-ray analysis was then carried out to determine the lattice parameter of γ and β phase in the alloy. The concentration profiles were measured on a JEOL 8600 EPMA operated at 15 kV.

RESULTS AND DISCUSSION

Microstructures of Diffusion Couples after Annealing. It can be noted in Figure 1 that most of the couples annealed at 1200°C revealed limited microstructural changes as a result of interdiffusion. Several couples (especially couples 1/7 and 2/8) contained Kirdendall porosity in the diffusion zone. It should also be noted in Figure 1 that a single phase layer of β formed at the initial interface of diffusion couples 1/7, 2/8 and 3/9 after annealing at 1200°C for 100 hours. The reason for this phenomenon is not well understood, but rapid diffusion through the continuous β phase may have been a factor.

Scatter in the Measured Volume Fraction of β. Significant interdiffusion is observed in the $\gamma+\beta$ region of couples 2/8, 3/9 and 4/10 as compared to that of couples 1/7 and 5/11 [8]. An example of the results of the measurements using image analysis is shown in Figure 3. The scatter in these measurements is mainly due to the random statistical error associated with sample size. However, the scatter around the interface of couple 3/9 can be explained by the single phase layer formed on the interface of couple 3/9 which leads to an extremely high volume fraction at that point. When the measuring frame is centered on the adjacent region where β is depleted, an extremely low volume fraction is obtained.

Analysis of γ and β Phase Boundaries. From the concentration versus distance data obtained by EPMA, the phase diagram boundaries for the two phase $\gamma+\beta$ region were determined. For simplification of the model, these phase boundaries were fit to straight lines as shown in Figure 4. The γ phase boundary determined in this work agrees with that determined by Nesbitt and Heckel [2] within 1 at.%. However, the β phase boundary had not been established until now.

Prediction of Concentration Profiles and Diffusion Paths of $(\gamma+\beta)_1/(\gamma+\beta)_2$ Diffusion couples. A computer program called "Profiler" was used to determine the concentration profiles of Al and Cr in each diffusion couple. As shown in Figure 5, all the diffusion paths except for couple 6/12 follow a zig-zag path. Since couple 6/12 is on a tieline, there is no interdiffusion between the terminal alloys and its diffusion path is just a straight line.

Comparison of calculated volume fraction profiles to experimental image analysis values. The average concentration profiles that were predicted by "Profiler" can be compared with the image analysis volume fraction data. To do this the average concentration profiles were converted to volume fraction profiles by applying the Lever Rule to the measured phase diagram in order to obtain atomic fractions of each phase. The atomic fractions were then changed to volume fractions with lattice parameter data [8]. Since the phase boundaries are linear, the method developed by Homayoun and Morral [7] for linear interpolation between tie-features in multicomponent phase diagrams was applied. The results agreed within experimental error in most cases (see Figure 3 for example).

CONCLUSION

From the experimental portion of this study on two phase diffusion couples, it can be concluded that: (1) Phase diagram boundaries can be established by performing electron probe microanalysis on two phase γ+β diffusion couples. (2) The ability to predict interdiffusion in $(\gamma+\beta)_1/(\gamma+\beta)_2$ diffusion couples with small concentration differences was demonstrated by accurately predicting volume fraction profiles of these couples. (3) The validity of the zigzag diffusion path theory developed by Hopfe and Morral was supported by the experimental results. (4) Diffusion in a two phase region with the assumption of constant diffusivity can be modeled similarly to diffusion in a single phase region using "Profiler".

ACKNOWLEDGEMENTS

The work at The University of Connecticut was supported by NSF Grant No: DMR-9025122. The work performed at Sandia National Lab was supported by U.S. DOE under contract no. DE-AC04-94AL8500. Paul Hlava's EPMA analysis is sincerely appreciated.

REFERENCES

1)   Nesbitt, J.A. and Heckel, R.W.: Metall. Trans. A, 18A: 2061 (1987).
2)   Nesbitt, J.A. and Heckel, R.W.: Metall. Trans. A, 18A: 2075 (1987).
3)   Nesbitt, J.A. and Heckel, R.W.: Metall. Trans. A 18A: 2087 (1987).
4)   Roper, G.W. and Whittle, D.P.: Met. Sci., 14: 541 (1980).

5)   Hopfe, W.D. and Morral, J.E.: (Submitted to Acta Metall. Mater., 1993.)
6)   Fujita, H. and Gosting, L.J.: J. Am. Chem. Soc., 78: 1099 (1956).
7)   Homayoun, A.S. and Morral, J.E.: Scripta Metall., 24: 1 (1990).
8)   Yeung,C.W.: M.S.Thesis, The University of Connecticut, 1992.

Table 1. Alloy composition prior to arc-melting.

Alloy #	At.% Cr	At.% Al	At.% Ni
1	17.5	23.5	59.0
2	18.7	23.2	58.1
3	19.7	22.2	58.1
4	20.0	21.0	59.0
5	19.7	19.7	60.6
6	18.7	18.8	62.4
7	17.5	18.5	64.0
8	16.2	18.8	64.9
9	15.3	19.7	64.9
10	15.0	21.0	64.0
11	15.3	22.2	62.4
12	16.2	23.2	60.6

Table 2. Volume fraction of β in each alloy from image analysis.

Alloy #	Composition	Volume% of β
1	Ni-23.1Al-17.1Cr	63
2	Ni-22.2Al-18.7Cr	58
3	Ni-21.4Al-19.3Cr	53
4	Ni-21.6Al-18.4Cr	54
5	Ni-21Al-17.7Cr	48
6	Ni-20.5Al-16.4Cr	42
7	Ni-17.6Al-17.6Cr	20
8	Ni-19.1Al-15.4Cr	29
9	Ni-19.9Al-13.5Cr	32
10	Ni-21.6Al-12.8Cr	43
11	Ni-22.5Al-14.6Cr	55
12	Ni-23.7Al-13.8Cr	63

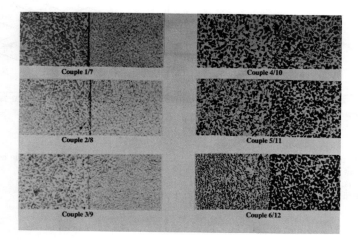

Figure 1. Microstructures of diffusion couples after annealing at 1200°C for 100 hrs.

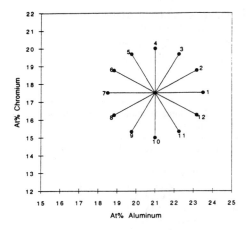

Figure 2.   Orientation of the initial diffusion couple design.

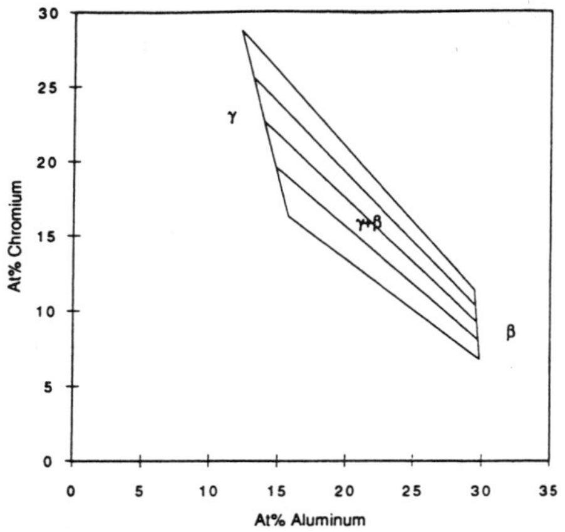

Figure 4. Curve fitted γ and β phase boundaries and tielines.

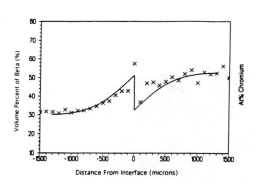

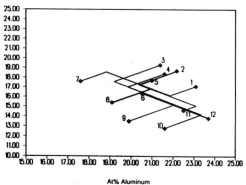

Figure 3.   Volume percent of β phase in couple 9/3: the crosses represent the experimental data measured using image analysis; the solid line represents the theoretical calculated values.

Figure 5. Diffusion paths of all the couples.

IV. CARBURIZING AND CARBONITRIDING

Materials Science Forum Vols. 163-165 (1994) pp. 197-202

OVER-CARBURIZING KINETICS AND MICROSTRUCTURE OF Z38CDV5.3 TOOL STEEL

V. Woimbée, J. Dulcy and M. Gantois

Laboratoire de Science et Génie des Surfaces (URA CNRS 1402)
I.N.P.L., Ecole des Mines, F-54042 Nancy Cédex, France

ABSTRACT

In a previous work, carburizing treatments to low alloyed or plain carbon steels have been applied, using direct cracking of an hydrocarbon such as propane at atmospheric pressure. This atmospher has allowed high speed of carbon transfer and avoids oxidation which is an important feature for treatment of alloyed steels. Short and low cost treatments yielding good wear resistance of plain carbon steel have thus been obtained: The present approach extends this treatment to alloyed steels such as Z38CDV5.3 (5% chromium content). Kinetics of weight increase versus carbide precipitation are examined. The microstructure is investigated by mean of microhardness curves, microprobe scanning and metallographic studies. Finally, subsequent post treatments are discussed.

KINETICS OF WEIGHT INCREASE

The specifications of carburizing treatments by a mean of an hydrocarbon at atmospheric pressure are reported in previous papers [1] and won't be described in this paper. Nevertheless, it is fondamental to mention that the key element in process control is a thermobalance for on line material transfer measurements to the specimens. Indeed, on line thermogravimetric measurements allow to track the process, conditioning the ultimate microstructure. **Figure 1** introduces some thermogravimetric curves obtained from the thermobalance, i.e. weight increase versus durations of treatment, specimens (1), (2), (3), and (4). A continuous microprobe scanning of the both elements C and Cr, analysed from the surface towards the core, is presented **Figure 2** for specimens (1), (2) and (3). The connection between the C and Cr intensity peaks reveals that these elements are present partly as carbides.

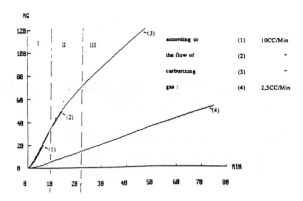

Figure 1 : Thermogravimetric curves of Z38CDV5.3 (chromium content : 5%) at 1000°C.

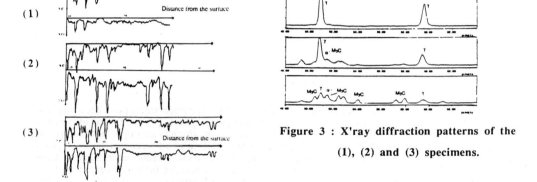

Figure 3 : X'ray diffraction patterns of the (1), (2) and (3) specimens.

Figure 2 : Continuous microprobe scanning from the surface to the core of tne (1), (2) and (3) specimens.

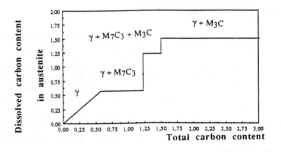

Figure 4 : Variation of total carbon content according to dissolved carbon content in austenite at 1000°C in equilibrium conditions.

These experiments exhibits three stages :

- first stage (I) : the slope of the curve is constant and this data depends on the kinetics of the carburizing gaz (comparaisons beetween the (1) and the (4) curves), no carbide precipitation is occurring,

- second stage (II) : the slope of the curve decreases slightly, the curve inflects. From X-ray patterns (**figure 3**), T.E.M. and optical observations, we noticed that this kinetics decrease was due to carbides precipitation,

- third stage (III) : onset of a new way of weight increase, the curve exhibits a low weight increase. A thorough study allows us to conclude on the increase of carbides layers of M_3C close to the surface.

    Figure 2 shows a deficiency of chromium content due to precipitation in the region close to the surface. A important precipitation occurs during the second stage and increases drastically during the third stage.The specific nature of the studied steel is demonstrated by revealing the modification of carbon diffusion, carbon surface concentration and precipitation of M_7C_3 and M_3C carbides by the alloying elements in the near surface region to be treated. M_7C_3 can be demonstrated by a mean of a Transmission Electronic Microscope. At this third stage of weight increase, only M_3C precipitation occurs, differences beetween alloyed and plain carbon steels are obtained because of the decrease in carbon diffusion. In order to clear the complexity of this carbides growth , we need to utilize data from equilibrium diagram. Also, we control carbon content in austenite according to bulk carbon content and carbides precipitation occuring in near surface region to be treated.

RELATION BETWEEN PRECIPITATION AND MICROSTRUCTURE

    Figure 4 gives the dependence between carbon content in austenite, bulk carbon content and precipitation at 1000°C. Precipitation of M_7C_3 occurs first at 0,6% content until 1,25% content. In the range 1,25 - 1,50%, both M_7C_3 and M_3C carbides precipitate. Below 1,50%, only M_3C carbide can be present. Itnhas to be kept in mind that carbides precipitation which holds chromium, decreases the percentage of chromium content dissolved in austenite. So, another diagram must be considered. A decrease of chromium content, decreases the area of M_7C_3 precipitation. A 2% chromium content alloy steel can no more precipitate M_7C_3. So, a small influence of subsequent 3% molybdenum content has to be taken into account.

    Figure 5 introduces the microprobe scanning according to depth for every stage of weight increase Kinetics of the four quenched specimens. We must connect those latter results with photographs from **figure 6**, showing the amount of precipitation. M_3C and M_7C_3 carbides appear light dark, respectively

    In the first stage, no precipitation occurs. Neverthless, Transmission electronic microscope carried out on thin specimens of (4), points out a great number of small particles M_7C_3.

Those can't be observed in specimen (1) for two reasons :

- carbon content is lower than in case of (4),

- treatment duration of (1) is too short to permit a effective diffusion of chromium to elaborate M_7C_3.

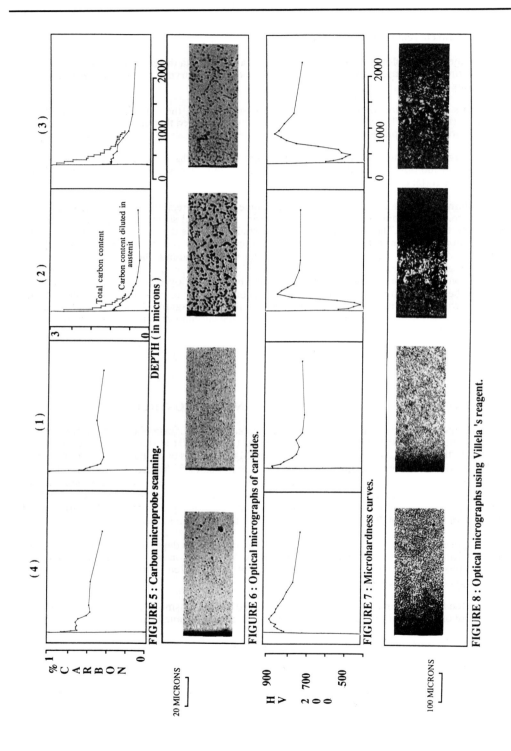

FIGURE 5 : Carbon microprobe scanning.

FIGURE 6 : Optical micrographs of carbides.

FIGURE 7 : Microhardness curves.

FIGURE 8 : Optical micrographs using Villela 's reagent.

In the second stage, both carbides precipitate. **Figure 5** represents the carbone content dissolved in austenite and the bulk carbon content, taking into account precipitated carbone. The carbone content dissolved in austenite (1,15%) is lower than the value provided by the Fe-5%Cr-C equilibrium diagram (1,25%). This discrepancy results from lowering chromium content in austenite due to precipitation and molydenum presence.

In the third stage, only intergranular precipitation of M3C occurs. The carbon content dissolved in austenite (1,25%) is lower than the value given by the equilibrium diagram (1,50%). Bungardt, Preisendaz and Lehnert [2] put in evidence the dependence of chromium content on carbon solubilities and areas of M7C3 precipitation in equilibrium conditions. The given value of 1,25% corresponds to the precipitation boundary of M3C for a Fe-3,5%Cr-C which is in good agreement with microprobe scanning of dissolved chromium content.

Figure 7 gives microhardness curves of quenched specimens (1), (4), (2) and (3). The shape is caracteristic of the microstructure shown **figure 8**, which presents a minimum and a maximum hardness value. Microstructure is constituted of martensite, carbides and residual austenite, induced by alloying elements during the quenching. The minimum value hardness is due to the amount of residual austenite. The maximum is connected with a low content of dissolved carbon. The surface, which is the carbone richest region, does not exhibit the poorest microhardness value because of precipitation of hard carbides and the lack of dissolved Cr and Mo in austenite due to precipitation. On increase the treatment duration, the minimum of hardness value is displaced towards the core, where a lower precipitation than in surface induces more Cr and Mo elements present in solution. The maximum hardness value spreads to higher depth, where carbon content in austenite, equals 0,7% about. This range of carbon content corresponds to M7C3 precipitation

POST TREATMENTS

The bad microstructure resulting from the intergranular precipitation of M3C and the large amount of residual austenite, can be improved by substential post treatments

After an enrichment of specimens with carbon, diffusion lowers the carbon content dissolved in austenite up to 0,7%. Precipitation of M3C is replaced by M7C3 precipitation and residual austenite is limited. The M7C3 carbides insure better impact mechanical properties. Carbides morphology becomes round and well ditributed, improved with a subsequent temperature lowering.

A double tempering strenghtens a stabilized microstructure of softened martensite, an absence of residual austenite and a secondary hardening by precipitation of small MC carbides, but tempering cancels residual strength due to the quenching.

An ultimate nitriding treatment elaborates compressive stress in the surface region, which improves fatigue resistance and raises microhardness value close to the surface. The ultimate microhardness curve and the ultimate microstructure are given **figure 9 and 10** respectively.

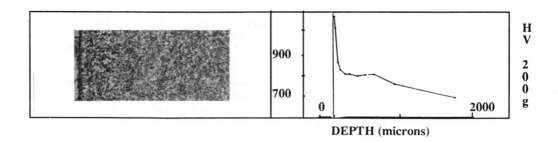

Figures 9 and 10 : Ultimate microhardness curve and ultimate microstructure, Nitriding at 500°C during 4 hours.

CONCLUSION

Mechanical properties and tool steel life service have been optimized by a duplex surface treatment :

- enrichment with carbon and associated diffusion to achieve the transformation of M3C in M7C3, final quenching.double tempering to achieve elimination of residual austenite and elaboration of second hardening

- nitriding to obtain compressive stresses together with a higher microhardness value in the surface region.

References

1) J. Dulcy et M. Gantois, "La cémentation accélérée", Journée A.T.T. - 1991, Internationaux de France du traitement thermique, Toulouse 26-28 juin 1991.

2) R. Benz, J.F. Elliot and J Chipman, "Thermodynamics of the carbide in the system Fe-Cr-C", Metallurgical transactions, vol 5, October 1974, P2235.

Materials Science Forum Vols. 163-165 (1994) pp. 203-208
© *1994 Trans Tech Publications, Switzerland*

CARBURIZATION OF HIGH-CHROMIUM STEELS

B.M. Khusid [1], E.M. Khusid [2] and B.B. Khina [2]

[1] A.V. Luikov Heat and Mass Transfer Institute, Byelorussian Academy of Sciences
15, P.Brovka St., 220 072 Minsk, Republic of Belarus

[2] Byelorussian State Polytechnic Academy,
65, F. Skorina Ave., 220 027 Minsk, Republic of Belarus

ABSTRACT

The influence of pack, paste, and gas carburizing media upon the structure of the carburized layer on hypo- and hypereutectoid high-chromium (0.3 and 0.8 wt% C, 3-17 wt% Cr) steels was examined. The optimum case structure is produced by a pack mixture containing 2 wt% γ-Fe$_2$O$_3$; the case is practically disposed of an oxidized zone and contains a thick layer of fine-grained carbides. The influence of the carburizing temperature and duration on strength and wear resistance of chromium steels is determined.

INTRODUCTION

High-strength chromium steels do not possess the required level of wear resistance for specific applications. Wear resistance can be significantly improved by carburization [1-3]. However, the advantages of the carburization of high-alloy steels, *viz.* the process simplicity, a large thickness (about 1 mm) of a hardened case, a gradual change in the carbon content and carbide volume fraction from the surface to the bulk are combined with substantial disadvantages: (a) typical carburizing media promote intensive oxidation of chromium thus lowering the surface properties; (b) the formation of a carbide zone increases the article brittleness due to local stress concentration in the subsurface layer.

The purpose of this work is to examine the effects of the carburizing media compositions and regimes on the carbide zone structure, wear properties, tensile, bending, and impact strength in order to obtain an optimum combination of the high surface hardness and wear resistance with a desired level of the bulk strength of carburized high-chromium steels.

EXPERIMENTAL PROCEDURES

For experiments, 12 high-chromium steels were smelted in an induction furnace: hypereutectoid, 80Cr4, 80Cr7, 80Cr9, 80Cr12, 80Cr14, and 80Cr16; hypoeutectoid, 30Cr3, 30Cr6, 30Cr9, 25Cr13,

25Cr15, and 25Cr17. The first number denotes the average carbon content of a steel (in 10^{-2} wt.%) while the second one is chromium concentration (in wt.%). The variation of carbon and chromium contents of these steels envelope the compositions of major high-strength tool and stamp steels [2,3]. The commercial 40Cr13 steel, 0.36-0.45% C, 12-14% Cr, 0.8% Mn, 0.8% Si (by wt.%), was also used in tests for the sake of comparison.

For pack carburizing, either charcoal or the commercial carburizator (granulated charcoal - 20 to 25 wt.% $BaCO_3$ - 3.5 wt.% $CaCO_3$) was used as a carbonaceous base. Various substances were used as energizers. Carburization was performed in sealed containers that were afterwards cooled to room temperature and unpacked.

The microstructure of the specimens was studied using optical and scanning electron microscopy (SEM), and the computerized image analysis. Phases were identified by X-ray diffraction analysis. The composition of the matrix and particles was examined by energy dispersive spectroscopy and electron-probe microanalysis (EPMA).

Tensile and bending tests were carried out on a universal testing machine Instron-1195. The specimen shape and sizes were cylindric (diameter, 6 mm, length, 50 mm) for tensile test and prismatic (10×5×55 mm) for bending test. Impact strength was studied by the Izod impact testing. For non-carburized heat-treated steels, 10× 10×55 mm samples with U-notch were used, the notch radius being 1 mm. For carburized steels with subsequent heat treatment, 10×10×55 mm samples without a notch were used. The impact strength value, KCU, was estimated by the absorbed energy per the original cross-sectional area.

Unlubricated sliding wear tests were performed in a ring-on-disk apparatus under dry ambient conditions. The sliding speed was 0.05 to 3 m/s and the normal pressure was 1 to 12 MPa. Pin abrasion tests were carried out in a pin-on-disk machine under the normal pressure of 1 MPa and the disk rotation speed of 80 1/min. The sample was gradually shifted from the periphery towards the disk center at a rate of 1 mm per revolution to provide a contact with a fresh abrasive paper. In both tests the linear amount of wear and mass loss were measured.

CHOICE OF THE CARBURIZING MEDIA COMPOSITIONS

The relationship between the carburizing ability and the oxidizing ability of an atmosphere formed in a container is determined by the CO_2-to-CO mole ratio which is governed by the carbon gasification rate. This ratio and thus the oxidation of a steel may be significantly lowered by the addition of a catalyst. In this work, a number of substances known as catalysts were tested as energizers for pack carburization: carbonates, hydrocarbonates, and hydroxides of lithium, sodium and potassium; carbonates of magnesium, calcium and barium; fluorides and chlorides of lithium, sodium, and potassium; potassium bromide and iodide; transition metal oxides, namely TiO_2, V_2O_5, Cr_2O_3, CrO_2, MnO_2, FeO, Fe_3O_4, α-and γ-Fe_2O_3, Co_3O_4, NiO, Nb_2O_5, WO_3; copper and lead oxides. The traditional energizers, viz. NH_4Cl, AlF_3, CCl_4, $KMnO_4$, Fe, and FeS were also used.

A carburized case on high-chromium steels consists of four typical zones: (1) an oxidized ("dark") surface layer; (2) a layer of coarse-grained, almost joined together carbide particles; (3) fine-grained globular carbide particles; and (4) a transition zone. The first and the second zones should be as thin as possible. The experiments performed have shown that $KHCO_3$, $NaHCO_3$ (proposed in [1]), NaF, $\alpha-Fe_2O_3$, Co_3O_4, NiO, and $\gamma-Fe_2O_3$ may be effectively used as energizers. Charcoal - 2 wt.% $\gamma-Fe_2O_3$ is the best carburizator [4]. It produces a case disposed of an oxidized zone and the interlayer of the coarse-grained joined-together carbide particles. A combination of two or more energizers does not improve the case structure due to intensive interaction of additives under heating.

The experiments on gas carburization were performed in an endothermic atmospheres: (mol %): CO, 1.96 to 20.7; CO_2, 0.05 to 0.20; CH_4, 0.15 to 0.75; O_2, 0.05 to 0.1; H_2, 40.2 to 41.4; the carbon potential being 0.89 to 1.02 wt.%. In order to improve the carburized case structure, ammonia was added to provide nitrogen potential of 0.022 to 0.062 wt.%. These atmospheres were developed for carburizing plain-carbon and low-alloy steels [5]. The temperature was 900 to 940 °C, and the duration was 15 h. It appears that the ammonia-containing atmosphere may be effectively used for carburizing steels with up to 7-9 wt.% Cr. The surface Rockwell C hardness of the latter attains 64-66 HRC.

For paste carburizing, the optimum $\gamma-Fe_2O_3$ concentration also appeared to be 2 wt.%. The use of $\gamma-Fe_2O_3$ in a paste instead of Na_2CO_3 (proposed in [1]) provides a significant decrease in the oxidized film thickness for steel containing 16-17 wt.% Cr and the elimination of this film for steels with 3-4 wt.% Cr.

STRUCTURE OF CARBURIZED CASES ON CHROMIUM STEELS

The variation of the carburizing medium composition primary affects the sizes of the case, oxidized layer, and zone of coarse-grained carbides; the morphology and thickness of the zone of fine-grained carbides are approximately the same. Increase in the chromium content of steels raises the volume fraction of carbide particles and their mean size. A rise in the carburizing temperature causes the carbide particles to grow while the process duration does not significantly affect their size. The carbide $(FeCr)_3C$ is the prevailing phase in the narrow subsurface zone while the carbide $(CrFe)_7C_3$ dominates in underlying layers.

The carbide zone thickness in hypereutectoid steels exceeds that in the hypoeutectoid ones by about 40%. Increasing carburizing temperature and duration enlarges the carbide zone. Alloying a high-chromium steel with Mo, Ti, V, and W (up to 2 wt.%) lowers the sizes of the case and the carbide zone. The Vickers microhardness in a carburized case gradually lowers with moving off the surface in the carbide zone and steepens in the transition zone. The microhardness of the carbide layer is about 1100-1300 HV.

To study the effects of a cooling rate, carburization in the charcoal - 2 wt.% $\gamma-Fe_2O_3$ was run in a thin-shell 25 mm diameter cylinder at 1000 °C for 6 h. The subsequent furnace, air, oil, and water - 10 wt.% NaCl cooling of the container were used. Both water and oil cooling resulted in the similar case morphology determined by the carburizing regime. EPMA demonstrated that the total chromium content (of both carbides and solid solution) was about uniform across the sample, *i.e.* chromium did not redistribute during carburization and subsequent cooling. Oil and water cooling induced the formation of isolated microvoids and the network of microcracks near the surface. Air cooling appeared to lower the size of the oblong-shape coarse-grained carbides and to raise the amount of fine globular particles. Furnace cooling induced the coarsening of carbide particles. These results show that the optimum morphology of a carburized case is formed under cooling at an intermediate rate (in air).

The following regimes of the post-carburization heat treatment have been proved to produce a high value of the surface hardness of a carburized steels and a high level of the tensile and impact strength of its core: quenching from 950 - 970 °C for steels containing 3 to 12 wt.% Cr, and from 1000 to 1020 °C for steels with 13 to 17 wt.% Cr; tempering at 180 °C for 2 h. A special test for tool and stamp steels is widely used [2,3]: the room temperature Rockwell C hardness, which a steel retains after it has been exposed to elevated temperature for 4 h, should exceed 45 HRC. The ultimate temperature determined by this test is an important property of such steels. Post-carburization heat treatment raises this ultimate temperature up to 600-640 °C for all steels. This treatment does not influence the microhardness of a carbide layer but increases that of a transition zone; the carbide particles become finer and acquire the shape of a globe.

OPTIMIZATION OF PROPERTIES OF CARBURIZED STEELS

Carburizing temperature and duration were varied in the ranges 950 to 1050 °C and 4 to 8 h because the enveloped the carburizing regimes necessary to provide the high wear resistance and surface hardness of chromium steels [6]. The charcoal - 2 wt.% γ-Fe_2O_3 carburizator was used. To describe the data on wear resistance and strength, the second-order regression model was used

$$y = b_0 + b_1x_1 + b_2x_2 + b_{11}x_1^2 + b_{12}x_1x_2 + b_{22}x_2^2 \qquad (1)$$

where y is the value of the property; $x_1 = (T-1000)/50$; $x_2 = (\tau-6)/2$; T and τ are the carburizing temperature (in °C) and duration (in h). The experiments were performed at nine different points. The coefficients determined by a least-squares method are listed in Table 1. For comparison, Table 1 summarizes the properties of steels carburized at 1000 °C for 6 h in charcoal - 2 wt.% γ-Fe_2O_3 and in charcoal - 15 wt.% $NaHCO_3$ (proposed in [1]).

In both tensile and bend tests, the carburized samples demonstrated an elastic behaviour, the plastic part of load-deformation graphs being small. Therefore, only the values of ultimate strength, σ_{UTS} and σ_{UBS}, were measured. A carburized layer occupied 20 to 40% of the total specimen half-thickness in all samples. σ_{UTS} of carburized steels is 1.2 to 1.7 times less than that of non-carburized ones. The values of σ_{UTS} and σ_{UBS} are lowered with increasing carburizing temperature. An increase in the carburization duration exhibits the same effect on strength which is more pronounced for σ_{UBS}. Varying chromium content from 6 to 17 wt% steeply decreases the strength of hypoeutectoid steels while σ_{UTS} and σ_{UBS} of hypereutectoid steels attain maximum values at about 12 wt.% Cr. These results are similar to those demonstrated by non-carburized heat treated steels.

The impact strength decreases with raising temperature. The KCU value is highly sensitive to carburization parameters; it can be varied by the factor of up to 10 under the examined conditions. Increase in the chromium content of a steel lowers the KCU; its value is greater for hypereutectoid steels. The effect of steel composition on the impact strength of carburized and non-carburized steels is different. For a carburized sample, the most part of the lost energy is connected with the crack initiation in the carbide layer, thus the role of bulk properties becomes weaker.

The fracture surfaces of heat-treated and surface hardened steels were studied by SEM. For both carburized and non-carburized samples, heavily dimpled surfaces were observed. A rise in the carbide phase volume fraction causes the plastic deformation of the carburized case to decrease.

Abrasive wear tests have demonstrated that carburization raises the wear resistance in comparison with that of heat-treated steels by a factor of 1.5. Wear losses are caused by microploughing and microcutting [6]. For non-carburized heat-treated steels the wear losses depend on the steel composition [6]. The coefficients of equation 1 were calculated for wear losses under a stable wear regime (wear distance was 10 to 30 m). Minimum wear losses for all steels correspond to carburizing at 1000 to 1030 °C for 5 to 7 h. The difference in the wear losses for carburized steels does not exceed 15% despite the difference in the carbon and chromium contents. Abrasive wear resistance depends mainly on the volume fraction, shape and size of carbide particles in the case.

The heat-treated steels without surface hardening cannot be used under heavy conditions of sliding wear (0.5 m/s, 2 MPa) due to intensive adhesion [6]. Carburization significantly reduced adhesion and thus dry wear losses [6]. The coefficients of equation 1 were calculated for wear loss at 1 km sliding distance at speed of 0.5 m/s and normal pressure of 2 MPa. Minimum weight losses correspond to carburizing at 1000 to 1040 °C for 5.5 to 7.5 h. The steel composition does not markedly influence the weight losses; for all steels the difference in values is less than 11%. because of similarity among the case structures. The oxidation-dominated wear mechanism partially accompanied by adhesion operates under the testing conditions (0.3 to 0.7 m/s, 1 to 3 MPa). An increase in normal load and a decrease in sliding speed promote transition from mild wear to adhesive one.

In comparison with the $NaHCO_3$-containing medium, the carburizator with $\gamma-Fe_2O_3$ provides 7 to 24% increase in σ_{UTS}, 20 to 74% increase in σ_{UBS}, 10 to 90% rise in impact strength, and an improvement of the abrasive (10 to 70%) and sliding (2 to 50%) wear resistance.

The results obtained enable one to estimate the optimum regime of carburization and subsequent heat treatment in order to produce a desired combination of wear and strength properties of high-chromium steels.

Table 1. Properties of carburized steels

Steel	Coefficients of equation 1						Energizer (wt%)	
	b_0	b_1	b_2	b_{11}	b_{12}	b_{22}	2% γ-Fe$_2$O$_3$	15% NaHCO$_3$
Ultimate tensile strength (MPa)								
30Cr6	1180	-113	32	0	-57	-119	1140	970
25Cr13	851	-53	-37	0	0	0	910	810
25Cr17	569	-60	-68	0	-30	0	540	500
80Cr7	914	-80	-129	0	0	-42	930	750
80Cr12	988	-113	-115	118	0	-66	900	770
80Cr16	767	-190	-52	158	-57	93	840	750
40Cr13	941	-90	-28	68	0	-41	870	740
Ultimate bending strength (MPa)								
30Cr6	1890	-389	-200	0	-63	0	1840	1043
25Cr13	1230	-239	-199	112	-75	126	1120	940
25Cr17	1050	-241	-159	0	0	0	1020	820
80Cr7	1500	-475	-161	67	0	176	1580	1080
80Cr12	1920	-417	-199	-134	-71	-129	2030	1050
80Cr16	1750	-304	-180	0	-87	0	1790	1030
Impact strength ($\times 10^{-2}$ MJ/m^2)								
30Cr6	9.8	-5.4	0.	6.2	0.	4.4	9.2	8.4
25Cr13	10.4	-2.0	1.4	0.	0.	0.	9.7	8.2
25Cr17	8.8	-1.0	0.	0.	0.	2.1	8.9	5.2
80Cr7	21.5	-9.7	4.3	0.	0.	0.	20.3	10.7
80Cr12	6.5	-8.1	-5.9	5.2	5.1	5.1	7.7	6.3
80Cr16	3.5	-4.4	0.	5.8	-3.7	4.3	4.2	3.2
40Cr13	5.4	-3.0	-1.8	3.7	-7.2	5.8	7.9	5.8
Abrasive wear loss ($\times 10^{-2}$ kg/m^2)								
30Cr6	70.6	-9.3	0.	11.0	0.	12.5	79.	88.
25Cr13	79.5	0.	0.	8.0	1.9	7.7	83.	90.
25Cr17	67.3	-10.6	-9.6	10.9	-2.2	10.1	71.	89.
80Cr7	75.1	-4.7	-4.0	3.5	0.	5.6	79.	96.
80Cr12	71.3	0.	4.4	9.0	0.	8.6	74.	92.
80Cr16	65.7	-3.4	-5.3	5.5	4.3	6.9	67.	115.
40Cr13	75.5	-3.1	-3.9	4.5	0.	3.6	80.	89.
Sliding wear loss ($\times 10^{-3}$ kg/m^2)								
30Cr6	180	-5.9	-7.5	7.6	0.	5.0	1.9	1.9
25Cr13	165	-3.0	0.	15.1	0.	14.6	1.7	1.8
25Cr17	155	-15.1	18.2	18.7	0.	25.4	1.5	1.5
80Cr7	113	-34.0	-36.6	52.7	-22.3	53.2	1.1	1.7
80Cr12	137	-37.1	-44.7	23.3	14.2	48.4	1.4	1.6
80Cr16	138	-32.7	0.	30.3	0.	38.8	1.4	1.7
40Cr13	121	-26.2	0.	36.0	-7.1	36.6	1.1	1.5

REFERENCES
1) Lyakhovich,L.S. (ed): *Thermochemical Treatment of Metals and Alloys. Handbook,* Metallurgiya, Moscow, 1981.
2) Geller,Yu.M.: *Tool Steels,* Metallurgiya, Moscow, 1983.
3. Poznyak,L.A., Skrynchenko, Yu.M. and Tishayev, S.I.: *Stamp Steels,* Metallurgiya, Moscow, 1980.
4) Aleshkevich,V.I., Voroshnin,L.G., Fraiman,L.I. and Khusid,E.M.: U.S.S.R. Patent SU 1,574,680 (C1.C23C8/64), 1990.
5) Pozharskiy,A.V., Pegisheva,S.A., Rozina,O.N., Semenova,L.M. and Korotkii,V.V.: Metalloved. Term. Obrab. Met., 1984, 4, 8.
6) Khusid,B.M., Khusid,E.M. and Khina,B.B.: Wear, 1993, in press.

Materials Science Forum Vols. 163-165 (1994) pp. 209-214

HIGH TEMPERATURE (1050 °C) CARBURIZING OF SAE 1020 STEEL AND ITS EFFECT ON FATIGUE FAILURE

A.C. Can

Pamukkale University, Faculty of Engineering, 20017 Denizli, Turkey

ABSTRACT

In this work, the mechanism of fatigue failure and the method of increasing fatigue strength have been summarised. Also, it has been dwelled on the surface hardening treatments. The work has aimed to find out whether the fatigue strength is increased by the application of carburizing in liquid salt bath (NaCN + NaCO3 + NaCl), widely used in practice. In particular, the carburizing at high temperature (1050 °C) at which carburizing time can considerably be reduced, and the residual stress on the case has been evaluated.

In the experiments, SAE 1020 steel was used. As a result, the carburizing at l050 °C has reduced the heat treatment time by 55 % for 12 % case depth/section thickness ratio in 6 mm diameter specimen, and has increased the fatigue strength by 10 % for 6% ratio of case depth and section thickness.

FATIGUE FAILURE AND INCREASING FATIGUE STRENGTH

Fatigue cracks are initiated at the free surface of ductile and homogeneous metals. Because:

a. A grain having a free surface will be able to deform plastically more easily than the body of metal that is surrounded by the other grains.
b. Surface grains are in intimate contact with the atmosphere.
c. Loading stresses are highest at the surface ·
d. Changes of section and surface roughness produce local "stress concentration" of greatest intensity of the surface.

Other materials may also exhibit fatigue characteristic but this is due to the propagation of a crack from some initial defect or flaw.

An important structural feature that is observed only in fatigue deformation on the surface ridges and grooves called slip-band extrusions and intrusions. Fatigue cracks initiate at intrusions and extrusions. Fatigue crack propagation occurs by a plastic blunting process [1].

Surface hardening improves the fatigue strength due to the micro structure formed and to development of a compressive surface residual stress during the hardening process. Liquid carburizing treatments are widely used to improve the wear resistance and fatigue strength of low carbon steels. A survey of literature reveals that when liquid case carburizing treatment is used fatigue strength can be improved up to 100% . The tempering of case hardened specimens, however, even at low temperatures of about l00 °C, results in a lower fatigue strength than that obtained in the fully hardened condition [2]. Liquid carburizing is a case hardening method for steels. In this method, steels are held above Ac_3 temperature in a molten cyanide bath so that carbon diffuses from the bath in to the metal and produces a case. The cyanide bath can be considered as the light case cyanide baths and the deep case cyanide baths that are arbitrary terms associated with liquid carburizing applications. The light case cyanide baths are usually obtained in the temperature range from 840 °C to 900 °C while the temperature range for deep case is between 900 °C and 950 °C. These baths consist of NaCN, NaCl, $NaCO_3$, $BaCl_2$, etc.

The light case baths are usually operated at higher cyanide contents than that of the deep case baths. The preferred operating cyanide contents are 10-23 % for light case and 6-16 % for deep case. Temperature range may be extended somewhat, but at temperatures higher than about 950 °C deterioration of the bath and equipment is markedly accelerated. However, rapid carbon penetration can be obtained at temperatures between 980 °C and l040 °C [3].

Because of diffusion of carbon in to the surface of the parts and the volume increase due to phase transformation, a residual compressive surface stress is generated. In quenching of a specimen from high temperature, the case is cooled first, then as the core cools and contracts, it develops a tensile stress which "pulls in" the case, introducing there a compressive residual stress [4].

Excessive quenching, like excessive carburizing, should be avoided, because high residual compressive stress in surface layers results in high residual tensile stress in the core. The latter may lead to fatigue cracks originating in subsurface of materials [5].

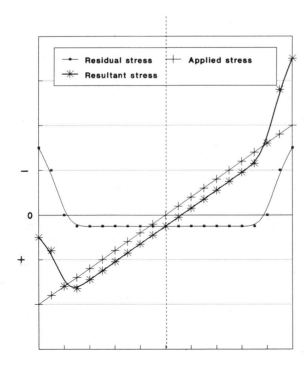

Figure 1. Stress distribution in case hardened specimen

The residual stress pattern in carburized specimens is shown in figure 1. The residual stress in the core occurs to balance the residual compressive stress in the case.

Figure 1 also shows the bending stress induced by the external load and the sum of these two stresses. Bending stress is decreased in the surface layer by the residual compressive stress as shown with bold line.

The magnitudes of residual stresses can be determined by calculations and experimental methods. The two most common laboratory methods are the x-ray and the dissection methods. The x-ray method that is non destructive, is limited presently to determine residual stress to a depth of two or three atoms below surface. The required laboratory equipment for this method is more expensive [4].

EXPERIMENTS

Specimens having a test section diameter of 6 mm and manufactured from SAE 1020 steel were used throughout this work. Shape of specimen and some data of the steel are shown in figure 2 and table 1 respectively.

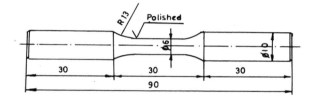

Figure 2 Shape of specimen

Table 1. Technical Properties of SAE 1020

Tensile strength N/mm^2	Upper yield strength N/mm^2	Lower yield strength N/mm^2	Elongation %	Hardness HV 71
462	291	280	25	127

Heat treatment of specimens carried out as shown in Table 2. The quenching medium used mineral oil. The fatigue tests were carried out on a rotating bending fatigue machine. The frequency of the load cycles was maintained at 50 Hz throughout the tests. Ten specimens were tested for each set of experiments. Results of fatigue tests are shown in Figure 3 and 4.

Table 2. Heat Treatment of Specimens

Experiment no	Normalizing temperature °C	Carburizing temperature °C	Quenching temperature °C	Carburizing time hour	NaCN concentration %
1	910	950	900	0.50	22
2	910	950	900	2.00	22
3	910	950	900	4.50	22
4	910	950	900	8.00	22
5	910	1050	900	0.25	15
6	910	1050	900	1.00	15
7	910	1050	900	2.00	15
8	910	1050	900	3.50	15

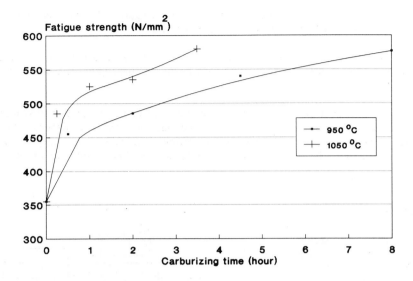

Figure 3. Effect of carburizing temperature on fatigue strength (N=10⁶)

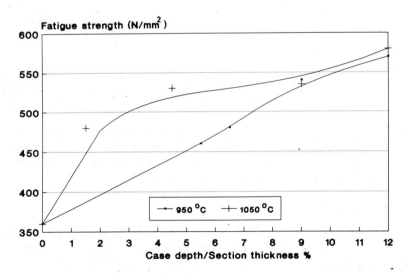

Figure 4. Effect of case depth/section thickness ratio on fatigue strength (N=10⁶)

Residual compressive stresses in the case were measured by using the dissection method and results are shown in figure 5.

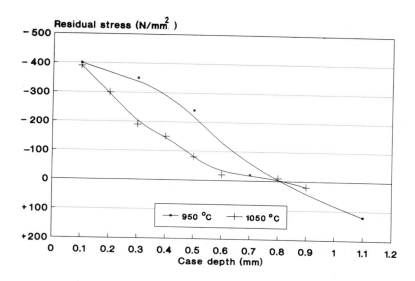

Figure 5. Residual compressive stress in the case

DISCUSSION

The case carburizing at 1050 °C has not decreased fatigue strength as shown in figure 3 and 4. Whereas, high carburizing time increases grain size as shown in figure 6. Especially, with shallow case, higher fatigue strength has been observed. This can be explained with residual stress distribution [6]. The residual stress pattern in the case that carburized at 1050 °C is concave and the residual stress pattern in the case that carburized at 950 °C is convex as shown in figure 5. If the residual stress pattern is concave, it will reduce the tensile residual stress in the core because the compressive residual stress at the surface is balanced with plateau of tensile stress in the core. Concave residual stress pattern decreases resultant stress that will initiate crack. This effect is shown in figure 7

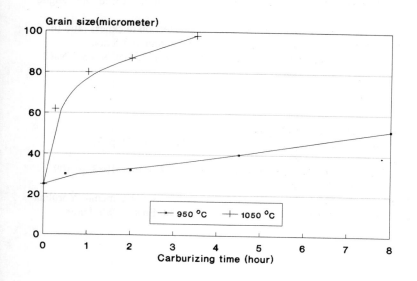

Figure 6 Effect of carburizing temperature on grain size

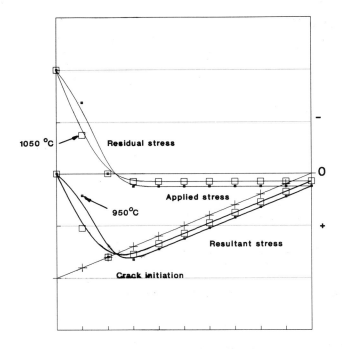

Figure 7. Effect of residual stress pattern on resultant stress

In addition, crack initiation point for 1050 °C carburizing temperature is near the surface. The strength of case increase near the surface. For these reasons high temperature (1050 °C) carburizing may increase the fatigue strength for SAE 1020 steel.

CONCLUSION

1. Carburizing of SAE 1020 case hardening steel at 1050 °C has not decreased the fatigue strength.
2. An 8 hour carburizing period is required to obtain 12 % case depth/section thickness ratio for 6 mm diameter at 950 °C, whereas at 1050 °C the required carburizing time is 3.5 hours.
3. The suitable bath composition for carburizing treatment at 1050 °C is 15% NaCN, 85% NaCl.

REFERENCES

1. Laird, C., : Fatigue Crack Propagation, ASTM STP 4l5, 1967, p. 131.
2. Luther, R.G., Willams, T.R.G. : Metallurgy and Metal Forming, 41, 1974, (3), p. 72.
3. Metals Handbook, American Society for Metals, (1964).
4. Black,P.H. and Adams, Q.E., Machine Design, Mc Graw-Hill Book Co. Inc., 1983, p. 109.
5. Parrish, G, : Heat Treat. Met, 4, 1977, 2, p. 45
6. Can, A.Ç.: Yüksek Sıcaklıkta (1050 °C) Sementasyonla Yüzey Sertleştirmenin Yorulma Mukavemetine Etkisi, Ph. D. Thesis, Dokuz Eylül University, Fen Bil. Enst., 1986, İzmir.

Materials Science Forum Vols. 163-165 (1994) pp. 215-220
© 1994 Trans Tech Publications, Switzerland

THE INFLUENCE OF CARBURIZING AND HEAT TREATMENT ON THE MECHANICAL PROPERTIES OF LH15 GRADE BEARING STEEL

M. Kulka, M. Przylecka, W. Gestwa and M. Piwecki

Technical University of Poznan, Poland

The problem of increasing working time of bearings elements from LH15 (52100) steel by the modification of its chemical composition, was a matter of many authors' interest. Bearing steel carburizing process and its influence on utilizable properties was also subject of penetrating investigations. It was stated, that carburized layer formed on elements made of LH15 steel, essentially increases these properties. In this paper, investi-gation results of carburizing and heat treatment influence on limited volumetric fatigue strength and hardness of diffusion layers formed on LH15 steel, were presented.

The problem of low-carbon steels carburizing was completely presented in the paper [1].

The influence of chemical and phase composition on utilizable properties of hyper-eutectoid bearing steels was extensively presented in the paper [2].

It results from investigations, that the highest utilizable properties can be obtained for 1.8 to 2.0 % of carbon content in the layer of upper zone. For higher contents, utilizable properties decreasing in diffusion layers formed on LH15 steel is observed.

In presented paper, investigation results of carburizing parameters and heat treatment after carburizing influence on limited volumetric fatigue strength and on hardness of diffusion layers formed on ŁH15 steel are presented. Investigation results are related to chemical composition of used ŁH15 bearing steel, according to PN-74/H-84041.   The parameters of carburizing and heat treatment – see table 1.

Dependence of total carbon content variations upon the distance from surface of carburized bearing steel is presented on the fig.1. On the base of this drawing, samples for following investigations were prepared, obtaining on theirs surface a different total carbon content. Samples prepared in this manner were heat treated at parameters according to table 1. Fatigue strength tests were carried out according to PN-76/H-04326, and hardness tests (Rockwell and Vickers methods), according to PN-78/H-04355 and PN-78/H-04360 respectively.

The results of these tests are shown on the fig.2, 3 and 4. From these figures follows, that maximal fatigue strength is obtained for 1.6 – 1.9 % of total carbon content on the surface of sample. It applies to variant 1 and 8 of heat treatment, and is compared to non-carburized bearing steel (fig.2)

Table 1.

Treatment variant	Carburizing		Hardening						Tempering	
			Austenitizing		Cooling		After-cooling			
	Tn [°C]	tn [h]	Ta [°C]	ta [h]	Tch [°C]	tch [h]	Td [°C]	td [h]	To [°C]	to [h]
1	900	15	850	0.3	120 oil	0.1	10 water	0.5	180	2
2			850		120 oil	0.1	10 water	0.2	150	1
3			800							
4			850	.25	20 oil	0.3				
5			800				---	---		
6			850		20 water	0.3				
7			800							
8			850	0.3	120 oil	0.1	10 water	0.5	180	2

Hardness measured as well with Rockwell as with Vickers methods
show a rapid increase for carbon content of 1.6 %. Over this
content, the increase is slight and for some heat treatment
variants it is stabilized at a constant level for 6 mm and 12
mm sample thickness (fig.3 and 4).

Variant 2 (fig.3) at carbon content 1.6 to 2.2 % on the surface is the best method of heat treatment for samples 6 mm
thickness. The same effect can be obtained for sampels of 12 mm
thickness (fig.4).

Moreover it was stated, that in the range of hardening temperatures from 800 to 850°C, the hardness is invariable and
independent of foundation material thickness. Foundation material thickness increase gives only insignificant decrease
of hardness. From view-point of used cooling medium, the oil
at 120°C appears the best, and is convenient to traditional
way of bearing steel cooling.

On the base of presented results it is possible to state,
that carburizing process considerably improves volumetric fatigue strength and hardness in comparison with non-carburized
bearing steel after traditional treatment.

The results we have obtained hitherto from complex investigations of carburized bearing steels utilizable properties
show, that the best effects were given after carburizing at
carbon surface content from 1.8 to 2.0 %. In these conditions
the smallest contact fatigue (fig.5) and the lowest abrasive
wear (fig.6) were abtained. From phase analysis it results,
that at these carbon contents, in carburized bearing steels
there exists about 20% of retained austenite and 17 % of cementite [4]. It is related as well to the lowest martensitic
transformation beginning temperature (Ms), to critical cooling
rate (fig.7), as to the most beneficial compressive stresses
(fig.8).

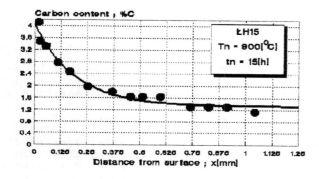

Fig.1. Carbon content
variations vs.
distanse from
surface of car-
burized ŁH15
bearing steel.

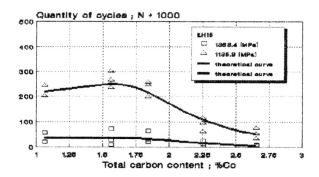

Fig.2. Fatigue life vs. carbon concentration at different stress levels of LH15 steel.

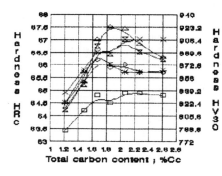

Fig.3. Changes of hardness HRc and HV30 regarding to car bon content for carburized LH15 steel samples of 6 mm thickness, after different heat treatments (according to table 1).

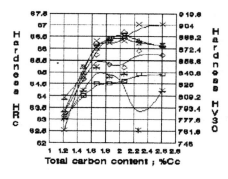

Fig.4. Changes of hardness HRc and HV30 regarding to car bon content for carburized LH15 steel samples of 12 mm thickness, after different heat treatments (according to table 1).

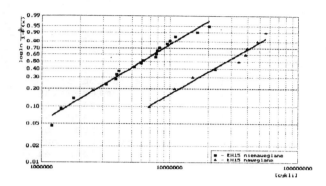

Fig.5. Results of con-
tact fatigue
life investiga-
tions for ŁH15
bearing steel
[2].

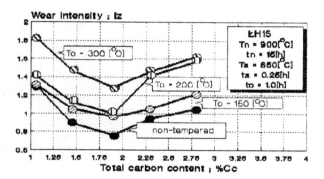

Fig.6. Abrasive wear in-
tensity vs. sur-
face carbon con-
tent of carburi-
zed ŁH15 steel
[3].

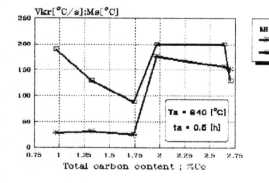

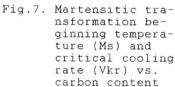

Fig.7. Martensitic tra-
nsformation be-
ginning tempera-
ture (Ms) and
critical cooling
rate (Vkr) vs.
carbon content
for carburized
ŁH15 steel [4].

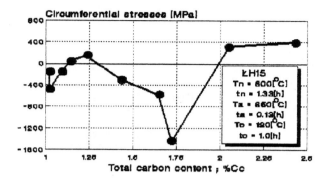

Fig.8.Variation of —
      circumferential
      stresses   with
      sub-surface la-
      yer carbon con-
      tent for carbur-
      ized ŁH15 steel
      [5].

REFERENCES
1. Krauss G. : ASM International, Materials Park, Ohio, 1990;
2. Przyłęcka M. : Mechanic Review , 1983 , 5 , s.7÷11;
3. Przyłęcka M. : Technical University of Poznań , Dissertation,
   1988 , 202 ;
4. Przyłęcka M. , Gęstwa W. , Kulka M. : Proceedings of VI
   Conference " Termoobróbka'92 " , 1992 , 150÷159;
5. Przyłęcka M. , Pertek A. : Proceedings of International
   Conference - XIVth Netional Days of Heat Treatment ,1992.

Materials Science Forum Vols. 163-165 (1994) pp. 221-226
© *1994 Trans Tech Publications, Switzerland*

PULSE PLASMA CARBURIZING OF STEEL WITH HIGH PRESSURE GAS QUENCHING

F. Schnatbaum and A. Melber

Leybold Durferrit GmbH, Hanau, Germany

ABSTRACT

Dependent on the plasma parameters used during pulse plasma carburizing a constant carbon mass flow to the surface occurs. The mass flow densitiy can be influenced by varying the plasma parameters. This allows a very close control to the process. The combination pulse plasma carburizing with high pressure gas quenching offers several other advantages. Parts from the automotive industry were pulse plasma carburized and quenched in oil and under high pressure gas using helium. To check the uniformity of carburization and the hardness after oil quenching and high pressure gas quenching respectively hardness profiles are measured. The distortion of components quenched in oil and under high pressure gas is determined. Results will be presented.

INTRODUCTION

The development in technology demands a continuous increase in corrosion, wear and fatigue resistance of parts. At the same time the saving of energy plus added emphasis on the economic and ecologic aspects is becoming more and more important in the creation of new products. The increasing demands to many parts are often only to be fulfill by heat treatment. The choise of heat treatment process is also being governed to a growing extent by the ecological and economical aspects. Plasma carburizing combined with high pressure gas quenching largely meets both demands. The benefits of this combination of processes are demonstrated by treatment results of components from the automotive industry.

PLASMA CARBURIZING

Plasma carburizing is performed in vacuum furnace at pressures between 10 Pa und 3000 Pa. The treatment gas consists of hydrogen, argon and a carburetted hydrogen gas as carbon supplier. Due to the composition of the treatment gas CO_2 cannot formed. This is likely to be of interest regard to the discussion on CO_2.

The walling of the furnace is connected as an anode. The work pieces to be treated are electrically insulated from the furnace walling and connected to the cathode. By applying direct voltage between 350 V and 1000 V a glow discharge is generated between the furnace wall (anode) and the work pieces (cathode). In the glow discharge the gases experience various reactions. Determinative for the process is that the hydrocarbon molecules are dissociated, ionized and/or set to excited states. Thereby large numbers of active carbon species are produced in the plasma, reach the surface and are embedded there. According to confirmaty statements quoted in literature this leads to a saturation of the component surface with carbon in just a few minutes.

PULSE PLASMA CARBURIZING

To avoid oversaturation of the surface with carbon and an unacceptable carbide formation associated therewith, the carbon concentration at the surface must be reduced. Hitherto a carburization phase is followed by a diffusion phase. To fulfill a certain specification a number of carburizing and diffusion stages of varying duration have to be carried out one after the other.

Investigations on carburizing in a pulsed direct current (DC) plasma have shown that a temporal constant mass flow carbon to the surface of the components take place. By varying the plasma parameter very high mass flows can be obtained which cause a saturation of the surface with carbon in just a few minutes. However, very small mass flows can also be produced. Consequently only the amount of carbon reaches the surface as diffuses into the component. Thereby oversaturation of the surface with carbon can be avoided.

As an example fig. 1 shows the dependence of the mass flow density carbon on the ratio of pulse duration / pause duration as determined. All other parameters were kept constant.

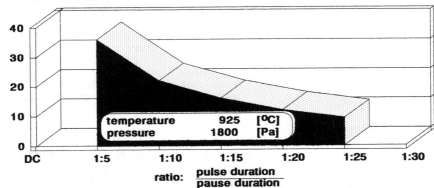

Fig. 1: carbon mass flow densitiy as a function of pulse duration/pause duration

The longer the voltage pulse in relation to the pause duration between the voltage pulses the greater the mass flow density. This means, by shortening the pauses in relation to the pulse duration such large mass flow densities of carbon can be produced that after a treating time of some minutes very high carbon contents can be obtained. By extending the pauses the mass flow can be reduced to such an extent that only the amount of carbon is supplied as diffuses into the work piece.

Thus, with a high mass flow at the beginning of the process a high surface carbon content to just below the saturation point can be obtained by choosing suitable parameter combinations. A steep carbon gradient into the material occurs.

Thereafter the mass flow is reduced to such a level that the surface carbon content remains just below the saturation point, but at the same time carbide formation in the surface zone is avoided. By limiting the mass flow density to the required level no other pauses without plasma for diffusion have to be made during the treatment. A steep carbon gradient is maintained and process times are as short as possible.

The mass flows of carbon, i.e. the process parameters, which must be chosen for the individual process stages are simulated and calculated on a computer and specified before the process starts.

HIGH PRESSURE GAS QUENCHING

The materials used in case hardening are mainly unalloyed and alloyed steels with carbon contents up to 0.25%. Hardening higher alloyed steels with high pressure gas at pressures up to 10 bar in nitrogen is a well established industrial process and is used for example for hardening tools made from high speed stell or hot work steel. However, the quench intensities obtainable at pressure up to 10 bar nitrogen are not generally adequate for the hardening of case hardening steels.

The process developed and patented by Leybold Durferrit for quenching under high pressure gas at pressures greater than 10 bar and using helium for example opens up further groups of material which can be quenched in high pressure gas because similar quenching intensities as are possible with oil hardening can be obtained.

After high pressure gas quenching the components are clean. As plasma carburizing is a "dry" process, quenching the components under high pressure offers a further process advantage. Post cleaning to remove oil residues adhering to the components is not necessary and the problems associated otherwise with disposal of such washing media do not arise.

PULSE PLASMA CARBURIZING WITH OIL QUENCHING AND HIGH PRESSURE GAS QUENCHING

For investigations on pulse plasma carburizing and quenching in oil and under high pressure gas two components from the automotive industriy were selected. Ball cages, made of material 21 NiCrMo 2 and injection nozzles, made of material 20 NiCrMo 2.

INJECTION NOZZLES

The treatment instruction of the injection nozzles prescribes several treatment steps. After carburizing and quenching the nozzles have to be deep frozen to -80°C to transform residual austenite to martensite. After the freezing step the injection nozzles are tempered. The specification requires a hardness of > 655 HV 1 in a depth of 0.2 mm and an effictive hardening depth of 0.4 - 0.5 mm at a hardness of 650 HV 1. Fig. 2 shows the result of hardness measurements after pulse plasma carburizing, oil quenching, deep freezing and tempering.

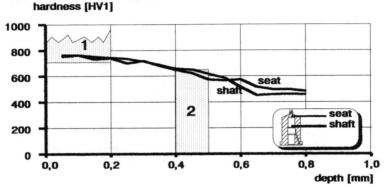

Fig. 2: hardness profiles measured after oil quenching

The hardness profiles were measured at the seat and at the shaft of the nozzle. As the hardness profiles show the carburization of the injection nozzles is very uniform. The specifications are fulfilled all hardness values are in the required range.

Fig. 3 shows the results of hardness measurements after pulse plasma carburizing, quenching under high pressure gas in helium at 10 bar, deep freezing and tempering. All hardness values are also in the required range, the specifications are fulfilled.

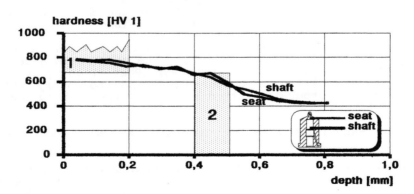

Fig. 3: hardness profiles measured after high pressure gas quenching using 10 bar helium

BALL CAGES

The specification of the ball cages requires a surface hardness of 670 - 790 HV 5 after tempering, a case depth of 0.8 - 1.1 mm at a hardness of 520 HV 1, a core hardness of 300 - 500 HV 30 and an approximately horizontal hardness profile within the first 5/10 mm. Fig. 4 shows the result.

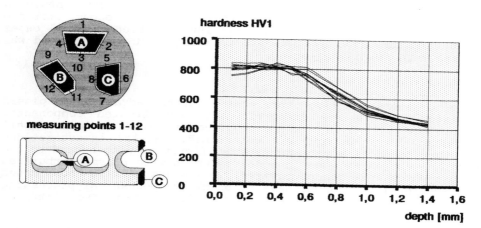

Fig. 4: hardness profiles measured after oil quenching

To check the uniformity of carburizing the ball cages were cut open, parts of the stem removed and the hardness measured at the points marked. All values are within a close range and all hardness values measured are within the required specification.

Fig. 5 shows the hardness profiles after high pressure gas quenching using helium at 15 bar and before tempering.

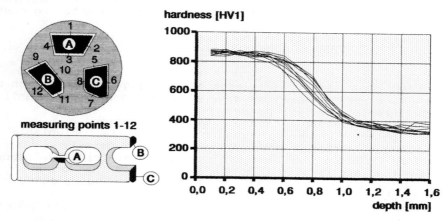

Fig. 5: hardness profiles measured after high pressure gas quenching using helium at 15 bar

All specifications are also fulfilled. Surface hardness, effective hardening depth and core hardness are within the required range. Also the plateau-like hardness profile within the first 5/10 mm is obtained without difficulty.

METALLOGRAPHIC INVESTIGATIONS

Because plasma carburizing is done in vacuum and no oxygen containing gas is used an internal oxidation of the parts cannot occur. This can be confirmed by several metallographic investigations of ball cages and injection nozzles. The surface of every part investigated was free from internal oxidation.

DIMENSIONAL STABILITY

It is often stated in literature that a further advantage of plasma carburizing is the low distortion of such treated components. Therefore the distortion of gas carburized and plasma carburized ball cages after oil quenching was determined. Initial investigations showed that there were no significant differences with regard to the dimensional change between those parts which were plasma carburized and those which were gas carburized. This means that in those cases when dimensional tolerances have to be close, mechanical processing is necessary, even though the carburization is free from internal oxidation.

Investigations on components plasma carburized and quenched under high pressure gas however showed that dimensional stability was greater when using this process than after oil quenching. In addition it is apparent that the scatter range is much smaller after high pressure gas quenching than after oil quenching. This is of considerable importance to the manufacture of components.

Nonvarying dimensional changes can be taken into account in the manufacturing process. By keeping the scatter band in dimensional change to a minimum in many cases it is possible to disperse with post mechanical processing even on components with close tolerances. This means that components can be finish mashined in the unhardened condition which gives considerable savings in manufacturing costs.

These investigations will be continued.

SUMMARY

Is has been shown that pulse plasma carburizing in combination with high pressure gas quenching has several advantages over conventional processes like
-    plasma carburizing prevents internal oxidation
-    the pulse plasma carburizing process is calculable and reproducible with a knowledge of mass flow density
-    plasma carburizing is an environmental-friendly process. Emissions are reduced to a minimum, CO_2 does not occur
-    no problems with regtard to cleaning and disposal of the respective media after heat treating
-    first investigations indicate that high pressure gas quenching has a negligible effect on the dimensions.

Materials Science Forum Vols. 163-165 (1994) pp. 227-232
© 1994 Trans Tech Publications, Switzerland

ION CARBURIZING - APPLICATIONS IN INDUSTRY

D.E. Goodman and S.H. Verhoff

Surface Combustion, Inc., Maumee, OH, USA

ABSTRACT

Glow discharge plasma processes have been commercially used for many years. Ion carburizing has emerged as one of the latest extensions of these plasma processes and has gained acceptance within the last 10 years. Performed under a vacuum, the process has all of the inherent advantages of typical vacuum processes including cleaner surface appearance, improved case uniformity, lack of intergrannular oxides, ability to process difficult geometries and a potential for process time savings. This paper will focus on several applications where ion carburizing has shown process advantages in the automotive and aerospace fields.

INTRODUCTION

Ion carburizing has been demonstrated to show its biggest advantages in higher temperature carburizing (greater than 1750°F), shallower case depths (up to .050" effective) and complicated geometric shapes of parts (gear teeth, blind holes/passages, etc.). Test work and production work accumulated has primarily dealt with gears, shafts, fuel injectors, bearing races, constant velocity joints and hydraulic components.

PROCESS OVERVIEW

Ion carburizing is one of many glow discharge plasma processes. A plasma is defined as an electrically-generated gaseous mixture consisting of positively, negatively and neutrally charged particles. This plasma is created in ion processing equipment by evacuating the chamber, introducing trace amounts of process gas and electrically charging the workload (cathode) with a negative voltage upwards to 1000 volts DC.

In addition to the process advantages mentioned previously, ion carburizing also has the benefits of:

- A cold-walled vessel construction which allows the unit to be placed in-line on a factory machining or assembly floor.

- The ability to start-up and shut down the equipment on an as-needed basis without requiring idling or seasoning of the equipment.

- No need for an endothermic carrier gas (RX®) and less process gas (CH_4) is required to produce a carburized part.

EQUIPMENT OPERATION OVERVIEW

A typical ion carburizing facility (see Figure 1) can be a single or multi-chamber vacuum furnace with a DC power supply for creating the glow discharge, an electrically isolated work hearth, a carburizing gas manifold system, a heating element isolation system and sophisticated controls to monitor and adjust all the necessary process variables.

Figure 1

In a typical cycle, the vacuum furnace chamber is evacuated and preheated to the operating temperature. A load is assembled by placing parts in standard vacuum work baskets with layers separated by mesh liner grids. When completely loaded, the baskets and parts are prewashed in a washer, using an alkaline based soap and allowed to dry. The load is charged onto the ion carburizing furnace quench chamber and set on the internal fork loading mechanism. The quench chamber is closed and evacuation begins.

When the heat chamber and the quench chamber have equalized in vacuum level, the load is transferred into the heat chamber and placed on the isolated hearth. The load is heated to temperature and soaked to insure temperature uniformity. When the soak is complete, DC power is applied to the load and hydrogen is introduced into the chamber at a preset flow rate. The pressure in the heat chamber is controlled by a microprocessor. (The purpose of this hydrogen pretreatment is to insure cleanliness and surface depassivation before beginning the carburizing segment of the cycle.)

When the pretreatment is complete, the hydrogen flow ceases and natural gas flow begins with the DC power remaining on. Typical natural gas flow is 0 - 90 scfh. Occasional arcing is suppressed by an ion arc detection/suppression system (Surface Combustion) integrated with the power supply. At the

conclusion of the carburizing time, the natural gas flow ceases, the DC power
supply is shut off and diffusion time begins at temperature under a hard
vacuum. Following diffusion, the load is equalized, transferred to the quench
chamber and oil quenched. This complete ion carburizing process is patented
by Surface and includes the control of watt density, carbon mass flow and
pressure.

CASE STUDY #1 - CONSTANT VELOCITY JOINT CAGES

CV joint cages provide several unique carburizing challenges in case
uniformity and distortion due to the thin wall construction. Post heat
treatment machining must be kept to a minimum. Typical steels used are AISI
8620H or AISI 5120H.

Figure 2

Surface Combustion developed ion carburizing
parameters to allow for .80% surface carbon
with an effective case depth of .050" to 50
Rc using a carburize and diffuse temperature
of 1800°F.

The ion carburizing cycle performed in a
two-chamber ion carburizing unit with an
integral oil quench is listed below:

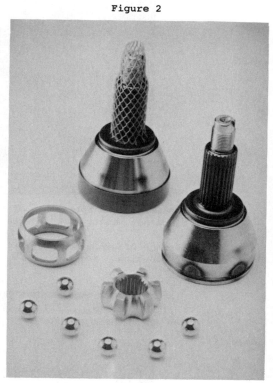

	Cycle Time (Minutes)
Load furnace, evacuate quench chamber and transfer load	15
Heat to 1800°F and soak	75
Hydrogen pretreatment	15
Ion carburize and diffuse at 1800°F	129
Cool to 1550°F	30
Equalize at 1550°F	30
Transfer load to quench chamber	10
Oil quench and drain	30
Total cycle time	334 Min
	or 5 Hr 34 Min

Conclusive process benefits included significantly reduced cycle time; minimal
process gas requirements for natural gas (no endothermic gas necessary;
reduced part distortion (see Figure 2) the CV joint cages (which enabled more
accurate machining before carburizing, plus ease of machining,

chucking/gripping of parts on flat edges, and finish grinding, etc. after carburizing); a system that operates in very close to a clean room environment in a machining/assembly plant.

The specific advantages of ion carburizing experienced over atmosphere carburizing were:

(1) Better case uniformity within the cage windows (due to the plasma providing an extremely uniform mechanism for transmitting carbon to the workpiece surface). See Figure 3.

Figure 3

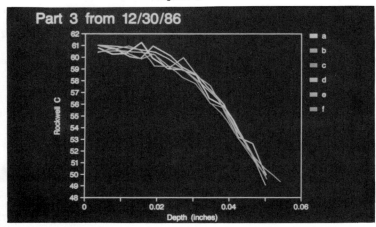

(2) Visually clean appearance of the workpiece surface (due to a lack of soot, since not nearly as much excess carbon is introduced).

(3) Absence of grain boundary oxides on the workpiece surface due to total vacuum processing. See Figure 4.

Figure 4 (1 of 2)

Figure 4 (2 of 2)

(4) Reduced total cycle time (by processing at 1800°F without experiencing alloy and brick life degradation associated with atmosphere carburizing).

(5) Improved distortion control of as-quenched parts in comparison of ion carburizing with atmosphere carburizing. (This benefit is most likely derived from quenching a more uniform case causing less distortional stresses within the part).

CASE STUDY #2 - FINE-PITCHED GEARING

Surface Combustion examined several samples of AISI9310 fine-pitched gear specimens to determine if ion carburizing could provide a more uniform case depth profile from the pitch to the root of the gear tooth.

The desired metallurgical properties were a .015"-.020" effective case depth with a .95% C surface carbon. Surface hardness range was 58-62 HRc.

Ion carburizing tests were conducted at 1650°F with a 30 minute ion carburize time and a 30 minute diffuse time at two partial pressure levels: 2 torr and 4 torr. The resulting carbon gradients are displayed. (See Figure 5)

The hardness profiles of the pitch and root of the gears showed that excellent case uniformity in fine-pitched gearing can be achieved. The samples processed at 4 torr are optimized and show that an effective case depth spread from .0155"-.0165" or .001" exists from the pitch to the root of the gear tooth. (See Figure 6)

CONCLUSION

Ion carburizing has been proven in many instances to have distinct process, productivity and environmental advantages over conventional carburizing processes. Significant growth will occur as more applications are demonstrated to show these advantages.

Figure 5

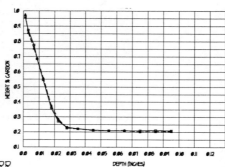

☐ - CURVE No.I - 2 TORR

△ - CURVE No.2 - 4 TORR

ION CARBURIZED AT 1650° F

ⓧ Surface ⓧ
Combustion, Inc.

Figure 6

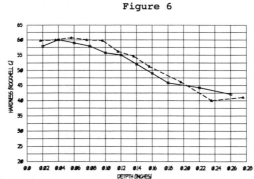

☐ - CURVE No.I - ROOT

△ - CURVE No.2 - PITCH

ION CARBURIZED
4 TORR AT 1650° F

ⓧ Surface ⓧ
Combustion, Inc.

REFERENCES

Schultz, T. J., Kuhn, T., Goodman, D. E., "Carburizing - Stepping Into The Future," Carburizing Processing and Performing Conference (July 1989)

Goodman, D. E. and Simons, T., "Constant Velocity Joints Being Produced With World's First Production Ion Carburizing Furnace," Industrial Heating (November 1987)

Verhoff, S. H., "Greater Uniformity Of Plasma Carburizing Rapidly," Industrial Heating (March 1986)

Materials Science Forum Vols. 163-165 (1994) pp. 233-238
© 1994 Trans Tech Publications, Switzerland

USE OF THERMOGRAVIMETRY FOR THE STUDY OF CARBONITRIDING TREATMENT OF PLAIN CARBON STEEL AND LOW ALLOY STEEL

M.S. Yahia, Ph. Bilger, J. Dulcy and M. Gantois

Laboratoire de Science et Genie des Surfaces (URA CNRS 1402),
I.N.P.L., Ecole des Mines, Parc de Saurupt, F-54042 Nancy Cédex, France

ABSTRACT

Carbon-nitrogen interaction and its consequences on the dissolution of these interstitiel elements as well as on their diffusion in austenite is investigated using an instrumented thermogravimetric apparatus developed in the laboratory. Carbon and nitrogen increase their mutuel thermodynamic activity which promotes their diffusion, but inhibits their dissolution.

INTRODUCTION

We apply the methodology of a carburizing process developed by J. Dulcy and M. Gantois[1] called "rapid carburizing" to study a new carbonitriding treatment. The final aim of this work is to investigate the effect of nitrogen on the mechanical properties obtained after quench hardening. It is then necessary :
- to set up carbon and nitrogen concentration profiles, and to
- characterize the carbon-nitrogen interaction in gamma iron.

We show in this paper some results obtained by thermogravimetry. Nitrogen transfer is done through ammonia decomposition on the surface of XC (plain carbon steel), CD4 (1%Cr, 0.2%Mo), and MC5 (1.2%Mn, 0.8%Cr) type steels at 870°C.

EXPERIMENTAL APPARATUS

In order to study the kinetics at the gas-solid interface we have used a thermogravimetric apparatus linked to a gas phase chromatograph developed in the laboratory (**figure 1**). The apparatus consists of:
- An instrumented thermobalance (Computer data aquisition) enabling an online access to the weight gain of the sample as a function of the operating parameters. The sample, hanged to the balance beam in the center of an alumina or a high temperature alloy tubular reactor, could be automatically unhooked to be oil or water quenched, so as to study its metallurgical structure.
- A set of mass flowmeters which regulate the inlet gas composition and flowrate.
- A gas phase chromatograph to qualitatively and quantitatively study the chemical species present in the atmosphere.
The computer data aquisition of the following: temperature, weight gain, inlet and outlet gas composition, enables the access in real time to the transfered mass flux, the mass balance, the nature of the chemical reactions in the gas phase and to their yield.

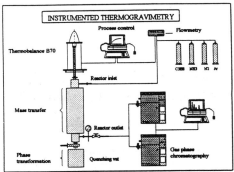

figure 1

CARBON-NITROGEN INTERACTION IN GAMMA IRON

     Thermodynamic data for the Fe-C-N system was obtained from literature[2,3,4]. This data is relative to the activity of carbon and nitrogen dissolved in austenite. The mathematical formalism is :

$a_C = \gamma_C(T, \%C, \%N). \%C$ $\%N$, $\%C$: dissolved nitrogen and carbon in weight percent
$a_N = \gamma_N(T, \%C, \%N). \%N$ where,
a_C and a_N are the carbon and nitrogen thermodynamic activity and,
γ_C and γ_N are the carbon and nitrogen activity coefficient, functions of T, %C, %N.
In this study we use the geometric exclusion model[5] applied to the ternary Fe-C-N system[6].
The curves of **figure 2** show that :
- The carbon activity increases as the nitrogen content of the solid increases for a fixed carbon content.
- The nitrogen activity increases as the carbon content of the solid increases for a fixed nitrogen content.

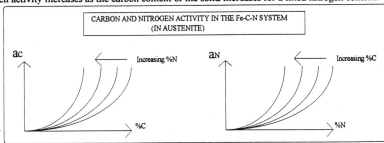

figure 2

     The notion of activity is correctly displayed from the experimental point of view by the thermogravimetric curves. Those of **figures 3a** and **3b** obtained during two nitriding treatments of an XC10 steel (0.1% nominal carbon content) and an XC65 steel (0.65% nominal carbon content), carried out under the same operating conditions, show that nitrogen intake is more important for the low carbon steel than for the high carbon steel. This means that for the same ammonia partial pressure we dissolve more nitrogen in XC10 steel than in XC65 steel.

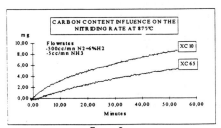

figure 3a

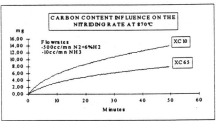

figure 3b

The ammonia partial pressure sets the nitrogen activity (mass action law), a higher carbon content (XC65 steel) inhibits nitrogen solubility in the austenitic phase (see **figure 3**).The nitriding reaction is:

$$NH_3 \Leftrightarrow [N]_{Fe} + \frac{3}{2}H_2, \qquad\qquad a_N = K(T).\frac{P(NH3)}{P(H_2)^{3/2}}$$

The notion of activity explains why, for identical operating parameters (same temperature and same ammonia flowrate), we have a difference in behaviour during nitriding treatments of a low carbon steel and a high carbon steel (samples having initially a homogenious carbon content).

SETTING UP OF CARBON AND NITROGEN GRADIENTS : PROPANE CARBURIZING FOLLOWED BY AMMONIA NITRIDING

In order to obtain the desired carbon and nitrogen concentration gradients, the carbonitriding treatments are made up of a carburizing stage followed by a more or less long nitriding stage (**figure 4**) during which carbon diffuses (zero propane flowrate, carrier gas 500cc/mn N2+6%H2, sample's surface = 1400mm²).

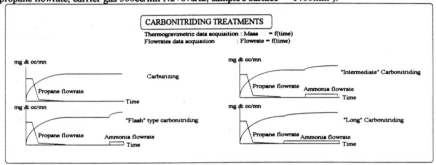

figure 4

"NITROGEN FLASH" TYPE CARBONITRIDING (27MC5 STEEL)

The aim of "nitrogen Flash" type carbonitriding (**figure 4**) is to obtain a steep nitrogen profile over a short depth (about 100 microns). The total treatment time is 110 minutes. 60 minutes are devoted to carbon intake and 50 minutes to its diffusion (zero propane flowrate). At the end of the carbon diffusion stage ammonia is introduced for 10 minutes. **Figures 5a and 5b** show the thermogravimetric curves obtained respectively for a carburizing and for a "flash" type carbonitriding treatments. The experimental diffusion profiles are shown in **figure 5c**. Surface carbon and nitrogen contents are respectively 0.6% and 0.3%.

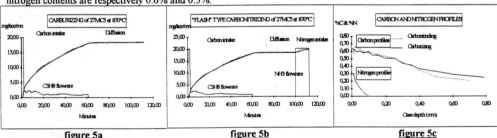

figure 5a   figure 5b   figure 5c

"INTERMEDIATE" AND "LONG" CARBONITRIDING (27MC5 STEEL)

In order to obtain various nitrogen profiles, we have made three carbonitriding treatments of the "long" and "intermediate" type. The total treatment time was 110 minutes, of which 60 minutes were affected to the carbon enrichment sequence. **Figures 6a, 6b** and **6c** show the thermogravimetric curves recorded for these three treatments. Nitrogen enrichment was carried out during, either the entire carbon diffusion stage (50 mn) using a constant ammonia flowrate (5cc/mn, **figure 6a**). 15 and 10cc/mn, **figure 6c**) or a fraction of this diffusion period using also a constant ammonia flowrate (10cc/mn, **figure 6b**). Thus we obtain three progressive nitrogen enrichment values (**figures 6a, 6b, 6c**) with identical carbon intakes.

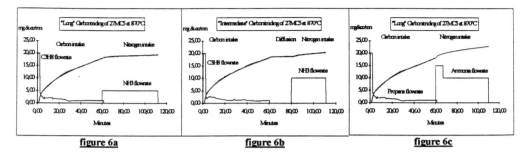

figure 6a   figure 6b   figure 6c

Analysis of the experimental diffusion profiles (**figure 7a**) confirms the thermogravimetric recorded curves :
- the three carbon diffusion profiles are sensibly identical.
- the three nitrogen profiles have increasing values.

Even though the samples, subjected to these three treatments, have different nitrogen profiles (but identical carbon profiles), their microhardness profiles (**figure 7b**) are sensibly the same. They show a constant hardness (800HV) over a depth of about 200 microns.

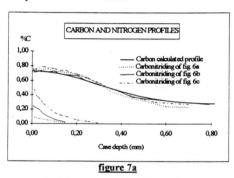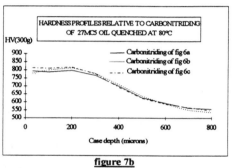

figure 7a   figure 7b

EFFECT OF THE PARAMETERS "INTERSTITIEL CONTENT" AND "CONCENTRATION GRADIENT"

Theoretical approach

If an element within a phase has a chemical potential gradient there exists then a transfert of this element by a diffusional process[7]. Carbon and nitrogen mutually increase their own thermodynamic activity (increase of the activity coefficient), which induces an increase of their chemical potential. Consequently we get :
- an inrease of the carbon diffusion coefficient in the presence of nitrogen atoms and conversely,
- an increase of the nitrogen diffusion coefficient in the presence of carbon atoms.

The formulation of the diffusion coefficients is:

$$D_C = D_1(T, \%C, \%N) : \text{carbon coefficient in } \gamma \text{ iron}$$
$$D_N = D_2(T, \%C, \%N) : \text{nitrogen coefficient in } \gamma \text{ iron}$$

These coefficients are functions of temperature (thermally activated phenomenon) and the respective carbon and nitrogen content. The carbon and nitrogen fluxes are then written as follows:

Carbonflux = $- D_1(T, \%C, \%N) . \nabla(\%C)$ **(1)**   Nitrogenflux = $- D_2(T, \%C, \%N) . \nabla(\%N)$ **(2)**

The transposition of the mass flux for carbon and nitrogen (equations 1 and 2) into the mass conservation expressions gives us two coupled equations :.

$$\frac{\partial \left(D_c . \frac{\partial C}{\partial x} \right)}{\partial x} = \frac{\partial C}{\partial t} \qquad\qquad \frac{\partial \left(D_N . \frac{\partial N}{\partial x} \right)}{\partial x} = \frac{\partial N}{\partial t}$$

The simultaneous numerical resolution of these equations gives the evolution of the carbon and nitrogen diffusion profiles with time.

Experimental approach

Using the thermogravimetric curves we demonstrate how the parameters "interstitiel content" and "concentration gradient" modify the diffusionnal processes in the Fe-C-N system.
To illustrate this point, we show how the value and the shape of the carbon profile affect nitrogen diffusion in low alloy steel. Three carbonitriding treatments were carried out :
- the first is a "long" type treatment of 27CD4 steel (**figure 8a**), called treatment #1
- the second is an "intermediate" type treatment of 27CD4 steel (**figure 8b**), called treatment #2
- the third is an "intermediate" type treatment of 27MC5 steel (**figure 8c**), called treatment #3

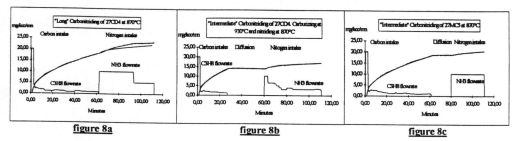

figure 8a   figure 8b   figure 8c

The comparaison of the weight gain during the nitriding stages following the carburizing sequences (**figure 9a**) shows that the nitrogen mass intake decreases in the following order: treatment #1, treatment #2, treament #3.
This may be explained by the character of the calculated carbon profiles at the start of the nitriding stage (**figure 9b**):
- for treatment #1, Carbon presents a concentration gradient (**curve 1, figure 9b**), whereas for treatments #2 and #3 (**curves 2 and 3** respectively) the profiles are "flat". Thus the nitrogen mass intake is more important for the first treatment.
- carbon content for treatment #2 is lower than for treatment #3, nitrogen mass intake is therfore higher for the second than for the third treatment.
The experimental and calculated nitrogen profiles are in agreement with the experimental weight gain curves.

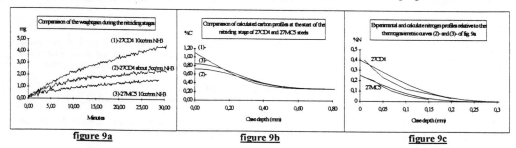

figure 9a   figure 9b   figure 9c

Treatments #2 and #3 are carried out on two different types of steel, namely and respectively 27CD4 and 27MC5. In order to eliminate any ambiguity associated with the alloying elements, we subjected two specimens of respectively 27CD4 and 27MC5 steel type to the same nitriding treatments (**figure 10, curve 2**). We observe that for these two steels, nitrogen enrichment is identical during the first 30 minutes of treatment.
The comparaison (**figure 10**) of the weight gain curves during the nitriding stage, with a prior carburizing sequence (27CD4, treatment #1, **curve 1 of figure 10**) and the nitriding stage without prior carburizing, confirms the improvement of the nitrogen enrichment by the presence of a carbon concentration gradient.

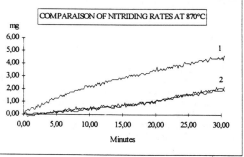

figure 10

EFFECT OF THE ALLOYING ELEMENTS

In order to show the effect of the alloying elements, namely chromium and manganese, we carried out three nitriding treatments under the same operating conditions (at 870°C with constant ammonia flowrate) of XC10, XC38 (0.38% nominal carbon), and 27MC5(0.27% nominal carbon) type steels **(figure 11)**.

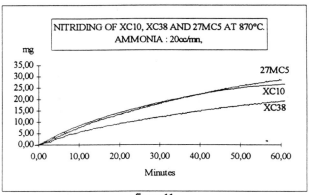

figure 11

From the three curves we infer the following:
- Confirmation of the effect of a homogenious carbon content : XC10 steel nitrogen intake is higher than XC38 steel (see comparaison of XC10 and XC65 nitidring treatments).
- Nitrogen intake is higher for XC10 than for 27MC5 during the first 40 minutes.
- After 40 minutes of treatment nitrogen weight gain of 27MC5 becomes higher than that of XC10. This might be explained by the presence of the alloying elements, namely chromium and manganese. which promote nitrogen dissolution by lowering its activity while carbon increases it and so disfavour nitrogen dissolution in austenite.

CONCLUSION

The use of instrumented thermogravimetry (data aquisition coupled with modelling) linked to gas phase chromatography shows results which explain on one hand the mass transfer into the solid phase and on the other hand the chemical composition of the atmosphere. In this study we are limited to the mass transfer into the solid phase. Some of the obtained results are:
- Nitrogen intake is more important for XC10 steel than for XC38 and XC65 steel.
- Nitrogen enrichment is promoted by a carbon concentration gradient comparatively to a "flat" carbon profile.

These experimental observations show how the nitrogen-carbon interaction in austenite modify the diffusion process of these elements and justify the repulsive nature of this interaction, which means that:
- carbon activity increases in the presence of nitrogen
- nitrogen activity is increased by carbon.

references

1) Dulcy, J. and Gantois. M. : "La cémentation accélérée", Journée A.T.T.T. -91, Internationaux de France du traitement thermique, Toulouse. 26-28 Juin 1991
2) Mori, T. et al : J. Jap. Inst. Met., 1967, 31, 887
3) Milinskaya, I. N. et al. : Russian Journal of physical chemistry, 1969, 43, 1318
4) Jarl, M. : Scandinavian Journal of Metallurgy, 1978, 7, 93
5) Kauffman, L. et al. : Decomposition of austenite by diffusion controled process, pp. 313-352, Interscience, New York 1962
6) Kurabé, H. : Transaction ISIJ, 1974, 14, 404
7) Darken, L.: Trans. AIMME, 1949, 180, 430

Materials Science Forum Vols. 163-165 (1994) pp. 239-244
© 1994 Trans Tech Publications, Switzerland

THE INFLUENCE OF CARBONITRIDING AND HEAT TREATMENT ON THE PROPERTIES OF LOW- AND HIGH-CARBON STEELS

M. Przylecka, W. Gestwa and M. Kulka

Technical University of Poznan, Poland

ABSTRACT

In this paper, the influence of carbonitriding atmosphere type, temperature and time, on carbon and nitrogen concentrations changes in case of single- and two-phase structure were defined. The influence of carbonitriding parameters on hardess and abrasive wear was examined. Changes of temperature Ms , retained austenite and internal stresses distributions were calculated. It was stated, that after carbonitriding, the single- and two-phase case have more beneficial properties than for carburized low-carbon steels.

INTRODUCTION

Carbonitriding is apart from carburizing , the most diffuzed thermo-chemical treatment process. The results of low-carbon and bearing steels carbonitriding are known from papers [2,3,6].   In this paper the investigations results of carbonitriding parameters influence on low- and high-carbon steels properties will be shown.

The influence of heat treatment on hardness and abrasive wear of steels 15,16HG and LH15 was examined.

Investigation results were connected with structural state, retained austenite quantity, martensitic transformation beginning temperature (Ms), and internal stresses.

Investigation results demonstrate, that for bearing steel it is possible to select suitable process parameters, and that bearing steel has better properties than low-carbon steels, that is connected with lower Ms tempperature, bigger retained austenite content and the lowest internal stresses.

METODOLOGY OF INVESTIGATIONS

Investigations of carbonitriding parameters influence were carried out for selected low-corbon steels and for hyper-eutectoid steel. Chemical constitution of tested steels (15, 16HG, LH15) - according to PN-72/H-84019 , PN-72/H-84030 , PN-74/H-84041 respectively. Carbonitriding processes were carried out with application of endotermic atmosphere produced from methanol and ammonia, and additionally with application of fluid bed. Carbonitriding process and heat treatment parameters are shown in the table 1. For investigation, the sampels that permit to point out carbon and nitrogen concentration changes for different distance from surface, were used (acording to PN-81/H-04040 and PN-82/H-36085).

Table 1.

Treatment variant	Carbonitriding		Hardening				Tempering	
			Austeni-tizing		Cooling			
	Twa [°C]	twa [h]	Ta [°C]	ta [h]	Tch [°C]	tch [h]	To [°C]	to [h]
WA1	900	3	900	0.5	120 oil	1.0	150	0.5
							200	
							250	
							300	
WA2	850	3	850	0.5	120 oil	1.0	150	0.5
							200	
							250	
							300	
WA3	850 fluid bed	3	850	0.5	120 oil	1.0	150	0.5
							200	
							250	
							300	

Samples for retained austenite, hardness and abrasive wear teasting were heat treated according to table 1. Argon was a protective atmosphere. According to carbon and nitrogen concentration change diagrams, the samples was ground, obtaining on the surface different carbon and nitrogen contents. Retained austenite in samples was measured by X-ray diffraction method. Hardness of hardened and tempered samples was determined by Rockwell "C" method according to PN-78/H-04355. Testing of abrasive wear was carried out with application of frictional couple composed of cylinder and plane. Tested cylindrical sampel played the cylinder, and sintered carbides plate played the plane.   Additionally in the paper, calculation of temperature Ms, retained austenite [3] and internal stresses [1,5] variations was made.   On the base of carbon and nitrogen concentration change diagrams, utilizing known process formulas, the run of temperature Ms changes, retained austenite and internal stresses of tested materials was calculated.

INVESTIGATION RESULTS

Fig. 1 to 4 present for carbonitrided steels 15, 16HG and LH15 carbon and nitrogen distributions regarding the distance from surface.   On the base of these distributions, investigations of abrasive wear, retained austenite and hardness (fig. 5 and 6) were prepared. On the fig. 5 tempering temperature influence on sampels hardness (HRc) is presented. From this drawing it appears, that the best results can be obtained for gas carbonitrided LH15 steel in all range of temperatures, in comparison with other tested materials and carbonitriding and hardening methods.   Also for this steel the lowest abrasive wear intensity coefficient for different tempering temperatures was achieved, apart from the temperature 300 [°C], at which 16HG steel demonstrated the lowest coefficient (see fig. 6).

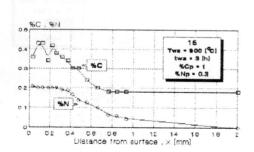

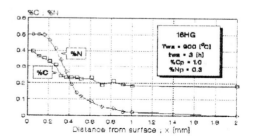

Fig.1.Carbon and nitrogen distribution vs. distance from surface of carbonitrided 15 steel.

Rys.2.Carbon and nitrogen distribution vs. distanse from surface of carbonitrided 16HG steel.

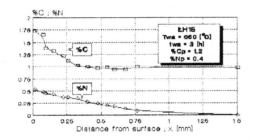

Fig.3.Carbon and nitrogen
     distribution vs. distance
     from surface of carbo-
     nitrided ŁH15 steel.

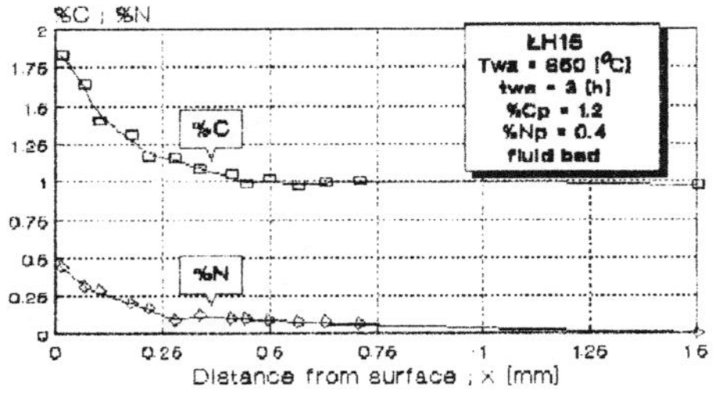

Fig.4.Carbon and nitrogen
     distribution vs. distance
     from surface of carbo-
     nitrided in fluid bed
     ŁH15 steel.

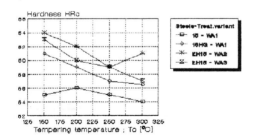

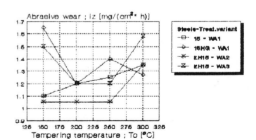

Fig.5.Hardness (HRc) of steels
     15, 16HG and ŁH15 tempered
     at different temperatures
     (according to table 1).

Fig.6.Abrasive wear intensity
     coefficient (Iz) vs. tempe-
     ring temperature of steels
     15, 16HG and ŁH15, heat
     treated according to table
     1.

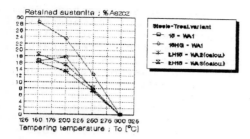

Fig.7.Retained austenite (%RA) distribution vs. tempering temperature of steels 15, 16HG and calculeted for ḼH15, heat treated according to table 1.

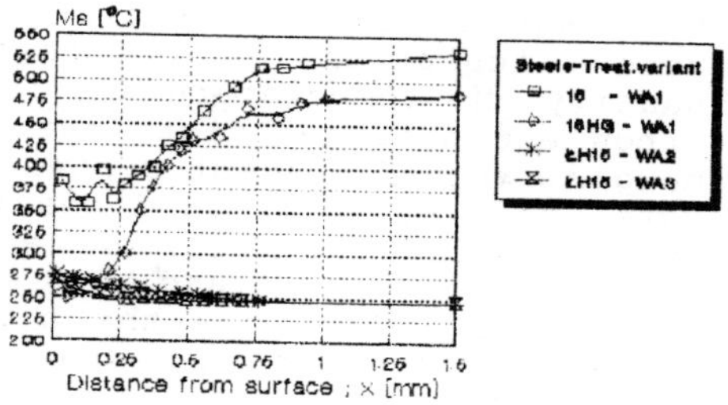

Fig.8.Temperature of Ms changes (calculated) vs. distance from surface for steels 15, 16HG and ḼH15 heat treated according to table 1.

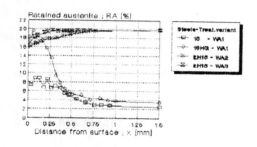

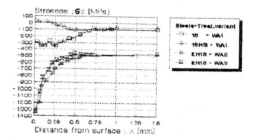

Fig.9.Retained austenite (%RA) changes (calculated) vs. distance from surface for steels 15, 16HG and ḼH15 heat treated according to table 1.

Fig.10.Internal stresses (σz) changes (calculeted) vs. distance from surfece for steels 15, 16HG and ḼH15 heat treated according to table 1.

Retained austenite investigations were carried out only for 15 and 16HG steels, and the results of these investigations are presented on the fig. 7. It arises from this figure, that retained austenite quantity decreases with tempering temperature growth, attaing for both steels the value close to 0 for tempering at 300 [˙C].

For comparison, the results of calculations were presented. From these calculations it is possible to reckon retained austenite quantity, Ms temperature and internal stresses on the strength of carbon and nitrogen distributions. Fig.8 presents Ms temperature changes, with different distances from surface. It is possible to state, that carbonitrided bearing steel has lower Ms values than low-carbon steels. Fig.9 presents retained austenite distribution in case, for 15, 16HG and LH15 steels. It results from this figure, that bearing steel has the biggest retained austenite contents. Next fig.10 presents internal stresses changes on the section of case. From stresses point of view, once more the best stresses distribution was obtained for bearing steel.

CONCLUSIONS

Presented results of invetigations authorise to the statement, that carbonitrided bearing steel has better properties, then low - carbon steels. It is connected with greater alloying elements content, influencing the lowest Ms temperature value. Also chemical composition influences carbon and nitrogen content in case properties, but just the opposite.

It is possible to ascertain that if temperature of process carbonitriding of bearing steel reduces in relation to temperatures of carbonitriding of low-carbon steel and carburizing of low-carbon and bearing steels, Working properties of layers growths.

The results obtained for bearing steel was compared to these presented in the paper [5] for carburized steel. Therefore it can be stated , that carbonitriding of bearing steel gives higher hardness, abrasive wear resistance and compressive internal stresses of material.

Results discussed in this paper, compose initial fragment of low-carbon and hyper-eutectoid steels carbonitriding process and properties investigations.

REFERENCES

1. G. Degallaix and others : Mémoires et Etudes Scientifigus Revue de Métallurgie, Février 1990, 113 ÷ 122;
2. G. Krauss : ASM International, Materials Park, OHIO, 1990;
3. J. Lesage & A. Iost : Advances in Surface Treatments, Pergamon Press, 1987, 5 , 45 ÷ 57;
4. W. Luty, J. Wyszkowski : Haerterei-Technische Mitteilungen, 27 (1972), 5 , 266 ÷ 271;
5. M. Przylecka : Technical University of Poznań, Dissertation, 1988 , 202 .
6. J. Slycke, T. Ericssen: J. Heat Treating, 1981, 2 .

V.   NITRIDING AND NITROCARBURIZING

Materials Science Forum Vols. 163-165 (1994) pp. 247-252
© 1994 Trans Tech Publications, Switzerland

ESTABLISHMENT OF QUASI EQUILIBRIUM STATES IN TECHNICAL SYSTEMS OF IRON AND NITRIDING GAS

Chr. Ortlieb, S. Böhmer and S. Pietzsch

Institute of Materials Engineering, Freiberg University of Mining and Technology, Germany

ABSTRACT

The Fe-N phase diagram is of particular practical importance in understanding and controlling the compound layer development in gasnitriding processes. Using selected methods of investigation it is possible to give a detailed description of establishment of quasi-equilibrium states between nitriding gas and iron and the approximation to equilibrium. It is demonstrated that the thermodynamic models, for example, can be applied in order to produce different nitrogen contents in ε-nitride depending on the nitriding conditions.

1. INTRODUCTION

A more exact process control of gasnitriding requires more reliable data of the equlibrium and the establishment of equilibrium between the nitriding medium and the nitrogen solved in the surface layer. In the nitriding practice equilibrium diagrams were used for orientation: the phase diagram experimentally determined by Lehrer (1) for gasnitriding and the results obtained by Naumann and Langenscheid (2). They were repeatedly confirmed experimentally and by calculation. Their transferability to technical conditions has particularly called into question by Edenhofer (3). Lehrer has performed its investigations under laboratory conditions, hence the equilibria established would change under different conditions. Special attention is paid to the equlibrium between the ammonia containing nitriding medium and the nitrogen dissolved in the ε-nitride phase. Calculations formally allowing a treatment analogous to the equilibrium between gasphase and carbon activity in γ-iron (4) - where a defined gas composition under constant temperature is assigned to each carbon activity - were suggested for discussion by Jentzsch, Kunze and others (5-8). The determination of the nitrogen content in ε-iron is based on the thermodynamic connection between the nitrogen partial pressure $p(N_2)$ of the nitriding gas and the nitrogen concentration and the thermodynamic models for regular or subregular solution of nitrogen in hexagonal iron (see fig. 1: Lehrer diagram with the isoconcentration lines according to the model of subregularsolution of nitrogen in hexagonal iron).

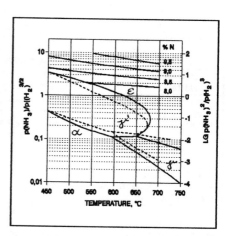

Fig. 1: Phase equilibria with NH_3-H_2 mixtures
solid lines: thermodynamically
calculated by Kunze
dashed lines: measurements by Lehrer

If sufficciently accurate and reliable experimental and calculated data are available the possibility of manufactoring of different surface layers in narrow tolerances is improved. A criterion of confidence in the reliability of determined data is the accordance of results obtained from nitrided iron foils and fom the surface-adjacent region of compound layers. The approximation to the equlibium from the outside gas as well as from the interior of solide to the gas-solid interface was studied too. A critical comparison with calculated data will there-fore guarantee the reliability of the registration of the real proportions and provide a basis for the practical use.

The main subjects of our research were investigations of foils and massive specimens gasoxynitrided in a chamber furnace which was equipped with solid electrolyte sensors. To limit error sources, the method of X-ray diffraction analysis and the methods of element concentration analysis were critically judged and combined with scanning electron microscopic investigations.

2. MANUFACTORING OF ε-COMPOUND LAYERS AND METHODS OF INVESTIGATION

Disc specimens ($\emptyset$ 30 $*$ 5 mm) of normalized steel Mk3Al were used for the examinations. Part of the specimen material was used as cold rolling and recrystallizing foils with a thickness of 10 µm. Informations on the chemical composition and heat treatment can be taken out Table 1.

Table 1: Chemical composition and heat treatment

chemical composition of steel Mk3Al						normalizing of material for massive specimen	heat tretment of material for foils
						950°C/20 min/aircooling	hot rolling at 1100°C to 4mm
C	Si	Mn	Cr	P	S		cold rolling to 0.8 mm
							rercystallizing 850°C/1 h
0.04	0.09	0.22	0.10	0.017	0.012		cold rolling to 0.01mm
							recrystallizing 720°C/2 h

Table 2: Methods of investigation used to characterize nitrided foils and massive specimens

method / kind of specimen	results
L M A :light mikroscopic analysis of massive specimens	- compound layer thickness; avereage of 40 thickness measurements in cross section - light microscopic structure evolution by etching
S E M: scanning elektron mikroscopic analysis of massive specimens	- compound layer thickness - optical distinction of different phases → see Pietzsch (9) - morphologic microananalysis of surface
C A: chemical analysis of nitrided foils	- total nitrogen concentration; integral and fractional temperature controlled method of melting extraction in inert gas
E P M A: electron probe microanalysis of massive specimens	- nitrogen content as a function of depth below the surface; obtained by measuring the nitrogen K_α intensities at points 2 µm apart on lines perpendicular to the surface; γ'-Fe$_4$N and ε-Fe$_2$N$_{(1-X)}$ standard specimens - maximum nitrogen content determined at the profile - nitrogen content as an average measured at points 10 µm apart along lines parallel to the surface after removing the surface rouhness by polishing
G D O S : depth profile analysis by glow discharge spectroscopy of massive specimens	- nitrogen concentration as a function of depth below the surface; only qualitative representation of voltage as a function of the sputtering rate
X R D: X-ray diffraction examination of nitrided foils and massive specimens	- phase concentration of the mechanically pulverized foils; Co-K_α radiation - phase concentration in the surface-adjacent region and as a function of depth of the massive specimens; Cu-K_α radiation - lattice parameters a and c of the ε-nitride - nitrogen concentration in ε-nitride using a functional dependence between the lattice constants and the content of intertitially dissolved nitrogen - real content of ε-nitride at the surface by applying radiation (Cu-, Co-, Cr-K_α) with different information depths and by interpolation to the immediate surface - macrostrains in ε-nitride at the surface of massive probes by applying the sin²ψ method; {223} reflection of ε-nitride

The nitriding treatment of foils and massive specimens was performed in a chamber furnace eqipped with solid electrolyte gassensors and conventional measurement technology (the furnace volume is about 150 l). The advantage of solid electrolyte measurement above all others is the possibility of directly determining the state of the nitriding gas in the furnace. The basis for the calculations was published in (10-12). The reactive gas atmosphere is characterized by the nitriding potential $r = p(NH_3)/p(H_2)^{3/2}$. In addition to the temperature it will be considered a main characteristic value controlling the growth of the compound layer.

Table 2 contains the methods of investigation used to characterize the nitrided iron foils and compact specimens, and the details concerning desired and actual values of the nitriding potential as well as nitriding temperature and duration of nitriding are included in Table 3.

Table 3: Comparison of nitrogen contents measured by different methods of investigation

nitriding conditions: temperature θ in °C / nitriding potential r / duration d in h		kind of specimen	% N (lattice constants)	% N (chemical)	% N (Electron Probe Microanalysis)		% N calculation by Kunze according to actual r values
desired values θ/r/d	actual values θ/r				parallel to surface	maximum value of profil	
550/ 3/ 4,8,16,24,48,96	550/ 3.80	foil massive	8.60 ± 0.2 7.90 ± 0.1	8.6 ± 0.2	8.9	8.9 ± 0,6	8.65
550/ 3/ 2,4,8	550/ 3.00	foil massive	8.30 ± 0.6 7.70 ± 0.1	8.8 ± 0.2		8.7 ± 0.2	8.43
550/ 3/ 1,2,4,8,16,32,48	550/ 3.40	foil massive	8.20 ± 0.3 7.80 ± 0.1	7.7 ± 0.02	9.7		8.55
510/ 3/ 12,36,96	510/ 2.95	foil massive	8.50 ± 0.1 8.00 ± 0.1				8.30
570/ 3/ 4,8,16	570/ 3.25	foil massive	8.50 ± 0.02 7.70 ± 0.02	9.1 ± 0.6		10.6 ± 1.3	8.55
570/ 3/ 1,2,4,8,24,48	570/ 3.60	foil massive	8.50 ± 0.04 7.60 ± 0.2	9.3 ± 0.4			8.65
590/ 3/ 4,8,16,24,32	590/ 2.50	foil massive	8.20 ± 0.1 7.45 ± 0	8.5 ± 0		9.7	8.30
590/ 3/ 1,2,4,8,16,24,32,48,96	590/ 2.95	foil massive	8.50 ± 0.15 7.55 ± 0.1	8.4 ± 0.2	8.0	9.7 ± 0.8	8.50
590/ 3/ ½,1,2,4,8,16,24,32,48	590/ 3.10	foil massive	8.50 ± 0.1 7.40 ± 0.1	9.0 ± 0			8.55
550/ 1.8/ 32,48,81	550/ 2.10	foil massive	8.00 ± 0.3 7.50 ± 0.2				8.10
550/ 6/ 4,8,16	550/ 5.50	foil massive	9.00 ± 0.1 8.20 ± 0.04	9.3 ± 0.1	8.9	10.4 ± 0.7	9.00
550/ 6/ 2,4,8	550/ 6.00	foil massive	8.90 ± 0.4 8.30 ± 0.05	9.5 ± 0.1		9.8 ± 0.4	9.10
550/ 9/ 1,2,4,8,16,24	550/ 8.40	foil massive	9.40 ± 0.2 8.70 ± 0.1	9.6 ± 0.1	9.0	10.5 ± 0.9	9.44
530/ 9/ 4,8,16	550/ 8.80	foil massive	9.20 ± 0 8.60 ± 0.1	9.6 ± 0.3	8.9	10.0 ± 1.2	9.35
570/ 6/ ½,1,2,4,8,24	570/ 6.00	foil massive	9.25 ± 0.35				9.15
570/ 9/ ½,1	570/ 9.00	foil massive	9.50 ± 0.1				9.50

3. METHODICAL RESULTS AND DISCUSSION

To improve the validity of chemical nitrogen analysis parallel examinations were made at different laboratories. The single values obtained at the different laboratories differed from each other by up to 50 percent. Due to the agreement of nitrogen concentrations measured at two laboratories, as well as the used conditions of analysis and standards, the nitrogen contents established at these laboratories were assumed to be relatively reliable. In addition to the integral method of melting extraction in inert gas the fractional temperature-controlled method was also used with the aim of distinguishing between different kinds of nitrogen bonding. Under the given analytical conditions no decomposition separated in time was observed of the different ε- and γ'-nitrides which were detected by the X-ray phase analysis.

The indicated error tolerance of the X-ray phase identification is 5 to 10 percent so that among our own investigations the nitrided iron foils and the surfaceadjacent region of the compact specimens, having a ε-fraction of more than 90 to 95 percent, were assumed to be single phase or to be at equilibrium, respectively.

The results of the nitrogen concentration measurements by chemical analysis and by X-ray lattice parameter determination (foils and massive specimens) are summarized in Table 3. For each nitriding technology the single values per nitriding time were combined into an average when equilibrium had established.

The information depth of the applied Cu-K_α radiation is about 3 μm. Therefore, the surface of a few massive specimens was investigated with radiation of different information depth so that it was possible to determine the fraction of the ε-nitride and the lattice constant in ε-nitride at the immediate surface by extrapolation. By this extrapolation it was concluded that measured ε-nitride fractions of about 85 to 90% amount to 100% at the immediate surface.

The chemically determined nitrogen concentrations as well as those from the lattice constants of the nitrided foils are in a good agreement but the nitrogen concentration at the surface of massive specimens - determined from the lattice constants as well - were found to be smaller in general than the foils data. The strains at the surface-adjacent region of ε-compound layers can be viewed as one reason. For a few specimen surfaces, measurements of macrostrains were performed. They amounted to about + 200 MPa. As a result of tensile stresses the lattice constants were measured too small and therefore the nitrogen concentrations were also determined too small (+ 200 MPa = 1 % nitrogen).

The comparison of measured phase concentration-depth profiles with scanning electron microscopic micrographs of the structure makes it clear that the representation so far does not corresponds to the structure of the micrograph and is not suitable to an exact description of approximation from the interior of solid to the gassolid interface. Therefore were made corrections represented in fig. 2. The maximum values of the nitrogen concentration determined from the concentration-depth profiles by EPMA are published in Table 3.

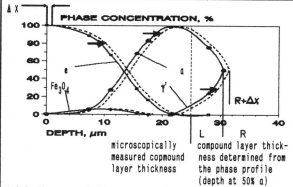

microscopically measured copmound layer thickness

compound layer thickness determined from the phase profile (depth at 50% α)

Originally measured phase concentrations
EFFEKTIVE PHASE CONCENTRATION (see Oettel (13))

1. correction according to the influence of the finite penetration depth of Cu-Kα radiation ; a simple shift by Δx of the abscissa yields a first approximation   TRUE PHASE CONCENTRATION (13)
2. determination of the proportion ε/γ' related to 100%
3. determination of the ratio (R + Δx)/L (=F)
4. correction of each depth below the surface multiplying by F

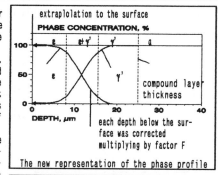

extraplolation to the surface

compound layer thickness

each depth below the surface was corrected multiplying by factor F

The new representation of the phase profile

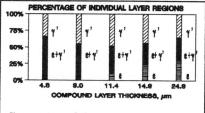

The percentages of the individual layer region were determined from the corrected phase profil

Fig.2: Correction of the measured phase profiles according to the structural informations obtained by SEM

They· are different to the results obtained with other methods of concentration measurements and cannot only be explained just by the local character of the electron-probe
microanalysis. Therefore the method does not yield accurate
results by investigations of equilibria establishment at
the surface of thin compound layers. Nitrogen contents
which were obtained by measuring parallel to the original
surface after removing the surface roughness are in a better agreement with the nitrogen concentrations of the nitrided iron foils. With the depth profile analysis by glow
discharge spectroscopy only quantitative nitrogen analysis
was done. They reasonably agree with phase profile and EPMA
investigations. Fig. 3 shows that reliable statements on
the structure of compound layers and the establishment of
gas-solid equilibrium can only be obtained if all informations resulting from electron microscopic structure investigations combined with methods of determining chemical
and phase composition were used.
The determination of the nitriding potential by utilizing
different methods of gas analysis -solid electrolyte (in-
situ) measuring the statement of nitriding gas in the furnace together with infrared analysis and gas chromatogra-
phy- guarantee the reliability of the measured nitriding
potential.

4. EXAMINATION OF THE ESTABLISHMENT OT AN EQUILIBRIUM BETWEEN NITRIDING GAS AND IRON WITH IMPROVED METHODS

In agreement with the selected nitriding conditions the investigated foils can be transformed into the ε-phase. The
nitriding times required for the establishment of defined
equilibrium (ε > 90-95 %) are 0.5 to 2 hours depending on
the used nitriding technology.
A comparison of nitrogen concentration and the fraction of
ε-nitride in foils and in the surface region of massive
specimens (values are not corrected to the immediate surface) plottet against the corresponding compound layer (Fig.
4) clearly shows that the establishment of the equilibrium
of the surface-adjacent region of massive specimens requires longer times and is controlled by the transformation γ'
into ε, although the nitrogen content at the ε-nitride
relatively fast attains a constant value which is equivalent to the equilibrium content. The percentages of the
individual layer regions as a function of the compound
layer (related to the nitriding time) are shown in Fig. 2.
When equilibium at the gas-solid interface is reached the
compound layer would have the same sequences of phases. The
outer part of the compound layer consists of ε-phase. At
intermediate depths, a two-phase structure of ε and γ'is
present. The bottom part consits of γ'-phase. The equilibrium between gas and the surface of iron will establish
faster with increasing nitriding potential and rising temperature. The alteration of the nitriding temperature from
550°C to 590°C has approximately the same effect as trip-
ling of the nitriding potential.
The investigation of compound layers (approximation to the
eqilibrium from the interior of solid) offers the possibility of extrapolation of phase and nitrogen content to the
immediate surface and increase the reliability of the statements together with structural investigations by SEM.

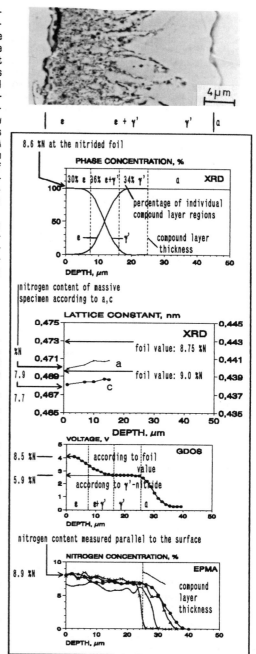

Fig. 3: combination of different methods of phase and
nitrogen concentration measurements with struc
tural investigations by SEM

Fig. 5 shows a sector of the Lehrer diagram with the isoconcentration lines calculated by Kunze and the nitrogen concentrations of foils obtained by X-ray analysis. They are reasonably agree together and thus they confirm the direct direct connection between nitriding potential, nitrogen partial pressure, temperature and nitrogen content established in the ε-nitride. The studies have shown that the nitrogen content in the ε-nitride can reliably precalculated by the thermodynamic model considering the ε-phase to be a subregular solution of nitrogen in hexagonal iron (see as well Tab.3) if the current nitriding parameters were exactly registered and reliable solid investigations were made.
It is possible to vary the nitrogen concentration in the ε-nitride even under nearly technical nitriding conditions. The methodical improvement of the gas and solid investigations should not only allow the establishment of defined nitrogen contents in ε-nitride according to the nitriding parameter, but also enable investigations in the region of the phase boundaries.

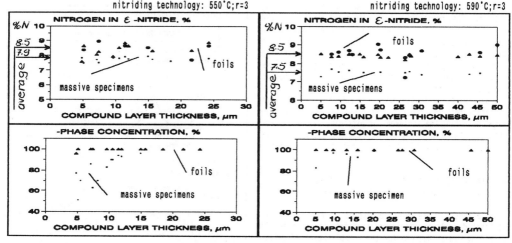

Fig. 4: Establishment of nitrogen and phase equilibria as investigated for nitrided foils and the surface-adjacent region of massive specimens

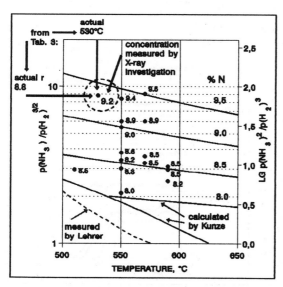

Fig. 5: Sector of Lehrer diagram: with our own results of nitrogen concentration determined by X-ray investigations (lattice parameter)

CONTENTS

1. E. LEHRER, Z. Elektrochem. **36** (1930),pp.383-392.
2. F.K. NAUMANN, G.K. LANGENSCHEID, Arch. Eisenhüttenwes. **36** (1965),pp.583-590; ibid. pp.677-682.
3. B. EDENHOFER, Nitrieren und Nitrocarburieren, Darmstadt 1990, pp. 165-185.
4. H.U. FRITSCH, H.W. BERGMANN 41 HTM (1986)1, pp.14-20
5. M. HILLERT, L.J. STAFFANSSON, Acta chem. scand. **24** (1970), pp.3618-3626
6. W.-D. JENTZSCH, S. BÖHMER, Kristall u. Technik 12 (1977), pp.1275-1283
7. J. KUNZE, steel research 57,(1986) pp.361-367
8. L. MADZINSKI, Z. PRZYLECKI, J. KUNZE , steel research 57 (1986),pp.645-649
9. S. PIETZSCH, S. BÖHMER, CH. ORTLIEB, paper to be published
10. H.-J. BERG, S. BÖHMER , Freiberger Forschungsheft B263, VEB Dt.Verl.f.Grundst.,Leipzig 1988
11. H.-J. BERG, Dr.-Ing. Dissertation, Bergakademie Freiberg,1985
12. H. ZIMDARS, Dr.-Ing. Dissertation, Bergakademie Freiberg,1987
13. H. OETTEL, Freiberger Forschungsheft B 265, VEB Dt. Verl.f. Grundst.,Leipzig 1989

Materials Science Forum Vols. 163-165 (1994) pp. 253-258
© *1994 Trans Tech Publications, Switzerland*

STRUCTURE OF POROUS AREAS OF NITRIDE-CONTAINING COMPOUND LAYERS

S. Pietzsch, S. Böhmer and Chr. Ortlieb

Freiberg University of Mining and Technology, Germany

ABSTRACT

In the present work specimens were gasnitrided and gasnitrocarburized in order to produce mono-phase ε and γ' layer areas as well as ε/γ' two-phase areas with a considerable variation of their chemical composition. Typical porosity phenomena are described and systematized in interdependence with the chemical, phase and crystalline composition of the compound layer. On the basis of a description of state, structural changes are discussed as a function of concentration dependent transformation processes in the compound layer together with the duration of existence of the layer areas.

1. INTRODUCTION

Porous surface areas of the compound layer influence the properties of nitrided components and specimens. Therefore the study of porosity has become an important aspect in the recent years. Results published so far do not allow a consistent understanding of the phenomena but give cause to different hypotheses about pore formation [1,2,3,4,5,6,7].

The objective of this work was the exact investigation of selected, typical structures by means of scanning electron microscopy, X-ray phase analysis and reliable methods of nitrogen and carbon determination. The aim was the determination of the relations between the chemical, phase and crystalline composition of the compound layer and the porosity phenomena as well as their changes during layer growth.

2. EXPERIMENTAL

In the present work specimens of normalized steel Mk3Al (0,04% C, 0,09% Si, 0,22% Mn, 0,017% P, 0,012% S, 0,1%Cr) were used. Nitriding and nitrocarburizing were carried out in a retort furnace (500 litre) with continuous monitoring of the oxygen potential and hydrogen content of the atmosphere and additional gas chromatographic analysis of the nitriding gas. The ratio $K_N = \Upsilon(NH_3)/\Upsilon(H_2)^{1,5}$ and the ratio $K_C = \Upsilon(CO)*\Upsilon(H_2)/\Upsilon(H_2O)$ were varied in order to produce mono-phase ε and γ' layer areas and ε+γ' two-phase layer areas with a considerable variation of their chemical composition (Table 1).

The following investigation methods were used for the characterization of the composition of the surface layer: scanning electron microscopy, X-ray phase analysis, X-ray determination of the lattice parameters and chemical analysis of iron foils [8].

The state of the nitriding and nitrocarburizing gas was additionally controlled by determining the equilibrium nitrogen and carbon content in iron foils (thickness ∼10µm). Extensive investigations [8] have shown transmission of these concentrations to the surface area of specimens to be possible.

Table 1: Nitriding parameters and composition of compound layer

K_N	K_C	T in °C	N-/C-content in the iron foil	phase composition of the compound layer
gasnitriding				
9	0	570	9,5[1] / 0	ε(γ')
3	0	570	8,5[1] / 0	ε+γ'
1,3	0	570	7,7[1] / 0	γ'(ε)
0,6	0	570	6,0[1] / 0	γ'
gasnitrocarburizing				
3	0,46	570	8,3[2] / 0,6[2]	ε(γ')
0,8	0,36	570	7,0[2] / 0,7[2]	ε(γ')
0,35	0,2	590	6,2[2] / 0,3[2]	ε+γ'
0,15	0,36	590	4,7[2] / 1,1[2]	ε(γ')

[1] determined by lattice parameters in the ε and γ'phase
[2] determined by chemical analysis

Scanning electron micrographs (magnification x 10,000) of sections areas parallel to the surface were used for the quantitative valuation of porosity . The photographs were evaluated by means of digital image processing (linear and area analysis).

3. RESULTS AND DISCUSSION

3.1. Characterization of porous states in dependence on the compositon of the compound layer

3.1.1. Gasnitrided layers
The phase concentration profiles and the SEM photographs of the compound layer (thickness ~12μm) are represented in dependence on the surface nitrogen content N_R in Figs. 1 and 2.

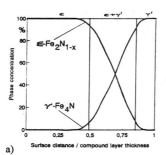 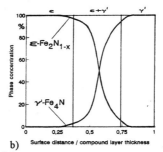 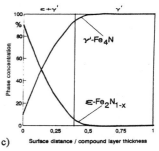

a)   b)   c)

Fig. 1: Phase concentration profiles of the compound layer, gasnitrided;
a) N_R=9.5%   b) N_R=8.5%   c) N_R=7.7%

For a surface nitrogen content of N_R=9.5% (Fig. 1a) the compound layer consists of three areas: ε nitride on the outside, two-phase ε+γ' in the middle and γ' nitride on the inside. With decreasing nitrogen content on the surface the thickness of the ε sublayer is reduced, while the thickness of the inside γ' sublayer is increased (Fig. 1b). If the surface nitrogen content is reduced further an ε sublayer is not observed and the compound layer consists of two areas: ε+γ' nitride on the outside, γ' nitride on the inside (Fig. 1c). For the variant N_R=6.0% the formation of the ε phase is suppressed and a mono-phase γ' compound layer is formed.

In dependence on the different chemical and phase composition of the compound layer clear differences can be shown in the porosity phenomena. The SEM photographs of variants $N_R=9.5\%$ (Fig. 2a) and $N_R=8.5\%$ (Fig. 2b) show that the surface near area (ε sublayer) is interspersed almost completely by pores. The pores occur inside the grains (grain porosity). The SEM photographs of the variants with reduced nitrogen content (Figs. 2c and 2d) illustrate that in the γ' nitride areas the pores are present along the grain boundaries as pore chains and pore channels (grain boundary porosity). The two-phase $\varepsilon+\gamma'$ areas indicate in addition to grain boundary porosity along γ' nitride crystals und grain porosity in the ε nitride also ε nitride areas still not showing grain porosity. Therefore a certain time is necessary until grain porosity is observed in the ε phase transformed from γ' nitride. This will also be demonstrated by the following investigations on layer growth. In addition to the dependence of the pore pattern on the composition of the compound layer, differences in pore size are also observed. Thus, the pores in the ε phase are smaller than the pores in γ' nitride (Table 2).

a) $N_R=9.5\%$ b) $N_R=8.5\%$

c) $N_R=7.7\%$ d) $N_R=6.0\%$

Fig. 2: SEM photographs of compound layer; gasnitrided
A pores in ε nitride   B pores in γ' nitride

3.1.2. Gasnitrocarburized layers

SEM photographs of the compound layers of nearly equal thickness which were produced by gasnitrocarburizing are represented in Fig. 3. The thickness of the porous area decreases in the case of reduction of the surface nitrogen content N_R. Contrary to this the thickness of the porous area of gasnitrided layers (equal layer thickness) is independent of the surface nitrogen content N_R (Fig.2).

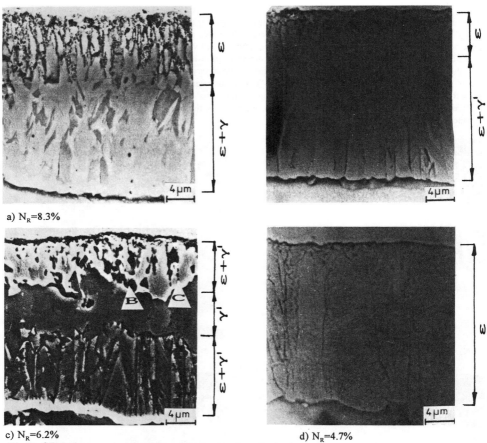

a) N_R=8.3%

c) N_R=6.2%   d) N_R=4.7%

Fig. 3: SEM photographs of compound layer; gasnitocarburized
B pores in γ' nitride   C pores in ε carbonitride

The external area of the compound layer of the variant with the relative high surface nitrogen content N_R=8.3 and a carbon content of C_R=0.6% is composed of ε carbonitride; the inner area consists of ε+γ' (carbo)nitride (Fig. 3a). In the case of reduction of the surface nitrogen content and a comparable carbon concentration the thickness of the ε carbonitride sublayer decreases (Fig. 3b). For the variant N_R=6.2% (medium nitrogen content) and reduced carbon content the compound layer is two-phase composed of ε+γ' (carbo)nitride with a nearly' monophase γ' area in the centre part of the compound layer (Fig 3c). For a surface nitrogen content of N_R=4.7% and a relatively high carbon content the compound layer consists of ε carbonitride with a small fraction (10%) of γ' nitride precipitated during cooling after nitrocarburizing (Fig. 3d). By this the experimental results of Naumann and Langenscheid [9] are confirmed which prove the existence of ε carbonitride with clearly smaller nitrogen content compared to the γ' phase.

For variants N_R=8.3% and N_R=7.0% the pores are present in the outer ε carbonitride sublayer along the grain boundaries as pore chains. The size of the pores is almost comparable with the pore size in ε nitride (Table 2). In the surface area of the two-phase ε+γ' compound layer (N_R=6,2%) fine pore chains in the ε carbonitride areas as well as greater pores in the γ' nitride areas are observed. No preferred presence of the pores along the phase boundary ε/γ' was observed (Fig. 3c). The compound layer with the low nirogen content (N_R=4.7%) does not show porosity (Fig. 3d).

3.2. Changes of porous states during compound layer growth (gasnitrided layers)

In Figure 4 the changes of the compound layer composition and the porosity phenomena during growth are shown.

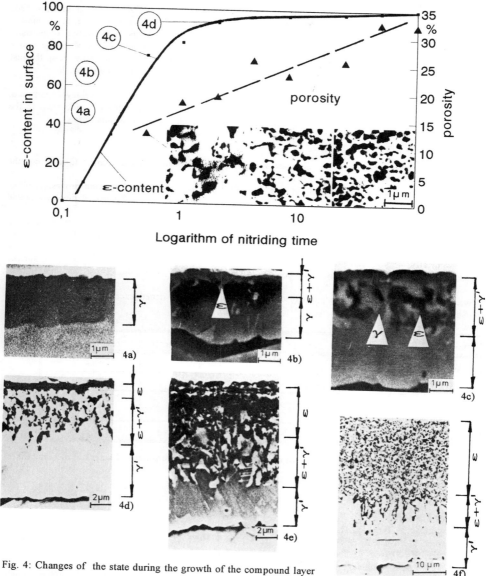

Fig. 4: Changes of the state during the growth of the compound layer

First of all a thin γ' layer without pores is formed (Fig. 4a). Then the transformation of the γ' phase into the ε phase follows in which the maximal nitrogen content is reached after a short time [8]. The compound layer then consists of two ares: ε+γ' nitride on the outside and γ' nitride on the inside. Fig. 4b) also illustrates that at first ε areas formed later do not show porosity in the same way. Before a continuous ε sublayer is formed, the pore formation begins in the surface near area of the compound layer (Fig 4c). At first the pores are present along the grain boundaries of the relatively "newly" formed ε crystals and the phase boundary of ε and γ' nitrides. The inner mono-phase γ' sublayer is free of pores.

If nitriding time increases a continuous ε sublayer is detectable on the surface (Fig. 4d). Simultaneously pore formation begins to increase inside the ε crystals. When the phase transformation γ'→ε in the surface near area is finished (continuous line) and the maximal surface nitrogen content is reached, the formation of pores in the surface area is further observed. This causes an inrease of the area fraction of porosity (interrupted line). In the deeper areas of the compound layer the transformation of the γ' nitride into ε phase continues. Therefore the two-phase area (transformation front) with the grain porosity along the just transformed ε nitrides and γ' nitride crystals is removed to the middle part of the compound layer.

The additional increase of nitriding time still causes a small increase of the ε sublayer part and, consequently, the area showing grain porosity (Fig 4e and 4f). Apart from this the ε nitride areas in the two-phase area of the layer also show inreasing grain porosity which is typical for this phase.

4. SUMMARY AND CONCLUSION

*A direct relation between the chemical, phase and crystalline composition of the compound layer and the porosity phenomena was demonstrated (Table 2).
* The changes of the porous areas during layer growth follow the phase transformation processes of the surface area and depend on the concentration conditions and the duration of existence of the investigated areas.
* In gasnitriding it is not possible to prevent pore formation with increasing layer thickness by a reduction of the surface nitrogen content (reduction of K_N). Porosity phenomena may be affected by the influence of the chemical and phase composition of the compound layer.
* In gasnitrocarburizing it is possible to restrict and even prevent porosity by a reduction of the surface nitrogen content in the range of the usual compound layer thicknesses of 10-20μm.

Table 1: Quantitative data of the porosity of the compound layer in dependence on chemical and phase composition [3]

Phase	porosity phenomena	area fraction of porosity	average pore size
ε nitride, N-rich	grain porosity	30 - 40 %	0,3 μm
γ' nitride	grain boundary porosity	15 - 20 %	0,6 μm
ε carbonitride, N-rich and middle N-content	grain boundary porosity	15 - 20 %	0,3 μm
ε carbonitride, N-poor	no porosity		

[3] determined by linear and area analysis of section areas parallel to the surface at a surface distance of 2-3 μm (compound layer thickness: 10-15 μm)

5. REFERENCES

1) Prenosil, B.: HTM, 1973, 28, 157.
2) Matauschek, J.; Trenkler, H.: HTM, 1977, 32, 177
3) Schröter, W.: Wiss. Z. d. Techn, Hochsch. K.-M.-Stadt, 1982, 24 , 795
4) Schröter, W.; Russev, R.; Ibendorf, K. : Wiss. Z. d. Techn, Hochsch. K.-M.-Stadt, 1982, 24 , 786
5) Somers, M.A.J.; Mittemeijer, E.J.: Surface Eng., 1987, 3, 123
6) Hoffmann, F.; Kunst, H.; Klümper-Westkamp, H.; Liedtke, D.; Mittemeijer, E.J.; Rose, E.; Zimmermann, K.: Nitrieren und Nitrocarburieren, Darmstadt, 1991, 103
7) Liedtke, D.: Nitrieren und Nitrocarburieren , Darmstadt, 1991, 114
8) Ortlieb, Ch.; Böhmer, S.; Pietzsch, S.: In this conference
9) Naumann, F.K.; Langenscheid, G.: Archiv f. d. Eisenhüttenwesen, 1965, 36, 6771

Materials Science Forum Vols. 163-165 (1994) pp. 259-264

GROWTH OF NITRIDE COVER LAYERS ON COMPOUND LAYERS OF IRON

S. Pietzsch and S. Böhmer

Freiberg University of Mining and Technology, Germany

ABSTRACT

In the present work the structure and growth of a special surface layer on nitride-containing compound layers which is designated as nitride cover layer is characterized by means of scanning electron microscopy. The results of the investigations show that the growth of the nitride cover layer on the original surface is governed by the parabolic law of diffusion controlled layer growth. Further more a relationship was observed between the growth of the nitride cover layer and the porosity of the compound layer below.

1.INTRODUCTION

Compound layers are the result of concentration dependent structural changes of the material surface layer [1]. About the formation of cover layers by growth processes on the surface there are contradictory data. Such processes were observed during the formation of the compound layer [2,3,4] and in connection with nitriding of iron oxide layers (designated for instance, as "iron film") [1,5]. Topographical investigations have shown structures which are simular to that of layers on surface of different substrates but not to the surface of compound layer. Therefore the current state of explanation of the different phenomena is unsatisfactory. The consequences for the obtained properties (for instance due to insufficient adhesive strength) cannot be calculated. This was the reason why scanning electron microscope investigations of selected compound layers had to be done.

2. EXPERIMENTAL

Normalized steel Mk3Al was used for the investigations (chemical analysis in [6]). The specimens were ground with SiC paper (R_z=0,06µm) and cleaned in alcohol before nitriding. The nitriding layers were produced by gasnitriding and gasnitrocarburizing. Table 1 contains the nitriding conditions.

The produced nitriding layers were investigated by means of a scanning electron microscope. The thickness of the nitride cover layer was measured at cross sections (magnification x 10,000). Digital image processing of scanning electron micrographs (magnification x 5,000) of cross sections was used for the quantitative evaluation of the porosity.

3. RESULTS AND DISCUSSION

3.1. Growth of the nitride cover layer

Fig. 1 shows the surface (1a), fracture (1b) and cross section (1c) of a γ' compound layer. A subdivision of the outer surface layer is possible into the following three areas: (1) nitride cover layer, (2) porous part of the compound layer and compact part of the compound layer (3). Directly on the surface a layer can be detected composed of nitride crystals (designated as nitride cover layer here). The thickness of this nitride cover layer is well discernible in fractures of the compound layer and carefully prepared cross sections.

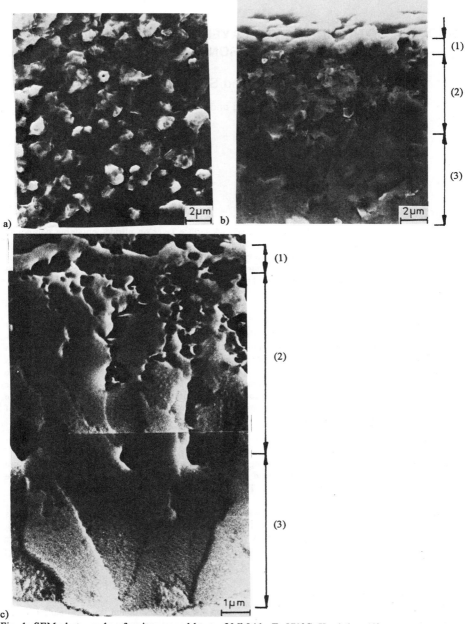

Fig. 1: SEM photographs of a γ'compound layer of Mk3Al; T=570°C, K_N=0.6, t=48h
a) surface   b) fracture   c) cross-section

As was mentioned above the existence of a surface layer on the compound layer which is chipping off easily is dicussed in the literature in connection with reduced oxid layers subsequently transformed into nitrides. Therefore a specimen was oxidized (450°C / 1 hour) before nitriding. By this an oxide layer was formed on the surface with a thickness of about 1μm. As Fig. 2a) shows the surface layer present after nitriding has only slight adhesive strength and chips off easily. The other specimens were ground before nitriding and were not preoxidized. Stronger oxidation at the beginning of nitriding can be excluded. The surface layer presented then does not chip off at the same value of strain (Fig. 2b), unlike the surface layer of the specimen which was preoxidized (Fig. 2a). Consequently the investigated nitride cover layers are not reduced iron oxide layers, as the following study of the growth of these layers will also demonstrate.

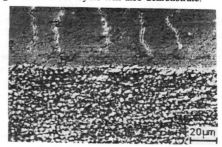

a) reduced iron oxide layer   b) nitride cover layer

Fig. 2: SEM photographs of a scratch ; F_N=100N; T=570°C, K_N=0.6, t=48h

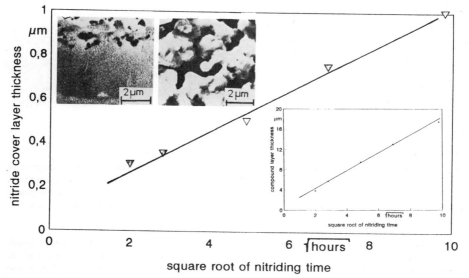

Fig. 3: Growth of the nitride cover layer and compound layer; T=570°C, K_N=0.6 (γ'-nitriding)

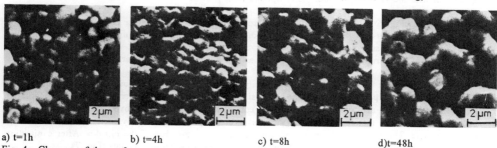

a) t=1h   b) t=4h   c) t=8h   d)t=48h

Fig. 4: Changes of the surface topography with inreasing nitriding time ; T=570°C, K_N=0.6

In Fig. 3 the growth of the nitride cover layer is illustrated; Fig. 4 shows the SEM photographs of the surface topography. At the beginning of nitride cover layer formation the isolated nitrides formed on the surface have not yet grown together into a continuos layer. On the other hand, a compound layer with a thickness of some micrometers can already be detected (Fig.3 - cross section of the compound layer). This phenomenon differs from that observed by Naumann [2] for polished and passivated specimens.

With increasing nitriding time a continuos nitride cover layer is formed. Apart from that the nitride crystals coarsen (Fig. 4) and an increase of the nitride cover layer thickness is observed (Fig. 3). A linear relationship between the nitride cover layer thickness and the square root of the nitriding time can be shown. This indicates a diffusion controlled growth of the nitride cover layer. Since after the removal of the nitride cover layer the structure underneath shows the topography of the original grinding state, the nitride cover layer is a layer growing on the original surface. Therefore the growth can be determined only by iron transport processes.

Table 1: Nitriding condition $^{1)}$ $K_N = \Psi(NH_3)/\Psi(H_2)^{1,5}$

T in °C	K_N $^{1)}$	t in h
Gasnitriding		
510	3,0	12, 36, 96
550	3,0	1, 2, 4, 8, 24, 48
570	0,6	1, 4, 8, 24, 48, 96
570	1,3	4, 8, 24, 48
570	3,0	1, 2, 4, 8, 24, 48
570	9,0	2
590	3,0	1, 2, 4, 8, 24
700	0,6	4
Gasnitrocarburizing		
590	0,15	8

3.1. Comparison of nitride cover layer and compound layer

- Growth: The growth of the nitride cover layer as well as the growth of the compound layer is determined by diffusion processes (Fig. 3 - parabolic law). The diffusion of nitrogen controls the rate of compound layer growth in the material while the growth processes of the nitride cover layer on the surface are determined by transport of iron. These processes are clearly slower than the growth of the compound layer which is governed by nitrogen diffusion.

- Factors influencing layer growth: The influence of the ratio K_N and nitriding temperature T on the nitride cover layer growth is illustrated in Fig. 5. Whereas the ratio K_N has only a relatively small influence on the layer growth, a strong influence of the nitriding temperature T is observed. The inrease of the slope k with increasing temperature is caused by an increase of the diffusion rate (Fig. 5b). These results are in good agreement with the knowledge about the compound layer growth.

- Composition: Fig. 6 shows the crystalline composition of the nitride cover layer. The crystal size is maximum 1μm. Because of the limited layer thickness (maximum 1,5μm) the nitride cover layer does not have a clear columnar grain structure like the compound layer (Fig. 1c). The chemical and phase composition of the nitride cover layer is the same as that of the compound layer.

- Porosity: Systematic investigations of the porosity of the compound layer have shown a direct relation between the chemical, phase and crystalline composition of the compound layer and the porosity phenomena [6]. Connections were also observed between the composition of the nitride cover layer and the porosity. Figs. 1a) and 4 show that the pores in a γ' cover layer are present essentially in the areas between the nitride crystals. The pores are the result of incomplete growth together of the nitrides. After ε nitriding especially with a high ratio K_N, porosity is also observed within the nitride crystals (Fig. 7).

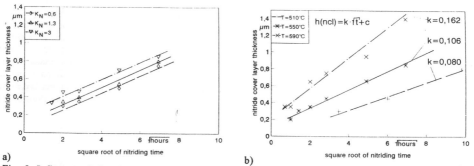

a)

b)

Fig. 5: Influence of a) ratio K_N (T=570°C) and b) nitriding temperature (K_N=3) on the growth of nitride cover layer ; h(ncl) - nitride cover layer thickness

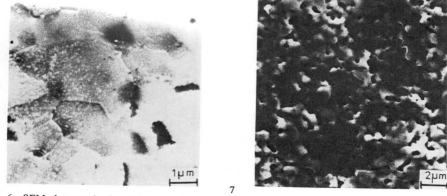

6   7

Fig. 6: SEM photograph of a nitride cover layer; section parallel to the surface
Fig. 7: SEM photograph of the surface on a nitride cover layer, T=570°C, K_N=9

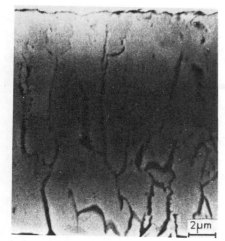
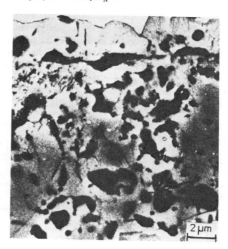

a) nitrocarburized, T=590°C, K_N=0.15, t=8h
Fig 8: SEM photographs of surface layer

b) nitrided, T=700°C, K_N=0.6, t=4h

In addition, the investigations have shown a direct relation between the appearance of the nitride cover layer and the porosity which is illustrated in Fig. 8. In the layer in Fig. 8a) porosity is not observed and no measurable nitride cover layer is detected either. Contrary to this the layer in Fig. 8b) shows a strong porosity and a "thick" nitride cover layer (also in comparison with the layer in Fig. 1). Moreover, a relatively good agreement between the area fraction of the pores and the nitride cover layer related to the compound layer is observed especially for γ' compound layers (Fig. 9). This can be assumed to be an indication that iron transport processes have importance for the pore formation in the compound layer. An exact atomistic interpretation of these connections is not possible because there are not exact informations about the iron transport processes in the iron nitride phases.

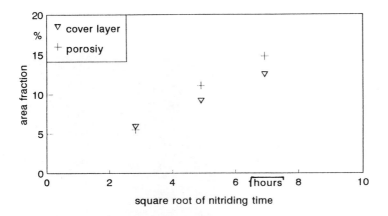

Fig 9: area fraction of the nitride cover layer and the porosity related to the compound layer; measured at cross sections of the surface layer

4. SUMMARY AND CONCLUSION

-It is proved by scanning electron microscopic investigations that cover layers on compound layers can be formed not only by the nitriding of reduced iron oxide but also by the growth of nitrides on the surface.
-It is shown by topographic investigations that the crystalline structure of the nitride cover layer differs characteristically from the structure of the compound layer. It can be concluded that the information about the surface of the compound layer often describes the surface of the nitride cover layer.
-The growth of the nitride cover layer is determined by a parabolic diffusion controlled law. It is probably controlled by iron transport processes.
-The adhesive strength of the layer on the surface is particularly disturbed if iron oxide is formed and reduced on the surface before the nitriding.

5. REFERENCES

1) Liedke, D.: in Nitrieren und Nitrocarburieren. Sindelfingen: Expert Verlag, 1986
2) Naumann, F.K.: Prakt. Metallogr. 1968 ,5 , 473
3) Dides, M.: Dissertation, Bergakademie Freiberg, 1975
4) Böhmer, S.; Jentsch, W.D.: Kristall und Technik, 1979, 14, 617
5) Edenhofer, B.: HTM, 1992, 47, 345
6) Pietzsch, S.; Böhmer, S.; Ortlieb, Ch.: In this conference

Materials Science Forum Vols. 163-165 (1994) pp. 265-272

THE EFFECT OF SURFACE FINISHING ON MECHANICAL AND MICROSTRUCTURAL PROPERTIES OF SOME NITRIDED STEELS

M. Boniardi and G.F. Tosi

Dipartimento di Meccanica, Politecnico di Milano,
P.zza L. da Vinci 32, I-20133 Milano, Italy

ABSTRACT
In this paper different nitrided compound layers are described in detail to evaluate the influence of variable initial surface roughness on mechanical and microstructural characteristics of nitrided alloy steels.
As base materials two nitriding steels (UNI 41CrAlMo7) with different chemical compositions are used; both gas and ion nitriding heat-treatments are performed modifying either time or temperature of the process. Initial differences in surface roughness are obtained by mechanical polishing.
The surface morphology of nitride layers and the microstructure of the samples are characterized using microhardness profiles, optical microscopy, x-ray diffraction and scanning electron microscopy.
Special attention is devoted to the variation of surface roughness before and after heat treatments: a correlation between nitriding processes and increase in roughness is found. The same roughness variation is also experienced on mechanical components. Lastly a possible mechanism of formation and growth of the compound layers is proposed.

INTRODUCTION

In recent years surface treatment technology has been involved in a great productive expansion due to investments in plants and "maturity" gained by the operators.
There is an obvious correlation between this development and the growing interest by scientists and engineers in new methods of mechanical and microstructural characterization of the nitrided layers which are able to describe more accurately the nitriding processes and the related phenomena at the metal surface [1, 2, 3, 4].
Metal surface is in fact a common site for nucleation of metallurgical and mechanical damage, such as fatigue, wear and corrosion, which usually compromise mechanical component life [5, 6].
Therefore this work is devoted to the investigation of the interaction between surface finishing and metallurgical characteristics in gas and ion nitriding treatments: this is to offer helpful information to the industrial user and to optimize properties for particular engineering applications.

EXPERIMENTALS
Materials and Samples
Two UNI 41CrAlMo7 nitriding type steels have been used and their
chemical composition is reported in table I.
This steel is one of the most popular of all nitriding steels and
it is used for a multitude of machine components such as shafts
and gear wheels or for any mechanical part subjected to fatigue
and/or wear conditions [7].
The tested steels were produced by electric arc furnace and,
after hot rolling operations, two round bars of 150 mm in
diameter were obtained. Then they were hardened at 910°C followed
by oil quenching and tempered at 600°C thus obtaining a surface
hardness between 320 and 300 HV.
Many samples (40x10x3mm in mean dimensions) were machined from
the round bars according to the diagram showed in figure 1. This
particular sampling method was performed in order to maintain a
constant radial distance from the bar center, that is a constant
hardness level on all samples.
After grinding, all samples were mechanically polished using
abrasive papers with various silicon carbide meshes (180, 320,
600, and 800 meshes) to gain different surface roughness. The
obtained values were similar to the one of lapped mechanical
components subjected to nitriding processes.

Treatments
All samples were subjected to standard industrial gas and ion
nitriding treatments as described in table II.
The nitriding treatments were performed in order to reduce as
much as possible the ε nitride layer thickness (thus gainig a
considerable advantage over γ' nitride) and to economize the NH_3
consumption.
Note: Each sample was identified by three parameters: steel type
(A or C), treatment number (1, 2, 3 or 4) and polishing mesh type
(180, 320, 600 or 800).

Testing
The following tests were carried out on the above mentioned
samples:
a) Vickers microhardness profile on transverse lapped section of
the nitrided samples (load: 50g, indentation time: 15").
b) Superficial x-ray diffraction analysis using Cu-Kα radiation
(λ=1.5405*10-1nm; 30°< 2Θ < 50°; scanning rate: 1°/min). Cu-Kα
radiation was used due to its low penetretion (about 5-15 μm) in
iron alloy to have a more precise determination of the phase
constitution of the surface layers [8].
c) Roughness measurements, before and after heat treatments,
considering the following parameters [9] in common use for
surface characterization [6]:
    Ra, roughness average value,
    Rq, roughness root mean square value,
    Rt, roughness peak-to-valley height.
d) Optical microscopy observations on transverse lapped section,
etched with Nital 2%, and scanning electronic microscopy (S.E.M.)
on the external surface to evaluate the nitride morphology of the
samples.

RESULTS AND DISCUSSION
a) Vickers microhardness
All samples were submitted to Vickers microhardness tests to evaluate any difference 1)between the various treatments and 2)within the same treatment (due to the sample surface finishings).
The obtained data were analyzed by the ANOVA as proposed in [10]. The main differences between the treatments are reported in figure 2, showing the Vickers microhardness profiles.
Within the same treatment, however, no statistical difference was discerned.
b) X-ray analysis
The same statistical analysis of a) was also performed on X-ray diffraction patterns. In table III the experimentals results, among the treatments, are reported.
In every case gas nitriding treatments showed a higher level of ε nitride; on the contrary, in ion nitriding the compound layer was found to be richer in γ' nitride.
When the treatment time is low, that is 10 or 12 hours, a (110) Fe-α peak was observed, which is more pronounced in ion nitriding.
No difference in the compound layer composition was discerned within the same treatment, considering the samples with various surface finishings.
c) Surface roughness
The experimental results of roughness measurements (Ra and Rt parameters) before and after heat treatments are reported in figure 3.
From the analysis by the ANOVA, the following considerations can be drawn:
-in each case there is a higher surface roughness after the treatments, with either gas or ion nitriding;
-roughness increase is higher when initial surface roughness is lower;
-roughness increase is higher in gas than in ion nitriding treatments;
Finally, the same measurements were taken on mechanical components before and after gas nitriding. The experienced roughness increase in the tested mechanical components was similar to that in the samples (see figure 4).
d) Optical and Electronical Microscopy
The metallographic aspects of the studied treatments are showed in figure 5. No particular difference was noticed within the same treatment.
The superficial nitride morphology was observed by S.E.M. and related micrographs are reported in figure 6.
In general the lower the surface finishing (that is a low value of Ra) the lower the mean nitride dimension. In addition, a growing in nitride dimensions and a progressive filling of the free surface can be observed with an increase in treatment time.
A particular consideration must be made for ion nitriding treatment. In this case, due to cathodic pulverization, a finer nitride morpholgy, with respect to gas nitriding, was observed. This, probably, affects the different tribological properties of the nitride layer.
e) Supplementary tests on cast iron
In addition to the above mentioned examinations, similar

measurements were extended to cast iron samples (nodular and malleable iron); they were gas nitrided and gas nitrocarburized to evaluate the roughness increase and the microhardness variation related to initial differences in surface roughness.
The experimental results confirmed a behaviour similar to steel samples: the Ra increase in cast iron was 0.2-0.4 μm after nitrocarburizing and 0.5-0.7 μm after nitriding.

f) Model of compound layer evolution

From the above mentioned experimental results, especially roughness and S.E.M. observations, a model of compound layer formation was elaborated and it is schematically presented in figure 7.

CONCLUSIONS

The effect of surface finishing on microstructural and mechanical properties of gas and ion nitriding samples was studied.
The experimental results show that there is a systematic increase in surface roughness after the studied surface treatments, which is higher in gas than in ion nitriding.
Within the same treatment, the superficial nitride morphology was influenced to a great extent by the imposed surface finishing of the samples. These results were confirmed both from samples and from mechanical spares.

REFERENCES

[1] R. Roberti, G.M. La Vecchia, G. Colombo, Proc. I° ASM Heat Treatment and Surface Engineering Conference, Amsterdam, 1991, p. 271.
[2] L. Sproge, J. Slycke, Proc. I° ASM Heat Treatment and Surface Engineering Conference, Amsterdam, 1991, p. 229.
[3] H. Du, J. Agren, Proc. I° ASM Heat Treatment and Surface Engineering Conference, Amsterdam, 1991, p. 243.
[4] M.A.J. Somers, P.F. Colijn, W.G. Sloof, E.J. Mittemeijer, Z.Metallkde, n°81, 1990, p. 33.
[5] S. Suresh, "Fatigue of Metals", Cambridge University Press, Cambridge, 1991.
[6] R.D. Arnell, P.B. Davies, J. Halling, T.L. Whomes, "Tribology - Principles and Design Application", Macmillan, London, 1991.
[7] K. E: Thelning, "Steel and its Heat Treatment", Bofors Handbook, Butterworths, London, 1975.
[8] D. Firrao, B. De Benedetti, M. Rosso, Met. It., n°11, 1981, p. 513.
[9] UNI ISO 4287, Surface roughness - Terminology - Surface and its parameters, 1991.
[10] G.W. Snedecor, W.G. Cochran, "Statistical Methods", The Iowa State University Press, Ames, Iowa, 1980.

Table I: Chemical compositions of the tested steels

Steel	C	Mn	Si	S	P	Cr	Mo	Al	Ni	Cu	Sn
A	0.39	0.61	0.23	0.009	0.026	1.56	0.24	1.01	0.14	0.20	0.014
B	0.38	0.60	0.19	0.014	0.017	1.57	0.24	1.13	0.14	0.21	0.015

Note: traces of Sb, Pb, B, Nb, Ti, B.

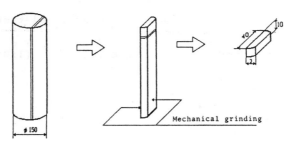

Figure 1: The used sampling method

Table II: Parameters of the thermochemical processes

N°	Process	Temperature	Time	Notes
1	gas-nitr. (10 hours)	505°C 525°C	5h 5h	K=20% K=70%
2	gas-nitr. (72 hours)	510°C 525°C	15h 57h	K=20% K=60%
3	gas-nitr. (120 hours)	510°C 525°C 535°C	30h 45h 45h	K=25% K=50% K=80%
4	ion-nitr. (12 hours)	530°C	12h	p=3.2mb H_2=80% N_2=20%

Notes: K=dissociation degree; p=pressure

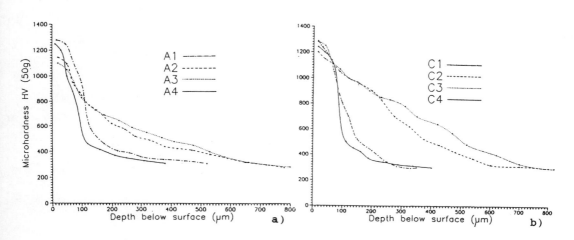

Figure 2: Microhardness profiles of the tested treatments
a) A samples
b) C samples

Table III: Phase analysis of the examined treatments

N°	ε (101)(100)	γ' (200)(111)	α (110)
1	+++	+	+
2	+++	+	−
3	++	++	−
4	+	+++	++

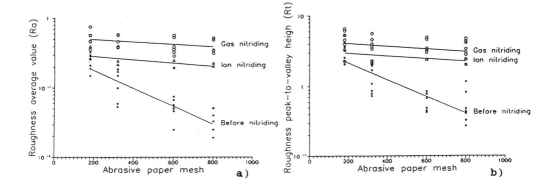

Figure 3: Roughness values before and after nitriding
    a) R_a values (in μm)
    b) R_t values (in μm)

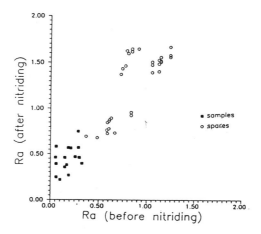

Figure 4: Roughness comparison between samples and spares
    (Ra values in μm)

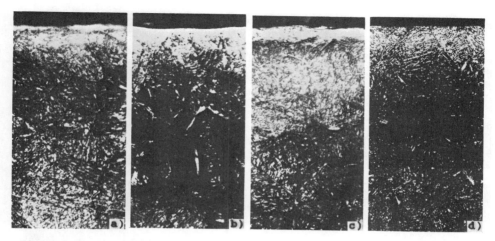

Figure   5: **Microstructural appearence of the nitrided layers of the tested treatments (mag. x 200)**
     a) treatment nº1   b) treatment nº2
     c) treatment nº3   d) treatment nº4

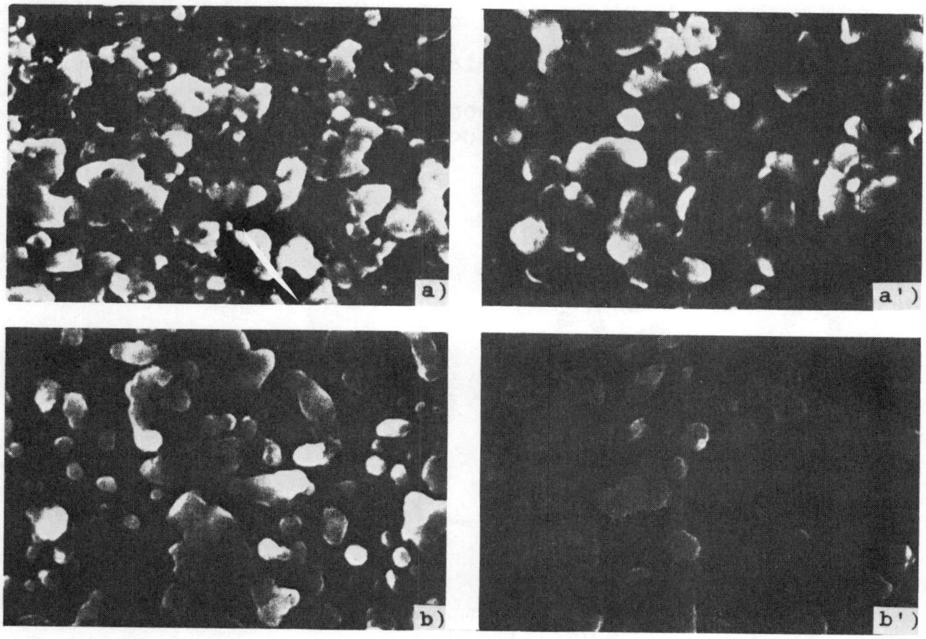

Figure 6: **Examples of the superficial nitride morphology of the tested treatments (mag. x 10.000)**
     a) A-1-180   a') A-1-800
     b) A-2-180   b') A-2-800

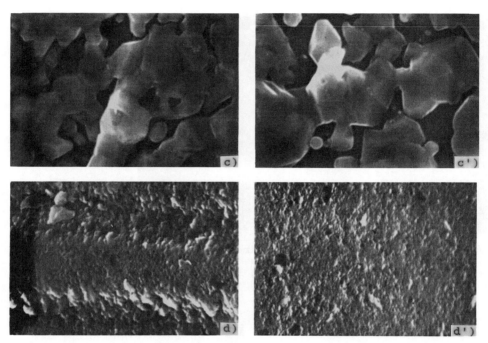

Figure 6: Examples of the superficial nitride morphology of the
          tested treatments (mag. x 10.000)
          c) A-3-180   c') A-3-800
          d) A-4-180   d') A-4-800

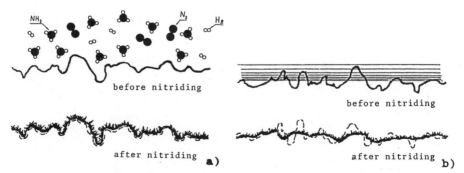

Figure 7: Model of compound layer formation
          a) gas nitriding
          b) ion nitriding

Materials Science Forum Vols. 163-165 (1994) pp. 273-278
© *1994 Trans Tech Publications, Switzerland*

THE KINETICS OF INTERNAL NITRIDING OF AN Fe-7at.% Al ALLOY

N. Geerlofs, C.M. Brakman, P.F. Colijn and S. van der Zwaag

Laboratory of Materials Science, Delft University of Technology,
Rotterdamseweg 137, NL-2628 AL Delft, The Netherlands

ABSTRACT

The kinetics of the internal nitriding of an Fe-7at.% Al alloy was studied using mass-increase data obtained in a thermobalance, for specimens nitrided in a NH_3/H_2 gas mixture at temperatures in the range of 803 - 848 K. Specimens were nitrided starting from both recrystallised and cold-rolled conditions. Contrary to expectations, the specimens did *not* exhibit strong nitriding interaction behaviour. Surprisingly, the recrystallised specimens displayed more tendency to strong interaction behaviour than the cold-rolled specimens. The nitriding behaviour of the latter varied from weak to intermediate.

1. INTRODUCTION

1.1 SCOPE AND PURPOSE

The nitriding of ferritic steels is used for enhancing the fatigue resistance and improving the wear and corrosion properties. The nitriding behaviour of alloyed steels containing many elements is difficult to analyse. The nitriding behaviour of binary Fe-X alloys where X is an nitride forming element, is more amenable to detailed characterisation. While the response to nitriding of Fe-Cr, Fe-Ti and Fe-V alloys is relatively well known, the nitriding behaviour of Fe-Al has not been studied in detail [1,2]. The present paper deals with the kinetics of internal nitriding of thin foil Fe-7at.%Al specimens, either recrystallised (REC) or cold-rolled (CR). Thermogravimetric analysis (TGA), microhardness determination and Electron-Probe Microanalysis (EPMA) were used.

1.2 NITRIDING INTERACTION BEHAVIOUR

The TGA method yields the nitrogen-mass uptake as a function of (nitriding) time. Each TGA data point represents the total nitrogen uptake of the specimen, averaged over all depths in the specimen. Therefore, straightforward analysis of the data is only possible in the case of *ideally weak nitriding interaction* or in the case of completely *strong nitriding* interaction behaviour [1,2]. In the first case nitrogen uptake is the same at all depths in the specimen. This situation occurs when the time required for nitrogen diffusion is short with respect to that necessary for nitride precipitation. *Ideally weak nitriding interaction* occurs when the free enthalpy of formation of the nitride is small or the energy barrier for nucleation is high. Strong interaction arises when the nitrogen arriving by diffusion at a certain depth in the specimen is immediately and completely bonded to the alloying element atoms. In this case a nitriding front progresses steadily inward. A sharp case/core transition occurs. The speed of the front depends on (i) the solubility and the diffusion coefficient of N in ferrite (ii) the concentration of the alloying element dissolved in the α-Fe matrix. The magnitude of the free enthalpy of formation of AlN suggests that Fe-Al would exhibit strong interaction behaviour as has been found for Fe-Ti [3], Fe-V [4] and also Fe-Cr [5,6,7]. This was not observed in Fe-2at.%Al. The observed *weak* behaviour is attributed to a high energy barrier for nucleation of AlN. The literature indicates that a change of the amount of alloying element content sometimes leads to changes in nitriding behaviour [5]. Therefore it was expected that Fe-7at.%Al would exhibit a different and stronger nitriding interaction.

2. KINETIC ANALYSIS

In the case of ideally weak interaction the precipitation of AlN progresses to completion at the same rate at each depth in the specimen. After nucleation particles grow by diffusion of Al while additional nucleation may still occur. The precipitation of AlN is carried to completion with *respect to time*. The kinetics of the process has been dealt with in the literature [2]. Precipitation of the nitride in the α-Fe matrix is slow. Nitrogen bonded to the alloying element is immediately replenished and saturation of the matrix with N is restored. Then the kinetics of nitriding is only related to the activation energy for diffusion of the alloying element and not to that for diffusion of N [1,2].

In the case of strong interaction the situation is different. Precipitation of AlN progresses to completion with *respect to depth in the specimen*. It is taken that all dissolved Al atoms at a certain depth in the specimen are instantaneously bonded to N. Provided the supply of N is sufficient, the nitriding kinetics can be described using internal oxidation theory [8]. Then the kinetics of the total process is strongly determined by the activation energy for diffusion of N.

3. EXPERIMENTAL

3.1 SPECIMEN PREPARATION

The Fe-Al alloy used in this study was prepared by melting iron and aluminium powder under an H_2 flow in a sintered Al_2O_3 crucible. Composition of the Fe-Al alloy has been determined by ICP-OES (Inductively Coupled Plasma Optical Emission Spectroscopy): [Al]=7.3at.% and a total content of Ti, Cu, Mg, Ca, V, Mn, P and Ni of less then 0.02at.%. The as-cast Fe-Al bar was cold rolled to a thickness of 0.32 mm, applying intermediate annealing treatments at 973K. In the final step, following annealing for 3h at 973K and pickling at 333K in a 50/50 mixture of HCl and water, the specimens were 60% cold rolled to a thickness of 0.13 mm. Specimens ($8 \times 14 \times 0.13$ mm^3) were cut from the rolled strip. The specimens to be used in the REC condition were again annealed 3h at 973K. Just prior to nitriding the specimens were chemically polished in Kawamura's [9] reagent obtaining a final thickness of 100 μm approx., etched 30 sec in 1% nital and ultrasonically cleaned in isopropanol. The hardness of the specimens prior to nitriding was 173 $HV_{0.025}$ for REC and 285 $HV_{0.025}$ for CR specimens.

3.2 NITRIDING

The CR and REC specimens were both fully and partially nitrided at 803, 818 and 848 K in a thermobalance (DuPont TGA 951) using an NH_3/H_2 gas mixture with a nitriding potential r_N (= $p(NH_3)/(p(H_2))^{3/2}$) of 2.45 x 10^{-4} Pa$^{-1/2}$ and a linear gas velocity of 15.0 mm.s^{-1}. The value of r_N was chosen such that no iron nitrides could develop during nitriding. Purification of the nitriding gases was performed as described earlier [10]. However, it was observed that the increased Al content of this alloy led to an extreme sensitivity to oxidation during nitriding. Traces of oxygen could lead to brown or even black specimen surface appearances. In these cases, the kinetics of nitriding was evidently disturbed and such specimens were discarded. The mass increase in the thermobalance was checked by weighing the specimens before and after the TGA experiments, using a Mettler mechanical microbalance. After nitriding, all specimens were annealed at 723 K in pure H_2 to remove the nitrogen not chemically bonded (denitriding).

3.3 METALLOGRAPHY AND COMPOSITIONAL ANALYSIS

The microhardness profiles of the cross-section of the specimens were determined using a Leitz Durimet Vickers hardness tester with a load of 5 g. Hardness values thus obtained were converted to typical hardness values for a 100g load, thereby eliminating the influence of elastic deformation. EPMA was employed for quantitative determination of nitrogen-concentration profiles across the specimens. Full details of the method are given in [1].

4. RESULTS AND DISCUSSION
4.1 THERMOGRAVIMETRIC ANALYSIS (TGA)
The TGA curves converted to fraction AlN precipitated are shown in figure 1a for the Fe-7at.%Al *REC* specimens. As in the case of internal nitriding Fe-2at.%Al [1,2] the TGA curves exhibited a plateau in the initial stage corresponding to saturation of the α-Fe matrix with nitrogen. In the present case the plateau was not so clearly distinguishable since nitride precipitation already started before saturation of the α-Fe matrix was completed. After an incubation period depending inversely on the nitriding temperature the nitrogen uptake increases rapidly until saturation is obtained. At saturation on average 8.8at.%N was adsorbed. Taking into account the initial Al content of the specimen and the amount of N required for matrix saturation (0.13at.%), an amount of 1.4at.% of *excess nitrogen* is obtained. This is by a factor of about 7 larger than found for Fe-2at.%Al [1]. After denitriding at 723 K still approximately 8at.%N remained. This indicates that the amount of *excess nitrogen* adsorbed at the AlN/matrix interfaces roughly equals the amount of nitrogen additionally dissolved in the strained matrix *plus* that adsorbed at dislocations [1].

The TGA curves of the *CR* specimens (figure 1b) also exhibited a behaviour similar to those obtained for the internal nitriding of Fe-2at.%Al alloy in the cold-rolled state [2]. Nitride precipitation starts at the beginning of the nitriding process, leading to *apparent* first order ('homogeneous') reaction kinetics. In contrast to the case of the Fe-2at.%Al alloys, almost the same amount of excess nitrogen was found in fully nitrided CR specimens as in the fully nitrided *REC* specimens.

Comparison of the present nitriding curves with those for Fe-2at.%Al showed a strong similarity in the shape of the curves but a reduction in the nitriding times for corresponding initial specimen conditions of a factor 6 ± 1. This reduction is larger than the reduction in the nitriding times predicted from a simple diffusion model taking into account the shorter distance between neighbouring Al atoms with increasing Al concentration: 2.54 $(= (7.3/2.0)^{2/3})$. The TGA curves did generally not reproduce very well: on repeating experiments significantly different nitriding rates could be found. In all these cases the specimens did not show any sign of visible oxidation after nitriding. However, applying ESCA analysis it was found that detectable amounts of Al_2O_3 were present on the surface of the specimen after nitriding. The corresponding signal vanished after sputtering off 15 nm of the top surface layer of the specimen. Nitriding kinetics may have been influenced.

4.2 NITRIDING INTERACTION BEHAVIOUR
The type of nitriding interaction behaviour can be determined by measuring N concentrations across the thickness for partially nitrided samples. The results are shown in figures 2a and 2b for CR samples nitrided at 803 and 848 K, respectively. At each temperature three specimens were used: 20% nitrided (curve a), 50% nitrided (curve b) and fully nitrided (curve c). Surprisingly, comparison of figures 2a and 2b shows that the nitriding interaction behaviour depends on the nitriding temperature. At a nitriding temperature of 803 K (figure 2a) the interaction behaviour can be classified as 'weak to intermediate'. At a nitriding temperature of 848 K (figure 2b) the nitriding interaction behaviour can almost be classified as strong.
This dependence of strength of interaction behaviour on nitriding temperature can also be deduced from microhardness measurements on the same CR samples (figures 3a and 3b). Good qualitative agreement between figures 2 and 3 is obtained. The hardness profiles for CR samples nitrided at 818 K (not shown) displayed an intermediate interaction behaviour, indicating that there is a continuous change in nitriding interaction behaviour with nitriding temperature.
The microhardness profiles for nitrided REC specimens are shown in figures 4a and 4b for nitriding temperatures of 803 and 848 K respectively. Again, at the lower nitriding temperature the nitriding interaction behaviour can be classified as weak, while the nitriding behaviour at 848 K is strong. The nitriding behaviour at 818 K (not shown) can also be classified as intermediate. At each nitriding temperature the degree of interaction is somewhat stronger for the REC specimens than for the CR specimens.

4.3 KINETIC ANALYSIS

Regarding the observed changes in nitriding interaction behaviour with nitriding temperature, no kinetic analysis can be performed since the nitriding mechanism changes with temperature. Essential for a kinetic analysis is that identical situations, but obtained at different temperatures, are compared. However, for those conditions where strong interaction is observed (at the highest nitriding temperature of 848 K), the observed nitriding kinetics can be compared to that predicted by the internal oxidation theory. Ignoring the effect of Al on the diffusion of N and assuming the nucleation barrier for precipitate formation to be zero, the kinetics of nitriding can be calculated. The nitriding kinetics determined experimentally is about a factor of 2 slower than the kinetics predicted. The discrepancy is attributed to the decelerating effects of the concurrent oxidation at the surface and the fact that some nucleation barrier for nitride formation still exists.

5. CONCLUSIONS

- The strength of the interaction behaviour in Fe-7at.%Al alloys depends on the nitriding temperature.
- For both recrystallised and cold-rolled specimens the degree of interaction increases from weak to almost strong with increasing nitriding temperature.
- At the same nitriding temperature recrystallised samples showed a slightly stronger interaction behaviour than the cold-rolled samples.
- The nitriding kinetics of CR and REC Fe-7at.%Al samples is rather different. The differences are similar to those observed in Fe-2at.%Al specimens.

ACKNOWLEDGEMENTS

The authors express their gratitude towards ir. W.G. Sloof and Mr. J. Helmig for EPMA determinations, to Dr. ir. M.A.J. Somers and ing. E.J.M. Fakkeldij for ESCA measurements and helpful discussions and to Mr. P. Hokkeling (Philips Research Laboratories) for manufacturing of the Fe-Al alloy. Thanks are also due to ir. M.H. Biglari for experimental assistance and discussions.

REFERENCES

1) Biglari, M.H., Brakman, C.M., Somers, M.A.J., Sloof, W.G. and Mittemeijer, E.J.: Z. Metallkde, 1993, 84, 124.
2) Biglari, M.H., Brakman, C.M., Mittemeijer, E.J. and Zwaag, S. van der: Proc. Int. Conf. Surf. Engng., P.Mayr ed., 1993, Bremen, BRD, DGM, Oberursel, BRD, 1993, p. 237.
3) Jack, D.H., Lidster, P.C., Grieveson, P. and K.H. Jack: Chemical Metallurgy of Iron and Steel, Proc. Int. Symp. Met. Chem., Sheffield, 19-21 July, 1971, The Iron and Steel Institute, London (1973), p.374.
4) Pope, M., Grieveson, P. and Jack, K.H.: Scand. J.Met., 1973, 2, 29.
5) Mortimer, B., Grieveson, P. and Jack, K.H.: Scand. J.Met., 1972, 1, 203.
6) Hekker, P.M., Rozendaal, H.C.F. and Mittemeijer, E.J.: J. Mater. Sci., 1985, 20, 718.
7) Wiggen, P.C. van, Rozendaal, H.C.F. and Mittemeijer, E.J.: J. Mater. Sci., 1985, 20, 4561.
8) Meijering, J.L.: Adv. Mat. Res., 1971, 5, 1.
9) Kawamura, K.: J. Japan Inst. Metals, 1960, 24, 710.
10) Somers, M.A.J., Pers, N.M. van der, Schalkoord, D. and Mittemeijer, E.J.: Metal.Trans.A., 1989, 20A, 15.

FIGURES

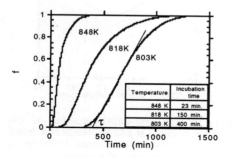

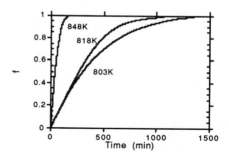

Figure 1a: Fraction AlN precipitated (f) for REC specimens as a function of nitriding time.

The value of the incubation time τ is taken as the part cut from the abscissa by the tangent in the point of inflection.

Figure 1b: Fraction AlN precipitated (f) for CR specimens as a function of nitriding time.

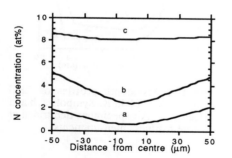

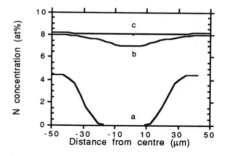

Figure 2a: Nitrogen-concentration (EPMA) depth profiles of CR specimens nitrided at 803 K. Symbols a, b and c: approx. 20% nitrided, approx. 50% nitrided and fully nitrided. All specimens denitrided at 723 K before EPMA measurements.

Figure 2b: Nitrogen-concentration (EPMA) depth profiles of CR specimens nitrided at 848 K. Symbols a, b and c: as in figure 2a. All specimens denitrided at 723 K before EPMA measurements.

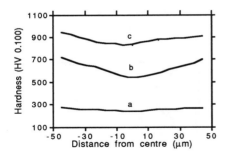

Figure 3a: Hardness-depth profiles of CR specimens nitrided at 803 K. Symbols a, b and c: as in figure 2a. All specimens denitrided at 723 K before hardness measurements.

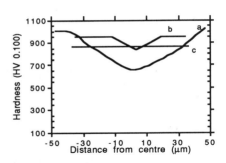

Figure 3b: Hardness-depth profiles of CR specimens nitrided at 848 K. Symbols a, b and c: as in figure 2a. All specimens denitrided at 723 K before hardness measurements

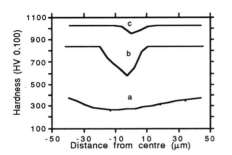

Figure 4a: Hardness-depth profiles of REC specimens nitrided at 803 K. Symbols a, b and c: as in figure 2a. All specimens denitrided at 723 K before hardness measurements.

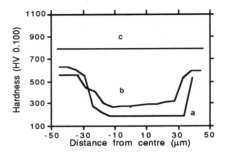

Figure 4b: Hardness-depth profiles of REC specimens nitrided at 848 K. Symbols a, b and c: as in figure 2a. All specimens denitrided at 723 K before hardness measurements.

Materials Science Forum Vols. 163-165 (1994) pp. 279-284

THE KINETICS OF NITRIDING LAYER GROWTH ON
Fe-Cr AND Fe-Ti ALLOYS

J. Ratajski, J. Ignaciuk, J. Kwiatkowski and R. Olik

The Solid State Physics Division at Technical University in Koszalin, Poland

ABSTRACT

This work presents some of the results on the diffusion zone growth kinetics in the Fe-Cr and Fe-Ti alloys. The results have been achieved in the temperatures of the process with help of constructed by the authors measuring system equipped with a magnetic sensor placed in the very furnace retort [1, 2].

The kinetics of the diffusion layer growth was tested as a function of:
- contents of Cr: 0.13; 1.12; 2.25; 3.18 and 5.23 wt%,
- contents of Ti: 0.15; 0.60; 1.10; 1.65 and 2.20 wt%

INTRODUCTION

A diffusion zone in nitrided iron alloys consists of solution of solid nitrogen in alloy matrix and alloying elements nitrides. The solid solution creation is accompanied by the development of residual surface macrostresses which are proportional to the differences in the surface concentration of nitrogen and its average contents in the solution [3]. Precipitations of the alloying elements nitrides cause a defined microstress state within the zone. Both macro- and microstresses being a function of the process time are the main factor influencing the magnetic properties of the nitrided material, its magnetic permeability mostly - the value of which may change up to several orders of magnitude during the process. The compound layer ($\varepsilon + \gamma'$) which is produced on the top of the diffusion zone is paramagnetic within the process temperatures and of magnetic permeability value of $\mu = 1$.

The above changes in magnetic properties are being used as the essence of our measuring system operation. This system can be applied to research the kinetics of the nitrided layer growth directly during the process. It also proves very useful for monitoring industrial nitriding processes.

This paper presents exemplary research results obtained by the application of this system, regarding the influence of alloying elements / Cr and Ti / contents in Fe upon the properties of the diffusion zone. The characteristics of changes in the induced voltage achieved from the process have been interpreted in reference to the own research as well as to literature data.

EXPERIMENTAL PROCEDURES
Specimen in the form of rectangular prism (30 mm x 5 mm, thickness 0.5 and 1 mm)
were prepared from Fe - Cr alloys (0.13; 1.12; 2.15; 3.18 and 5.23 wt% Cr) and Fe -Ti
alloys (0.15; 0.60; 1.10; 1.65 and 2.20 wt% Ti). Fe - Cr alloys specimen were annealed
prior to the nitriding process at 1223 K in a hydrogen atmosphere while those made of
Fe - Ti alloys were annealed at 1123 K and in an argon atmosphere.
During the nitriding process the specimens were placed in the magnetic sensor situated in
the retort of a laboratory vertical furnace. A gaseous nitriding atmosphere was flowing
through the sensor, with a linear flow rate of 16 cm/s. Values of the voltage signal induced
in the sensor were recorded all the time through the nitriding process. All the processes
were conducted at the temperature of 833 K and in the atmosphere of % NH_3 / % H_2 =
15/85; 40/60; 60/40 and 100 % NH_3.
After defined periods of the process the specimens were water quenched and then tests
were carried out:
- optical microscopy was performed with a Neophot-2 microscope (Carl Zeiss Jena),
- the microhardness profiles were determined using Vickers microhardness tester
  according to Hanemann, adapted to the Neophot-2 microscope (applied load 100 g),
- X-ray diffractometry was performed for phase identification and residual surface macrostress analysis. Co K_α and Cr K_α radiations were applied. X-ray diffraction line profiles
  were measured according to the present time method in steps on 0.01° or 0.05° 2Θ
  employing counting time of 100 s. For the determination of macrostress, the $\sin^2\Psi$
  method was applied.

RESULTS AND DISCUSSION
I. Fe -Cr ALLOYS
   Changes in the voltage induced in the sensor as a function of the nitriding time for the
Fe - Cr alloys specimens have been presented in figure 1.   In terms of quality these
courses have identical shape, i.e. within the first period of nitriding a decrease of the

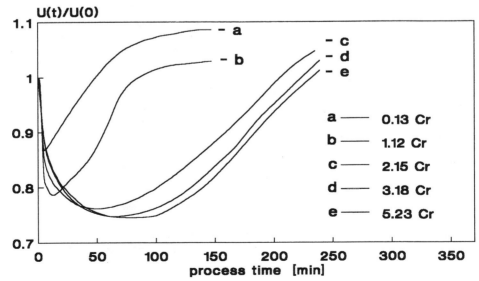

Fig. 1. Induced signal in magnetic sensor v. nitriding time (15 vol.% NH_3 / 85 vol.%
H_2 , T = 833 K) of Fe-Cr specimens of thickness 0.5 mm.

recorded signal is being observed which then, after having reached its minimum value grows up until it reaches its maximum. Both time needed for reaching these extreme points as well as their values depend on chromium contents in the alloy.

Earlier research [1, 2] conducted for samples made of Fe has been a reference point for the interpretation of presented courses. Some exemplary courses of recorded changes in voltage for these samples within the range of solid solution αFe(N) have been shown in figure 2. In this case both the moment at which the minimum signal is being observed as well as the values of the minimum are the function of the nitriding atmosphere composition. On the basis of calculated profiles of nitrogen concentrations in the diffusion zone (figure 3), at the assumption that the surface concentration of nitrogen achieves its equilibrium value with the nitriding atmosphere relatively slowly, as well as on the basis of research in distribution of microhardness along the cross-section and literature data [4, 5], following interpretation of observed changes in the measured voltage signal has been presented:

- recorded minimum of signal corresponds with maximum residual surface
  macrostress which is directly proportional to the difference between surface nitrogen
  concentration and its average contents in the specimen,

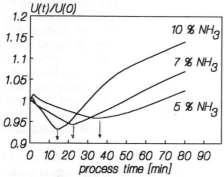

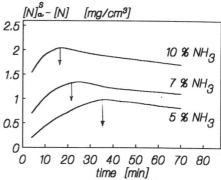

Fig. 2. Induced signal in magnetic sensor v. nitriding time (10/90; 7/93 and 5/95 vol.% NH$_3$/ vol.% H$_2$, T = 833 K, thickness 0.5 mm.

Fig. 3. Calculated differences between the surface nitrogen concencentration tration value and its average contents in Fe specimen v. time.

- an annihilation of macrostress occurs for the nitriding time which is longer than the time of recorded minimum signal,
- the voltage signal value bigger than its initial one is connected with the increase of magnetic moment per Fe atom growth in presence of atoms of nitrogen [6].

Diffusion zone of nitrided Fe - Cr alloys at the nitriding temperature consists of a solid solution of nitrogen in αFe lattice and chromium nitrides. The rate of nitrides creation depends on chromium - nitrogen interaction which changes from a weak to a strong one according to chromium contents in the alloy. Chromium nitrides do increase nitrogen solubility in the matrix net causing through this higher macro-stress in comparison to Fe and moreover lead to the development of residual microstress. These phenomena have been reflected in figure 1 where the courses of changes in measured voltage signal are presented. Among these courses, there are two families of characteristics visible which differ from each other due to the difference in the rate of voltage signal changes. Characteristics obtained for Fe - Cr alloys of 0.13 and 1.12 wt% Cr can be included into the first family while those obtained for 2.15; 3.18 and 5.23 wt% Cr into the second one.

The courses creating the first family of characteristics distinguish themselves by the short time of reaching the minimum value: 6 and 12 min. respectively for 0.13; 1.12 wt% Cr, and by a big differenciation between the minimum value and quick growth of signal in the time of process longer than the time of the registration of the minimum.

Times of reaching the minimum are definitely higher for the other family of characterisitcs; 50, 65 and 85 min respectively for 2.15; 3.18 and 5.23 wt% Cr. Moreover, a small differenciation between signal minimum value is observed here; also, the signal after time of the process corresponding with its minimum value grows relatively slowly.

The research on microhardness distribution along the cross-section of specimens indicates that for alloys of small chromium contents / 0.13 and 1.12 wt% - first family of characteristics /their case-core interface has been moderate - the diffusion front is ahead of the front of reaction. This interface is sharp in the alloys which courses have been included into the other family of characerictics; the reaction front here is ahead of the diffusion one /specimens of 2.15 wt% Cr though show some insignificant moderation of this interface/.

On the X-ray diffractogram of the samples of 0.13 wt% Cr nitrided for 6 min-up to the time of minimum signal registration and then dropped immediately down to water from the magnetic sensor, some reflexes from γ' phase / Fe_4N / have been observed. This proves that after this time of the process the concentration of nitrogen has achieved the maximum value at the surface of diffusion zone. At the same time it is also in equlibrium with the maximum nitrogen concentration in the zone. The biggest macrosterss should be thus expected to occur, at this time moment of the process. As it proves, residual surface macrostress measured by a $\sin^2\Psi$ method for these samples does have a value of - 240 MPa after 6 min and then, after 20 min of the process this value drops down to - 95 MPa. Such a violent drop of stress is caused by an equalization of nitrogen concentration profiles in diffusion zone. Thus, minimum at the voltage courses achieved from the process responds to the maximum value of residual surface macrostress in the diffusion zone of the Fe - Cr alloy of 0.13 wt% Cr.

The X-ray reflexes γ' phase of 1.12 wt% Cr sample are visible after 9-10 min of the process, but here this time is smaller than the time in which the voltage signal obtains its smallest value. An analysis of X-ray diffraction line broadening conducted for 0.13 and 1.12 wt% Cr samples nitrided in increased time interval indicates significant broadening of lines from the phase αFe after 12 min of the process for the samples of higher chromium contents. This is a result of chromium nitrides precipitation which in case of alloys of this contents of chromium is being reflected in the voltage courses registered with a magnetic sensor by shift of the minimum towards longer process time and essential decrease of its value in comparison to the 0.13 wt% Cr alloy. To sum up, in the alloy of 1.12 wt% Cr, a maximum state of stress was created within the process time corresponding with the minimum of the measured voltage and being a resultant of macro- and microstresses.

Considerable differences between the times of achieving the minimum signal for the I-st and II-nd families of characteristics are connected with the growth of nitrogen solubility in Fe - Cr alloys along with the growth of Cr contents. After 30 minutes reflexes from a γ' phase are still not being identified in any of the X-ray diffractograms of alloys of 2.15; 3.18 and 5.23 wt% Cr contents. Much slower growth of signal after the time of the registration of the minimum is evoked by the change in location of the reaction front in regard to the diffusion one. As for the 3.18 and 5.23 wt% Cr alloys the reaction front is ahead of the diffusioin zone, it combines the decrease of the nitrogen concentration gradient and annihilation of macrostresses with a moving of the reaction front.

II. Fe -Ti ALLOYS

Titanium is characterized by a strong interaction with nitrogen, independently to its contents in the alloy. This property of titanium implies the fact that the reaction front in Fe - Ti alloys is always ahead of the diffusion front. Titanium nitrides which are formed do not, in contradiction to chromium nitrides, practically coagulate during the time of the process. Moreover, a reaction of discontinuous precipitation does not take place in Fe -Ti alloys.

Figure 4 presents changes in the voltage induced in the sensor in the function of the process time of nitriding of Fe -Ti alloys of 0.15; 0.60; 1.10; 1.65; 2.20 wt% Ti contents. It is imposible, contradictory to Fe - Cr alloys, to distinguish the two families of characteristics among these courses. On the basis of results presented in figure 4 , there is shown in figure 5 diagram of relationship between the root square of processes duration for the intervals at which the minimum of the signal is observed in function of titanium contents in the alloys - figure 6. This diagram presents directly proportional relationship between these quantities .

At the same time an diffusion zone thickness was determined right after characteristic times the process; this was performed on the basis of microhardness distribution determined along the cross-section of specimens - figure 6.

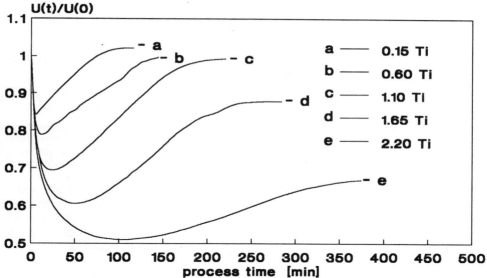

Fig. 4. Induced signal in magnetic sensor v. nitriding time (15 vol.% NH_3/ 85 vol.% H_2 , T = 833 K) of Fe-Ti alloys specimen of thickness 1 mm.

This diagram additionally presents thickness data calculated for this zone and for these times of the process. A model [7] describing the nitriding front advances at a rate dictated by internal nitriding theory has been used here:

$$X^2 = 2 \times [N]^s_\alpha \times D_\alpha \times t \ / \ [Ti] \tag{1}$$

where: - $[N]^s_\alpha$ - nitrogen concentration at the surface in the iron matrix,
  - $[Ti]$ - titanium concentration in the specimen,
  - D_α - nitrogen diffusion coefficient in pure ferrite,
  - t   - nitriding time.

Full interpretation of results presented in figure 4 still requires some precise measurements of microstress as function of the time process. First approximation however allows to draw following conclusions:
- identical shape of voltage - time courses for all the tested alloys confirms that interaction titanium - nitrogen is independent from the titanium contents in alloy,
- minimum of the recorded signal responds to the max. stress state in the diffusion zone
- maximum stress state was formed for the same thickness of the diffusion zone in all tested Fe - Ti alloys,
- discrepancies between the measured and calculated thicknesses of the diffusion zone (figure 6) for the 2.20 wt% Ti alloy are probably resulting from the assumption (equation 1) of constant surface nitrogen concentration for αFe and from the assumption of a constant coefficient of diffusion independent from nitrogen concentration.

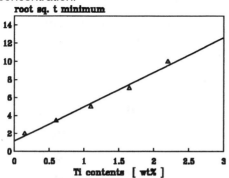

Fig. 5. The relationship between the root square of process duration for the intervals of which the minimum of the signal is observed.

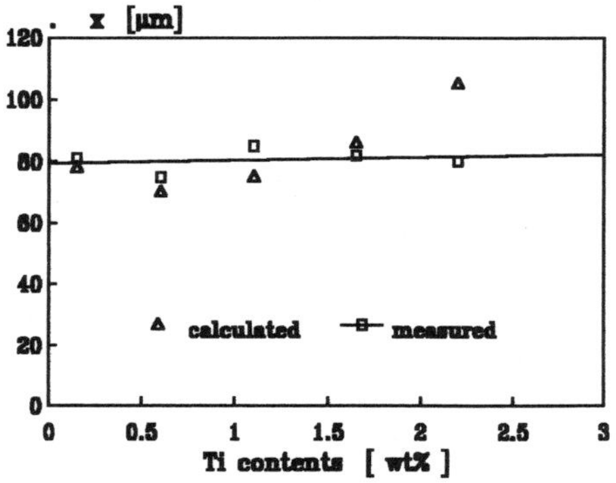

Fig. 6. The diffusion zone thickness after nitriding time at which the minimum of the recoreded voltage signal is observed v. titanium contents.

For Fe - Cr alloys, time of the process at which maximum stress in diffusion zone is created, also depends upon the contents of Cr. In this case though, due to changeable interaction chromium - nitrogen and due to the occurence of the discontinuous precipitation reaction, no linear relationship between the square root of minimum signal achievementtime and chromium contents in the alloy has been observed.

References:
1) Ignaciuk, J., Kwiatkowski, J., Olik, R., Ratajski, J., Zyśk, J.: Proc. 5th Internat. Cong. Heat Treatment of Materials, Budapest, 1986, 1030.
2) Ratajski, J., Ignaciuk, J., Kwiatkowski, J., Olik, R.: Proc. of HNS, 1988, 423.
3) Mittemeijer, E.J.: TMS/AIME meeting, 1983.
4) Rozendaal, H.C.F., Mittemeijer, E.J., Colijn, P.F., Van der Schaaf, P., J.: Met. Trans. A, 1983, 82A, 255.
5) Straver, W.T.M., Rozendaal, H.C.F., Mittemeijer, E.J.: Met. Trans. A, 1984, 15A, 627
6) Mitsuoka, K., Miyujima, H., Ino, H., Chikazumi, S.: J. Phys. Soc. Japan, 1984, 53, 2381
7) Spitzig, W.A.: Acta Metall., 1981, 29, 1359.

Acknowledgments

Financial support of the Scientific Research Commitee - (Komitet Badań Naukowych - KBN) is gratefully acknowledged.

Materials Science Forum Vols. 163-165 (1994) pp. 285-292
© 1994 Trans Tech Publications, Switzerland

SOME MORE DATA FOR THE EUTECTOID EQUILIBRIUM OF THE Fe-N-C SYSTEM

R. Russev and S. Malinov

Technical University, Varna, Bulgaria

ABSTRACT

Investigations and data referring formation and eutectoid transformation of γ-phase in Fe-N-C System are ambiguous. Siycke.. and Do.. give
proof for principally new phase equilibriums at 550-600°C in this
system formed by Naumann and Langenscheid in 1965, the system being
still considered the only reliably one. These phase equilibriums are
based on the principal assumption that there exists a limited in
temperature 550-570°C two-phase $\varepsilon+\alpha$-zone which determines changes in
the sequense of nonvariant solid phase transformations. As a result the
plane of ternary eutectoid at 575°C is higher than that in Naumann and
Langenscheid system(565°C). Furthemore, the ternary eutectoid in this
case is a mixture of $\alpha+\gamma'+\varepsilon$ while in Naumann and Langenscheid system it
is $\alpha+\gamma'+\theta(Fe_3C)$.
The present report deals with the examination powdered samples with
various concentrations of N and C. These examinations were mainli aimed
at explanation of the above-mentioned contradiction. Thermal effects of
solid phase transformations taking place at 550-650°C are obtained by
precise differential thermal analysis, specific temperatures of
eutectoid processes being determined as well. The type and the quantity
of phase are controlled by X-ray structural analysis.
The report contains some important facts and proofs contributing to
the analysis, complement and support of some controversial points
concerning eutectoid processes in Fe-N-C system. The most important
conclusions are:
-the temperature of the eutectoid equilibrium in Fe-N-C system is 570°C
-ho changes of the phase composition are observed before γ-phase
formation;
-γ-phase formation starts after a certain incubation period(4.5min. at
570°C) and decreased with the rise of isothermal soaking temperature;
-at isotermal heating alove 610°C ε-phase is formed with low content
of N and C corresponding to stoicnomentric $Fe_3(N,C)$. This transformation takes peace after the γ-phase formation at heating and after
complete transformation of the α-phase, which proves that the
equilibrium $\varepsilon+\alpha$ is impossible.

INTRODUCTION

A number of critical papers concerning phase equilibrium and transformations in the Fe-N-C - system proposed by Langenscheid and Neuman [1] have recently appeared. The significance of the system for studying a lot of problems in material science demands its thorough analysis. J. Slycke, L Sproge and J Agren [2] critisized Langenscheid's system mainly in terms of the type and sequence of non-variant equilibria:

according to Langenscheid:	according to Slycke:
$\gamma \rightleftharpoons \alpha + \gamma' + Fe_3C$ at $565°C$	$\alpha + \varepsilon \rightleftharpoons Fe_3C + \gamma'$ at $550°C$
$\gamma + \varepsilon \rightleftharpoons \gamma' + Fe_3C$ at $575°C$	$\gamma \rightleftharpoons \alpha + \gamma' + \varepsilon$ at $580°C$

In Slycke's model eutectoid transformation at heating is theoretically followed by another non-variant transformation: $\varepsilon + \alpha$ -- $\gamma + Fe_3C$.
Having analysed both models in previous papers [3,4], the authors of this report have specified this transformation and plotted relative isothermal and polythermal sections. Later, Du [5] also mentioned this eventual equilibrium in his substantial theoretical work.
Taking into account the complete coincidence of the models proposed by both Slycke and Du on the one hand, as well as our model, on the other, further discussion proves necessary. However, it should be mentioned that the new model is developed only of the base of theoretical data and conditions for phase transformations in ternary systems. Data supporting the existence of the two-phase $\alpha + \varepsilon$ equilibrium and proving the new model do not seem very convincing.
Evidently, experimental proving of the authenticity of one of these models is a task which requires not only precise accuracy of the experiment but also its correct setting and planning.
Of all possible four-phase equilibria in the two models the most interesting is the eutectoid one which occurs at $565°C$ previous to another non-variant peritectic reaction in Langenscheid's model while in Slycke's model it occurs between two peritectic reactions at about $576°C$. Phases taking part in the eutectoid equilibrium are different in both models, being $\alpha + \gamma' + Fe_3C$ in Langenscheid's model, and $\alpha + \gamma' + \varepsilon$ in Slycke's one. All these factors result in distortion and changing of the isothermal triangle of that equilibrium in the Fe-N-C diagram(Fig.1).
The authors of the present paper tried to prove experimentally by thermal analysis the occurrence of this transformation as well as of any other eventual non-variant transformations. The paper contains the most important results obtained which give reasons for conclusions to be made concerning the temperature of the eutectoid equilibrium and its phases.

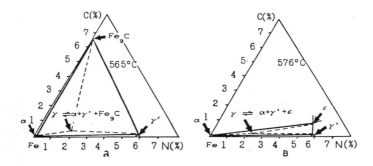

Fig.1. Non-variant eutectoid planes in the Fe-N-C system
according to Langenscheid (a) and Slycke...Du(b)

EXPERIMENT PROCEDURE

Preparation of samples is one of the problems in thermal   analysis   of
the Fe-N-C system. As it is well-known Fe-N-C alloys with   considerably
high pescentage of N can be synthesised only by diffusion.   To   obtain
the proper phases composition and homogeneity of   peculiar   ones   for
respective conditions of nitrocarburising, iron powder with particles of
<0.5 mm was used. It was saturated for a period of time sufficient   for
obtaining an equilibrium chemical and   phase   composition.   "Carbonit"
technology of gaseous nitrocarburising in NH_3 and CO_2 meduim [6],   developed in Bulgaria, was used for the purpose. This method provides low
carburising potential due to which   the   ternary   Fe-N-C   system   is
analised in zones close to the   peripheral   Fe-N system. Technological
parameters   of   the   saturation   such   as   temperature,   degree   of
dissociation   and NH_3 and CO_2 ratio   were   chosen   after   thorough
preliminary calculation aimed at obtaining   chemical   composition   and
phases lying within the eutectoid triangle.
Thermal analysis carried out in   a   specially   developed   installation
where the samples were heated and colled in a neutral atmosphere   (Ar)
with a controlled speed. The temperature of the   standart   iron   powder
was controlled parallel to the analysis of the nitrocarburised samples.
Thus, differential analysis was performed which allowed thermal effect
to be determined with high accuracy. Ceramic crucibles   with   powdered
samples weighing 7g were immersed into a bath of molten metals (Pb-Sn)
providing homogeneity of the temperature field and   the   same   heating
rates. Thermocouples of Chromel-alumel-type were used   to   control   the
temperature of both the analysed and   standart   samples. The   error   in
temperature measurements did not exceed $1^\circ C$ The type and   quality   of
phase transformations were controlled   by   X-ray   diffraction   (Co-Kα)
after hardening of the samples. The thermal analysis   of   the   samples
shown in Tabl.1   was   made   in   two   ways-at   continuous   heating   and
cooling   and   at   isothermal   heating. Two rates of continuous heating
called below "fast" and "slow" were used.

	slow	fast
heating	$3-3.5^\circ C/min$	$15-18^\circ C/min$
cooling	$4.8-5^\circ C/min$	$27-30^\circ C/min$

Table 1 Conditions of gas nitrocarburising, chemical and phase compositions of powdered samples subjected to thermal analysis

N°	T°C	α,%	NH/CO	N,%	C,%	phases*
1	600	80	100/10	2,9	0,44	α+γ'+ε
2	600	80	100/30	3,9	0,46	α+γ'+ε
3	650	80	100/10	3,3	0,17	α+γ'+ε
4	650	80	100/30	2,8	0,26	α+γ'
5	650	60	100/10	3,1	0,24	α+γ'+ε
6	650	60	100/30	3,3	0,25	α+γ'+ε
7	700	80	100/10	2,8	0,16	α+γ'
8	700	80	100/30	2,0	0,10	α+γ'
9nitr.	650	80	100/0	2,9	0,0	α+γ'

*Phase composition obtained after slow cooling in a flow of the NH_3 of the nitrocarburized samples. There are small amounts of the ε-phase.

RESULTS

Data obtained by thermal analysis of the nitrided samples were used as a comparison base. Fig.2 shows time-temperature and differential curves registered at continuous heating and cooling of a nitrided sample and of one of the nitrocarburised samples. Data for all the rest nitrocarburised samples are given in Table 2. Rather high temperature hysteresis even at very low rates (1°C/min) was observed, which may be caused by long incubation period. That was the reason for the thermal analysis to be performed at isothermal heating conducted through immersing the samples into a heated metal bath at temperatures close to the supposed eutectoid one.

The X-ray diffractional analysis of all the samples studied showed that the registered thermal effect is mainly caused by γ-phase formation. Critical temperatures shown in Table 2 should not be taken for final conclusions concerning the temperature of the eutectoid equilibrium because the effect of the incubation period is considerable and other phase transformations especially in nitrocarburised samples may well be possible. It is obvious, however, that this transformation in nitrocarburised samples occurs at lower temperatures than those for the same transformation in the Fe-N system. To estimate the temperature of the eutectoid transformation thermal analysis was made during isothermal heating at 540,550,560,570,580,590 610,630 and 650°C. Table 3 contains some of the results which prove that the eventual value of the temperature for that equilibrium lies in the interval of 560-570°C. While in nitrided samples no austenite is registered at temperature 590°C, this phase is registered at 570°C in all nitrocarburised samples at isothermal soaking. Control isothermal soaking at 565^{+2}°C during 35min was undertaken which showed neither thermal effect nor changes in the initial phase composition.

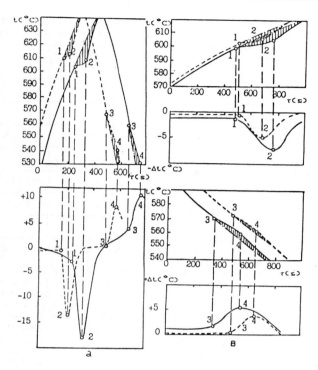

Fig.2 Time- temperature and differential curves obtained
by thermal analysis at continuous "fast" (a) and "slow"
(b) heating and cooling

―――――― nitrocarburized sample
― ― ― ― nitrided sample

1-onset of the reverse eutectoid transformation
2-end of the reverse eutectoid transformatiion
3-onset of the eutectoid transformation
4-end of the eutectoid transformation

Table 2 Temperatures at which thermal effect of the eutectoid
transformation starts and finishes at continuous heating and cooling

| | heating | | | | cooling | | | |
| | fast | | slow | | fast | | slow | |
	st.	fin.	st.	fin.	st.	fin.	st.	fin.
1	604	612	598	605	565	540	570	560
2	605	613	597	605	560	530	569	561
3	604	613	595	605	560	530	570	555
4	608	615	597	606	560	530	575	555
5	602	611	594	603	563	532	570	555
6	607	612	595	607	565	530	570	555
7	603	608	597	604	560	530	570	560
8	605	615	598	605	565	530	575	555
9nitr	610	618	603	610	568	540	575	560

Table 3. Phase composition of the nitrocarburised samples after isothermal soaking at various temperatures and subsequent hardening

$T^{\circ}C$ / N°	init.	550	560	570	580	590	610	630
1.	α+γ'+ε		α+γ'+ε	α+γ'+ε+γ	α+γ'+ε+γ			
2.	α+γ'+ε		α+γ'+ε	α+γ'+ε+γ				
4.	α+γ			α+γ'	α+γ'+γ	α+γ'+γ		γ'+γ+ε*
5.	α+γ'+ε		α+γ'+ε	α+γ'+ε+γ		α+γ'+ε+γ		
6.	α+γ'+ε	α+γ'+ε	α+γ'+ε	α+γ'+ε+γ	α+γ'+ε+γ	α+γ'+γ	γ+γ'+ε*	γ+γ'+ε*
8.	α+γ'			α+γ'+γ	α+γ'+γ	α+γ'+γ	α+γ+γ'	γ+α+α'
9-n	α+γ'					α+γ		

ε*-modification of the ε-phase resultes from the peritectic reaction with the composition coresponding to the Fe5(N,C)

Fig 3 represents time-temperature and differential curves at various temperatures at isothermal heating. The differential curves clearly show that phase transformation occurs at isothermal soaking at 570°C, X-ray diffraction analysis proving that transformation to be caused by γ-phase formation. This transformation started after an incubation period which depended mostly on the temperature of soaking and less on the composition of the sample studied.

A more detailed and thorough analysis of the difractograms showed that along with the γ-phase formation at 570°C, which is the most important result of the present work, there was another change of the phase composition observed at heating above 600°C-formation of ε-phase(its parameters being a=2,64A; c=4,31A; c/a=1,63) with low N and C contents corresponding to stoichiometric Fe5(N,C). Similar transformation was observed by both Langenscheid and Slycke. In the first case it resulted from the peritectic reaction γ'+Fe3C ⇌ ε+γ, while in the second it occurred before formation of γ as a result of the peritectic reaction Fe3C+γ' ⇌ ε+α. However, in both cases the ε-phase has low N and C contents and strongly differs from the ε-phase formed by diffusion. Our experiment has established that the ε-phase is formed at 610°C and it is in equilibrium rather with γ than with α-phase because at such temperatures the latter disappears while the γ-phase increases. Fig. 4 presents typical difractograms showing changes in phase composition after isothermal soaking and proving, to a great extent, the type and sequence of reactions which take place.

Certain denitriding is observed after isothermal soaking above 630°C especially in samples with high N content, as a result of which α+γ' phase composition is formed after slow cooling, chemical composition of C being unchanged -0.4-0.5%. This indicates the presence of Fe3C as well since α and γ'-phases cannot dissolve this amount of carbon. The quantity of Fe3C in that carbon content of the samples cannot be recorded on the difractograms.

The present work was not aimed at complete analysis of all phase transformations and equilibria in the controversial temperature interval. It presents only results related to the eutectoid equilibrium.

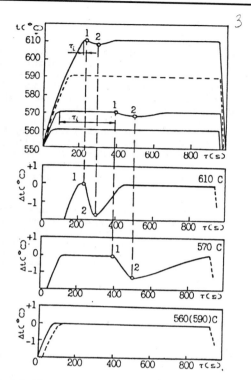

Fig. 3 Time-temperature and differential curves obtained by
isothermal analysis at various temperatures
   1 – onset of the reverse eutectoid transformation
   2 – end of the reverse eutectoid transformation
   τ_i – incubation period

——— nitrocarburised sample N°6
– – – nitrided sample

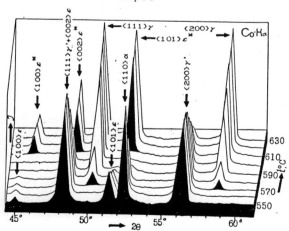

Fig. 4. Difractograms showing changes in the phase
composition of N°6 nitrocarburized sample after its
isothermal analysis.

CONCLUSIONS

The thermal analysis of Fe-N-C powdered samples with chemical and phase compositions lying in the eutectoid plane enables some conclusions concerning the eutectoid reaction to be drawn:

-the temperature at which that transformation occurs is 570°C

-no changes of the phase composition are observed before γ-phase formation at both continuous and isothermal heating;

-γ-phase formation starts after a certain incubation period(4.5min. at 570°C) and decreased with the rise of isothermal soaking temperature;

-at isotermal heating above 610°C ε-phase is formed with low content of N and C corresponding to stoichiometric $Fe_5(N,C)$. This transformation takes place after the γ-phase formation at heating and after complete transformation of the α-phase, which proves that the equilibrium $\varepsilon+\alpha$ is impossible.

As a result of the above-mentioned experimental analysis of phase transformations occurring with in the temperature interval and chemical composition which allow formation and decomposition of the γ-phase according to the familiar models of the Fe-N-C systems, the authors can draw the following conclusion: phase equilibria and transformations proposed by Slycke...Du are not confirmed. Thus, the experiments described have proved the validity of the Fe-N-C phase diagram proposed by Langenscheid.

REFERENCES

1. F.K.Naumann and G.Langenscheid:Arch.Eisenhuttenwes.,36,(1965),677
2. J.Slycke,L.Sproge and J.Agren:Scand.J.Metall.,17,(1988),122-126
3. R.Russev,MiTOM,6,1991,14-16,(russ)
4. R.Russev,S.Malinov,S.Grosdeva,J.Argirov,Proceedings of National Congress on Heat Treatment,Varna,(1991),3-8
5. H.Du and M.Hellert,Z.Metallkunde,82,(4)(1991),310-316
6. R.Russev,Konferencia XII Celostatne dni tepelneho spracovania (1988),115

Materials Science Forum Vols. 163-165 (1994) pp. 293-308
© 1994 Trans Tech Publications, Switzerland

THE SULPHONITROCARBURISING PROCESS SULF IONIC®

S. Deshayes, P. Jacquot and E. Denisse

Innovatique SA (HIT Group),
25 Rue des Frères Lumière, F-69680 Chassieu, France

ABSTRACT

Sulphonitrocarburation is a surface treatment that is essentially applied to steels, and which allows to strongly improve the tribological properties of mobile surfaces.
The new procedure that we have developed consists of sulphonitrocarburising steels i.e. enriching the iron sulphide and nitride content of the superficial microlayer of steels. We have chosen to inject a gaseous mixture made up of nitrogen, methane and dilute sulphurising gas in a plasma nitriding furnace.
Our study concerned the influence of parameters such as temperature, pressure and treatment time. We shall especially present the influence of the ionic sulphonitrocarburising treatment on the tribological properties using the results of the Faville test and we shall compare the micrographic structure as well as the morphology of the layers obtained by the salt-bath and plasma methods.
In addition, we shall emphasize the ease of application and the versatility of this procedure, and we shall then conclude by presenting some examples of the industrial applications of sulphonitrocarburised pieces.

I - INTRODUCTION

In the field of general mechanics, problems relating to friction, seizing and simultaneous wear are often resolved by a thermochemical treatment of the nitriding type carried out in molten salt baths. For the last few years, constraints relating to environmental protection have also become more stringent. In order to take into account these new constraints and with a view to improving working conditions in the thermal treatment workshop as well as to limit the washing operations of the pieces after treatment, we have developed a non-polluting thermochemical procedure, carried out under vacuum with the aid of plasma.

II- EXPERIMENTAL SET-UP
II - 1 The ionic sulfonitrocarburising furnace

The operation of the ionic sulphonitrocarburising procedure requires the use of a plasma vacuum furnace. This equipment was constructed by BMI, the French manufacturer of industrial vacuum furnaces (Fig. 1) and is made up of the following elements :

- a vacuum chamber (ø 500mm, H 500mm) in which the pieces to be treated may be suspended or placed on a semi-soft steel plate. This chamber (BMI type VI63) contains three interior stainless steel protective screens.
- a current and voltage-variable electrical power supply system equipped with a system permitting rapid exchange against arcing. This system allows creating a luminescent discharge between a cathode (lower platen where the pieces are placed) and an anode (furnace walls). A 20kW generator produces a direct current pulse.
- a gas distribution system consisting of gas cylinders linked to three Brooks mass flow meters calibrated with nitrogen, and a programmable automaton (model KN8) which uses one or other of these flow meters depending on the percentage of gas required.
- a temperature regulation and measurement system which consists of two K-type (chromel-alumel) thermocouples placed directly within the pieces : a thermocouple which measures the temperature of the pieces and an "overheat" thermocouple, which acts as a system security in case the allowed temperature is exceeded (Tmax= 600°C).
- a pumping installation consisting of an Edwards vane pump having a maximum pumping rate of 40 m^3/h, allowing to obtain a vacuum of 0.3 mbar before discharge is established.

II - 2 The specimens treated

The specimens tested are round in shape, 40mm in diameter, 10mm thickness, with roughness index (Ra) equal to 0.8μm and two different grades of stainless steel (annealed reference state), i.e. 35CD4 and XC38 (table 1). Cylindrical XC38-grade steel samples in the annealed state, of diameter 6.35mm and length 36mm, were used to carry out the Faville friction tests.
The specimens and samples were carefully cleaned in trichloroethane before treatment.

II - 3 Characterisation methods

For each of the sulphonitrocarburising methods studied, we used four different analytical techniques, to show respectively the structure, morphology, nature and tribological properties of the sulphonitrocarburised layers.
Metallographic examination carried out with the aid of a Zeiss optical microscope was principally aimed at showing the presence and structure of the material before and after treatment. This observation was done after cutting, mounting under heat, polishing and chemically attacking the surface (Nital 3%), which allowed us to measure the thickness of the different layers formed.

Vickers hardness distribution allowed to measure:

- the hardness of the layers formed at the surface 100μm from the edge under a 50g load,
- the hardness at the heart of the piece,
- and to determine the conventional depth of nitriding measured under a 100g load for a hardness equal to that of the heart plus 100HV (Synecot DN1 norm).

Scanning electron microscopy observations using a Philips SEM 525 were carried out as well to show the structure and morphology of the nitrocarburised layers, as well as the appearance of the outer surface of the sulphonitrocarburised samples.

The microanalyser (Kevex diode) allows us to identify the elements present in a microvolume ($1\mu m^3$), or on a much larger surface. This identification is based on the measurement of the X-ray photons emitted by the matter.

Analyses carried out on the one hand by X-ray diffraction at normal incidence using cobalt $K\alpha$ ray, and on the other, by grazing angle X-ray diffraction using copper $K\alpha$ ray, allowed us to determine the nature of the chemical compounds and the stoechiochemistry of the iron nitrides and sulphides formed on the first few microns of the sulphonitrocarburised layers by the molten salt bath or ionic methods.

In the FAVILLE test (Fig. 7) the samples are sollicited by gliding under a huge load. The test consists of rotating a cylindrical sample 6.35mm in diameter between two jaws cut into a 90° V shape, at a constant speed of 330 rpm.

The distance separating the two approaching jaws is a linear function of time, the jaws crush the sample with an increasing force that acts during the rotation at a speed of 0.11m/s. The test is considered to be over when the load begins to decrease, signifying that the sample has attained a temperature that is sufficient so that under the so-called plastic flow load, its flow rate is greater than the speed at which the jaws approach each other.

The test is carried out dry and the result obtained is the mean of 5 tests. This test is easy to carry out and allows to rapidly establish comparisons between different types of treatments at minimum cost.

III - COMPARISON BETWEEN TWO SULPHONITROCARBURISING PROCESSES
III - 1 Sulfonitrocarburising in salt (or liquid) baths

We especially decided to study two processes of nitrocarburising in salt baths, the first (salt bath process n° 1) containing a small quantity of chemically-bonded sulphur in small quantity, and the second (salt bath process n° 2) devoid of sulphur [1].

The two processes are similar, both from the point of view of the chemical nature of their components, and of their utilisation temperature, the latter is comprised between 570°C and 580°C.

EDS microanalysis spectra (Fig. 4) obtained on an XC38-grade steel sulphonitrocarburised in salt baths according to processes 1 and 2 show the presence of oxygen issuing from oxidised species, and from nitrogen, iron and sulphur, it being impossible to detect carbon, given its low molecular weight (12 g.mol^{-1}).

Nascent nirogen and activated carbon formed during the two salt bath processes lead respectively to the formation of a superficial layer of ε ($Fe_{2-3}CN$) and γ' (Fe_4N) carbonitrides containing a microporous zone that is on the hand rich in sulphur and on the other, a dense and compact nitrogen-rich zone, the latter preceding the nitrogen diffusion layer.

The sulphur used in the first process is not a constituent element of the nitride layer but remains present and occluded in the microporosities of the superficial layer, promoting its growth and improving its tribological properties.

III - 2 Ionic sulphonitrocarburising

Ionic sulphonitrocaurburising is carried out at low pressure (0-10 mbar) in a plasma issuing from an "anormal" type luminescent electrical discharge. The principal function that we expect from plasma of a gaseous mixture containing nitrogen (N_2), methane (CH_4), hydrogen (H_2) and a sulphurising gas is the creation of chemically active species that allow to control the reactions susceptible of developing at the surfaces of the pieces to be treated.

In this medium that is totally out of equilibrium since it exists at low pressure, at a temperature comprised between 450 and 600°C, and partially ionised, it is very difficult to characterise the active species that depend on the nature, concentration and energetic distribution of the electrons of the discharge.

The EDS microanalysis spectra (Fig. 6) obtained on an XC-38 grade steel that has been ionically sulphonitrocarburised shows the presence of nitrogen and iron with oxygen not being detected as the ionic process is carried out under vacuum.

III - 2-1 Creation of the plasma state under low pressure

The plasma is obtained by establishing a potential difference between the pieces to treated which constitute the cathode of the electrical system and the metallic walls of the furnace which act as the anode.

The regime of the luminescent discharge exploited is of the "anormal" type [2].

The active ions formed and accelerated in the vicinity of the cathode bombard the pieces which are heated up by the kinetic energy liberated. A thermo-ionic and thermochemical reaction is then produced between the ionised gases and the substrate. The different gas molecules are dissociated and partially ionised, creating chemically active species and atoms that can react chemically and diffuse at the solid surface.

The temperature attained by the charge depends on the density of the applied power (expressed in W/cm^2) and that dissipated at the surface of the pieces placed in the furnace.

III - 2-2 The treatment cycle

The ionic sulphonitrocarburising cycle is composed of five sequences (Fig. 2) :

1st sequence : pumping down the chamber to 0,05 mbar.

2nd sequence : etching of the pieces by ionic bombardment (hydrogen plasma) between 250°C and 500°C to eliminate all traces of contaminant elements at the surface of the pieces to be treated.

3rd sequence: ionic nitriding under a gas mixture containing nitrogen, dilute sulphurising gas and methane, thus forming a single diffusion layer with little or no formation of white layer.

4th sequence: ionic sulphonitrocarburising under a gas mixture containing nitrogen, dilute sulphurising gas and methane, allowing to form a layer of iron nitrides doped in sulphur.

5th sequence: cooling under 900 mbar with a gas mixture of hydrogenated nitrogen stirred by a turbine so as to increase convection phenomena.

IV - COMPARATIVE RESULTS OF THE SALT BATH AND PLASMA TECHNIQUES
IV-1 Colour of the pieces - surface roughness

Whatever the grade of the steel that is sulphonitrocarburised by the salt bath or ionic ways the pieces are of a mouse-grey colour, intense and uniform, they are slightly pulverulent to the touch and their surface roughness is unchanged after treatment (Ra=0.8μm).

IV-2 Morphology compared between the layers formed by the salt bath and ionic processes

** salt bath method (Figs. 3 and 4)

For the given treatment parameters (tables 2 and 3), the surface of a sample (of XC38 or 35CD4 grade steel) sulphonitrocarburised by the salt bath method, successively presents, from the surface to the heart of the piece :

- a microporous combination layer of 25μm mean thickness,
- a nitrogen diffusion layer of about 0.2mm.

Since both nitrogen and carbon have diffusion coefficients that are different in the steel, the combination layer obtained (also called nitrocarburised layer) possesses a complex microstructure, with different layer thicknesses of ε ($Fe_{2-3}CN$) and γ' (Fe_4N).

The nitrocarburised layer is microporous [3]. We note from the SEM image (Fig. 4) that the pores formed have a small diameter and are arranged in columnar fashion over two-thirds (for the salt bath process n°1) and one-third (for the salt bath process n°2), of the nitrocarburised layer thickness.

The appearance of microporosities at the surface of the nitrocarburised layer can be explained by a chemical aggressivity of the salts of the two baths which are corrosive and lead to dissolution of the matrix iron. As for the internal porosity, it is probably due to recombination of nascent nitrogen (N) into diatomic nitrogen (N_2) [3].

The Vickers hardness measured at 100μm from the edge of the sample and under a 50g load is 623HV for XC38 steel and 689HV for 35CD4 steel.

** ionic method (Figs. 5 and 6)

For treatment parameters that are identical to those announced for the salt bath method, the surface of a sample (of XC38 or 35CD4 grade steel) sulphonitrocarburised by the ionic method successively presents from the surface to the heart of the piece:

- a dark and microporous sulphurised layer (FeS) of thickness comprised between 1 and 5μm,
- a clear and microporous nitrocarburised layer of 20μm mean thickness,
- a nitrogen diffusion layer of about 0.2mm.

On the SEM image (Fig. 6), we note that the number of pores formed by the ionic method in the nitrocarburised layer is much smaller and that their distribution is more dispersed than those formed by the salt bath method. In addition, they only occupy half the thickness of the nitrocarburised layer.

The compactness of this nitrocarburised layer influences its superficial Vickers hardness measured at 100μm from the edge of the sample and under a 50g load the latter is 756 HV for the XC38 steel and 800HV for the 35 CD4 one.

Analysis by X-ray diffraction at normal incidence (Fig.6) allowed us to determine the nature of the iron formed in the nitrocarburised layer, i.e. ε ($Fe_{2-3}CN$) and γ' (Fe_4N) and an analysis by grazing angle X-ray diffraction allowed us to :

- identify the iron sulphide formed as being FeS, showing that the sulphur has high affinity for iron and destabilises the nitrides formed,
- determine that the adjacent compact nitrocarburised layer is composed of γ' (Fe_4N)

  nitride and that the surface microporous layer is composed of the ϵ ($Fe_{2-3}CN$) nitride.

IV-3 Tribological test: the FAVILLE test

Thanks to the FAVILLE test, we were able to reveal a value expressed in kgf (or Newtons per second) characterising the friction of a tested material couple surface until appearance of plastic flow or seizing (table 3).

The results obtained show that overall, the ionic sulphonitrocarburising treatment confers better tribological behaviour to the pieces than the salt bath sulphonitrocarburising one. In fact, with the first treatment, values of 700 to 740kgf were obtained, while for the second, these were only about 600kgf.

V - INDUSTRIAL APPLICATIONS

Having realised the preparatory stage of the process, we allowed it up by developing this treatment to industrial applications (Fig. 8).
These are specified as follows :

- plastic mould pieces (movable arms of the interrupter, ejectors, mould carriages, tubulars, etc,...) in Z 38 CDV 5,
- construction pieces (sliding rails, clutch wheels, etc,...) in low carbon XC18-type steel,
- anti-wear pieces in Z 160 CDV 12 tool-steel,
- screws in 35 CD4 steel, and so on , ...

The principal advantages of the SULF IONIC ® process are as follows:

- good resistance to seizing owing to the presence of a real layer of microporous iron sulphide (FeS),
- little dimensional variation,
- possibility of carrying out this treatment within the 500-570°C temperature,
- clean treatment without washing or sandblasting,
- programmable automatic procedure with reproducible results.

V I - CONCLUSIONS

The ionic sulphonitrocarburising treatment (SULF IONIC ®) that has been developed is a sulphur-based thermochemical process which allows to superficially enrich steels of various grades and to improve their friction properties by diminishing their tendency to seizing.

In addition, this procedure uses plasma and vacuum techniques which offer easier workshop conditions and totally eliminate the washing operations of the pieces which are necessary in the case of salt bath treatments.

At present the SULF IONIC ® treatment is industrially applied on pilot production (VI 63 type BMI) equipments that have specially been conceived for work in the presence of sulphurising gas. In addition, this treatment can be carried out in large capacity vacuum furnaces while respecting the environment at the same time.

REFERENCES

[1] Liquid Nitriding, ASM Handbook 1991, Volume N°4, p. 410 - 419, Heat Treating

[2] Gaseous and Plasma Nitrocarburizing, ASM Handbook, Volume N°4, p. 425 - 436, Heat Treating

[3] Internal and External Nitriding and Nitrocarburizing of Iron and Iron - Based Alloy, Thesis 1989, M.A.J. Sommers

AKNOWLEGMENTS

We would like to thank the following HIT group companies (Bmi, Partiot, Vide Adour, Vide Express, and Vide et Thermochimie de Normandie) who financed and participated in the development of this procedure and in its industrialisation.

We also thank the ANVAR who partially finance this development and Mr Dusserre who carried out the metallographic analyses.

FIGURE N° 1 : Ionic sulphonitrocarburising equipment (typeVI63)

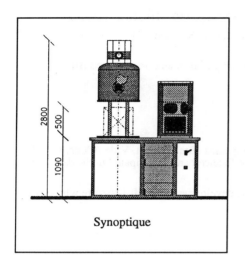

Synoptique

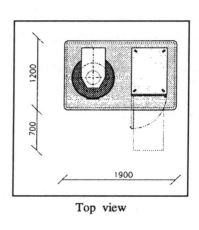

Top view

FIGURE N° 2 : Ionic sulphonitrocarburizing cycle (BMI type VI63 furnace)

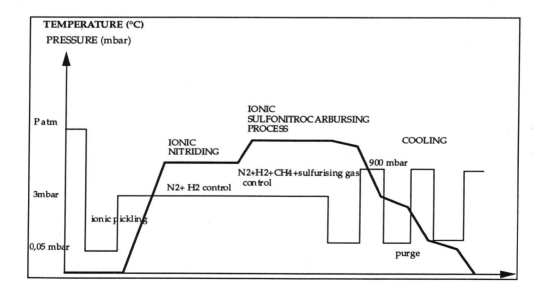

FIGURE N° 3 : Microstructures and hardness distribution of salt bath sulphonitrocarburised parts (salt bath processes n°1 and n°2)

a) XC 38 steel

b) 35CD4 steel

a)

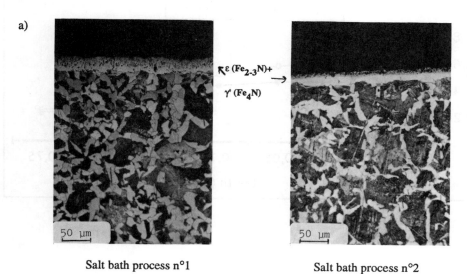

Salt bath process n°1   Salt bath process n°2

b)

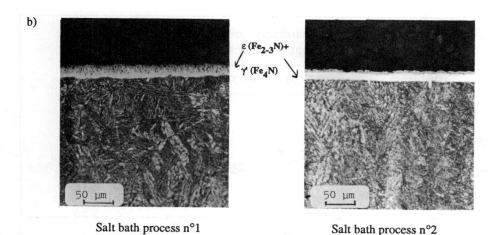

Salt bath process n°1   Salt bath process n°2

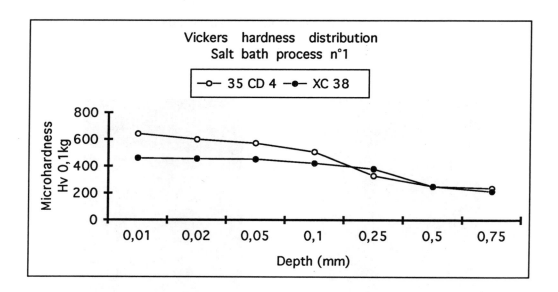

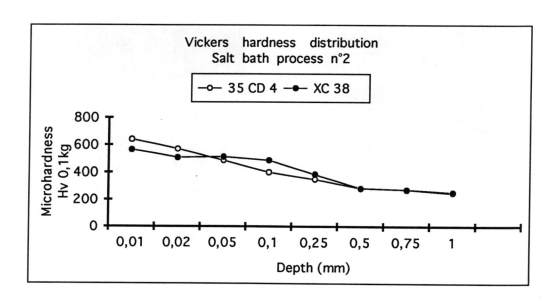

FIGURE N° 4 : SEM observations of the polished section and the surface of sulphonitrocarburised parts using salts bath processes (process n°1 on the left et n°2 on the right)

a) XC 38 steel - the polished section

b) XC 38 steel - the surface

a)

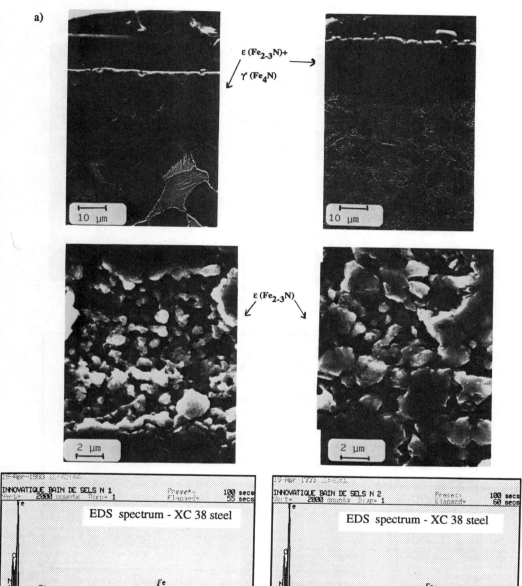

FIGURE N° 5 : Microstructures and hardness distribution of ionic sulphonitrocarburised parts
a) XC 38 steel
b) 35CD4 steel

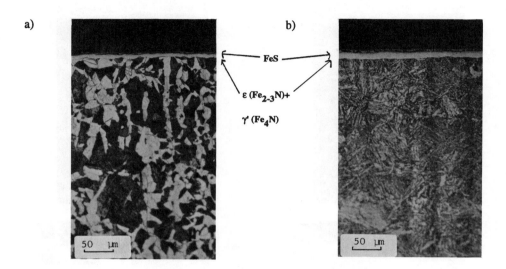

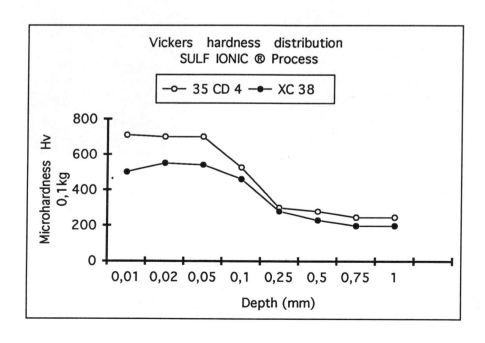

FIGURE N° 6 : SEM observations of the polished section and the surface of sulfonitrocarburised parts using the ionic process (process SULF IONIC ®)

a) XC 38 steel - the polished section

b) XC 38 steel - the surface

a)

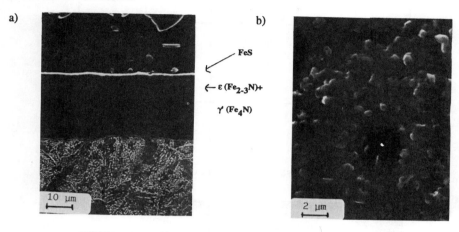

b)

FeS

← ε (Fe$_{2-3}$N)+

γ (Fe$_4$N)

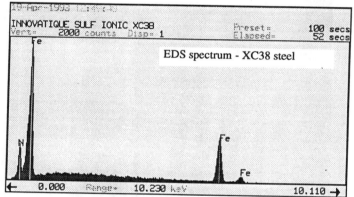

EDS spectrum - XC38 steel

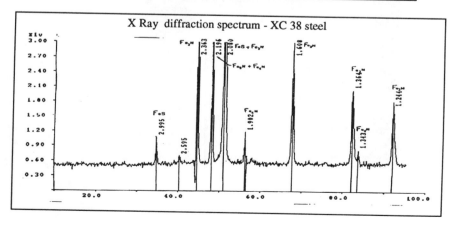

X Ray diffraction spectrum - XC 38 steel

FIGURE N° 7 : Principle of the FAVILLE test

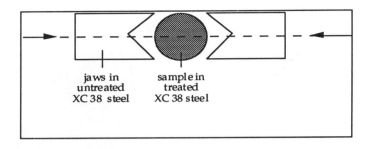

FIGURE N° 8 : Industrial application of the SULF IONIC ® process

TABLE N° 1 : Chemical composition of the steel grades used (% weight)

Steel grade	%C	%Si	%Mn	%P	%S	%Cr	%Mo	%Ni	%Cu	%Al	%Ti	% Sn
35CD4	0,36	0,25	0,75	0,02	0,026	1,1	0,21	0,2	0,15	0,032	0,003	0,008
XC 38	0,38	0,27	0,76	0,009	0,031	0,17	0,03	0,1	0,09	0,031	0,002	

TABLE N° 2 : Metallographic characterisation of sulphonitrocarburised layers prepared with salts bath or ionic processes for the following conditions of treatment : time : 3h, temperature : 570°C, pressure : 3 mbar.

Steel grade	Microporous white layer thickness (µm)	Nitriding depth (mm)	Surface hardness
35CD4 (1)	21-28	0,25	690 (HV 0,05 kg)
35CD4 (2)	18-23	0,22	670 (HV 0,015 kg)
XC 38 (1)	20-24	0,19	625 (HV 0,05 kg)
XC 38 (2)			

(1) Salt bath process n°1

(2) SULF IONIC®

TABLE N° 3 : Metallographic and tribological characterisations of sulphonitrocarburised 35CD4 et XC 38 steel parts with salt bath or ionic processes at 570°C

Steel grade	Time (h)	Microporous white layer thickness (µm)	Nitriding depth (mm)	Faville test (kgf)
35CD4 (1)	1h30	12-19	0,26	690 - 740
35CD4 (2)	1h30			600
35CD4 (3)	1h30	24		600
35CD4 (1)	3h	18 - 23	0,22	740
35CD4 (2)	3h	21-28	0,25	600
35CD4 (3)	3h	30		700
XC 38 (1)	1h30	10	0,26	670
XC 38 (2)	1h30	20-24	0,19	600
XC 38 (3)	1h30	22		600
XC 38 (1)	3h	9 -11	0,36	740
XC 38 (2)	3h			600
XC 38 (3)	3h	30		700

(1) SULF IONIC®

(2) Salt bath process n°1

(3) Salt bath process n°2

Materials Science Forum Vols. 163-165 (1994) pp. 309-314
© 1994 Trans Tech Publications, Switzerland

PRESENT STATE OF SALT BATH NITROCARBURIZING FOR TREATING AUTOMOTIVE COMPONENTS

G. Wahl

Leybold Durferrit GmbH, Hanau, Germany

ABSTRACT

Nitrocarburizing in salt melts is an economical heat treating process for improving the corrosion resistance, wear resistance, fatigue and rolling fatigue strength of work pieces. As there are a number of variants to the process it can be adapted to provide the improvements in component property required.The numerous and widely diversfied applications demonstrate its great popularity in all branches of industry.

Other developments such as high temperature nitrocarburizing or the recycling of nitriding salts and effluents can give added savings coupled with an environment-friendly process cycle.

INTRODUCTION

Salt bath nitrocarburizing plays an important role in the technical application of heat treating automotive components. In many cases salt bath nitrocarburizing is employed as an alternative to other surface layer processes such as case hardening or hard chrome plating with equally good or better results at greater economy.

Rising costs of energy and raw materials caused all branches of industry to look around for methods of reducing manufacturing costs. The process characteristics of salt bath nitrocarburiing makes it well able to meet this challenge because

- the treating temperature is low
- reproducibility of results is good
- operating costs are low
- and the process is not harmful to the environment.

A surface layer consisting of an outer compound layer and a subjacent diffusion layer is formed during nitrocarburizing. A number of component properties are changed by this nitride layer. Table 1 shows a list of major automobile parts which are being nitrocarburized by manu-facturers around the world in high volume production.

Table 1

Improved properties of TUFFTRIDE® treated parts						
Part	Material	Fatigue strength	Hot strength	Resistance to wear	Resistance to scuffing	Running behaviour
Crankshaft	SAE 1045	•	–	•	•	•
Inlet valve	X45CrSi9	•	–	•	•	•
Outlet valve	austenitic steel	•	–	•	•	•
Camshaft	chilled cast iron	–	–	•	•	•
Rocker arm	SAE 1045	–	–	•	•	–
Rocker shaft	unalloyed steel	–	–	•	•	•
Cylinder head	cast iron	–	•	•	–	–
Chain wheel	alloyed sinter. iron	–	–	•	–	–
Differential carrier	cast iron	•	–	•	•	–
Pinion shaft	SAE 8617	–	–	•	•	•

Looking at the Table we can see that a wide variety of materials ranging from un-alloyed medium carbon steel to high alloyed austenitic material, high speed steel, various qualities of cast iron and also sintered iron are salt bath nitrocarburized.

It is widely known that the compound layer produced by nitrocarburizing ist very resistant to abrasive wear and corrosion (1,2,3,4,5). Under combined operating conditions, such as wear and corrosion in particular, the durabilitiy of the compound layer shows its benefits. The diffusion layer improves the surface's resistance to pressure and strength under vibrational conditions.

Corrosion resistance can be further enhanced by an oxidative post treatment of the compound layer formed during nitrocarburizing.

NITROCARBURIZING AT HIGH TEMPERATURES

In addition to the oxidative post treatment, interest has grown in recent years for nitrocarburizing at higher temperatures. Higher treating temperatures produce much thicker surface layers at equal treating times or, based on the same layer thickness, enable treating times to be considerably reduced. In many cases the plant capacity and the economy can thereby be increased.

The use of various nitrocarburizing temperatures promises benefits both with regard to quality and economy. Nitrocarburizing at above 590°C and then reducing to 580°C gives improved properties such as wear and corrosion resistance and al so fatigue strength. Assuming that the two-stage nitrocarburizing enables the total treating time to be reduced, this gives a savings potential.

Fig.1

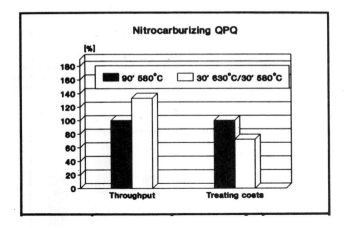

CHANGES IN COMPONENT PROPERTIES BY THE QPQ TREATMENT

CORROSION RESISTANCE

The most stringent corrosion test under DIN 50 021 conditions is the CASS test. Results obtained by a comparison between QPQ treated and hard chrome plated piston rods withs layers of 10 - 12 μm and 30 - 35 μm are given in Fig. 2. As can be seen, corrosion resistance after QPQ treatment is much better than after hard chrome plating. There was no difference in behaviour between the 10 - 12 μm layer and the 30 - 35 μm hard chrome layer. After being sprayed for 8 hours all chrome plated parts showed corrosive attack over the whole surface. Even after 16 hours however the QPQ nitrocarburized samples merely showed corrosive attack on 10 % of the surface.

Fig. 2

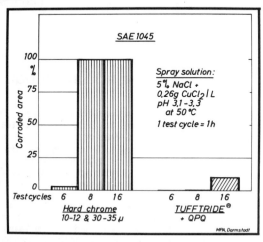

A number of Institutes and Technical Colleges have carried out investigations on the corrosion resistance of various types of surface layers. In the Thesis by Rauch galvanic, chemical and thermochemical surface engineering processes are compared with regard to their resistance. Nitrocarburized layers of unalloyed and low alloyed steels which were subsequently oxidized proved to be very good.

WEAR RESITANCE AND RUNNING PROPERTIES

The intermetallic structure of the compound layer reduced the friction and tendency to weld with a metallic counter partner. Nitrocarburized components are well known for their excellent sliding and running properties plus higher wear resistance. In comparisons with hard chrome plated layers the compound layer also shows its superiority. Everyday use has shown that nitrocarburized components such as transmission shafts, plus gauges, hydraulic aggregates etc. have a longer service life than hard chrome plated ones.

Fig. 3

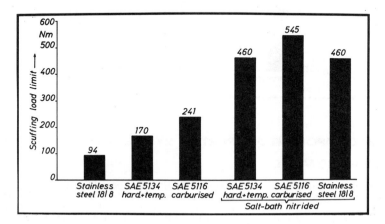

The suitability of a material or material pairing for given application depends on the wear they have to withstand. Salt bath nitrocarburized counter partners have proved to be very good against adhesive wear. Their tendency to scuff is much less pronounced than that of other surface layers. Fig. 3 shows the scuffing load limit of gears made from various materials (according to Niemann-Rettig).

INCREASING THE FATIGUE STRENGTHAND ROLLING FATIGUE STRENGTH

Increasing the fatigue strength of parts is also widely practised. To obtain the required component properties a treatment of 1 - 2 hours is usually sufficient.
An increase in fatigue strength of over 100 % was measured on unalloyed and low alloyed medium carbon steels after 90 minutes salt bath nitrocarburizing.
It is well documented in technical literature that, by nitrocarburizing at higher temperatures, an increase in fatigue strength compared with the normal treatment is obtained. Our investigations on notched rotating samples confirm this tendency.

The slower rate of cooling in the oxidative AB1 salt bath has an effect on the fatigue strength of unalloyed steels, as indicated in the two left-hand columns of Fig. 4. If parts are nitrocarburized at 630°C for 90 minutes the fatigue strength after salt bath cooling is similar to that of water-cooled samples after normal treatment. Depending on the treating time at 630°C, the values of samples treated by the two-stage process are also higher than those of samples treated normally for 90 minutes at 580°C and cooled in the oxidative salt bath.

Fig. 4

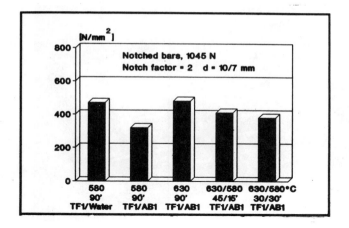

PRACTICAL APPLICATION OF SALT BATH NITROCARBURIZING

A major advantage of using nitrocarburizing processes is that the formation, structure and properties of the compound layer are only slightly influenced by the material being treated. This gives the design engineer the possibility of choosing material for wear parts from a manufacturing economic aspect. Due to the good wear behaviour of the nitride layer, timing gears of diesel engines which were formerly made from case hardened, hardened and tempered or cast materials are now nitrocarburized. Timing gears which are to be treated by this process are usually made from medium carbon steel C 45 in the normalized condition.

Gear wheels for oil pumps, crankshafts and chain wheels for camshafts as well as synchronizing rings are being nitrocarburized with great success. There is a growing tendency to manufacture such parts from sintered iron. Attention should be paid that the density after pressing and sintering is as high as possible of the sintered iron is below 7 g/cm³ the parts have to be specially cleaned in hot water and boiled in oil after treatment.

Heat treatment combinations, e.g. case hardening or induction hardening plus nitrocarburizing are also in common use. Fig.6 shows camshafts made for petrol and diesel engines. In most cases the treating time is 90 minutes at 580°C which gives the camshaft surface good sliding properties. The camshafts pictured are either made of chilled iron or tungsten arc remelted on the cams.

Fig.5

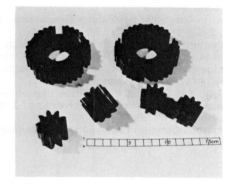

Fig. 6

Automobile manufacturers have been using the process for many years to increase the fatigue strength and running properties of crankshafts. Alloyed and unalloyed medium carbon steels, normalized or hardened and tempered are beeing employed.

Treating cylinder liners of diesel engines also has a beneficial effect. The compound layer formed during nitrocarburizing greatly improves running properties, wear and corrosion resistance. A treatment of 90 - 120 minutes duration is usually long enough to give the cylinder liners the required service life.

Reducing the wear on valves is one example of how the nitrogen retains its effectiveness even at high temperatures. Wear on the shaft is, of course, also reduced. In many cases valve shafts are totally or partially hard chrome plated to protect them against corrosion or wear. If these parts also have to withstand vibrational load it should be borne in mind that the strength is considerably reduced due to the very unfavourable effect of the chrome process (6, 7, 8).

REFERENCES

1. J. Müller,   Das Nitrieren von Eisenlegierungen und seine Auswirkung auf den Verschleißwiderstand, HTM 9 (1954, Heft 3, Seite 25-44
2. B. Finnern,   Entwicklung und praktische Anwendung des TENIFER-Verfahrens, ZwF 70(1975), Heft 12, Seite 659-664
3. G. Wahl,   Salzbadnitrieren EinwirtschaftlichesWärmebehandlungsverfahren mit vielseitigen Anwendungsmöglichkeiten, ZwF 75 (1980), Heft 9, Seite 414-418
4. H. Rettig,   Die kritische Grenzlast bei Zahnrädern, Maschinen-Markt 75 (1969), Heft 80, Seite 183-189
5. H. Kunst,   Eigenschaften von Proben und Bauteilen nach Badnitrieren und Abkühlen in einem Salzbad, HTM 33 (1978), Heft 1, Seite 21-27
6. M. Ohsawa   Der heutige Stand des Nitrierens im Fahrzeugbau in Japan, HTM 34 (1979), Heft 2, Seite 64-71
7. G. Wahl   Anwendung der Salzbadnitrocarburierung bei kombinierter Verschleiß- und Korrosionsbeanspruchung, ZwF 77 (1982), Heft 10, Seite 501 - 506
8. G. Wahl   Salzbadnitrocarburieren nach dem QPQ-Verfahren, VDI-Z 126 (1984), Heft 21, S. 811-818

VI.   BORIDING

Materials Science Forum Vols. 163-165 (1994) pp. 317-322
© 1994 Trans Tech Publications, Switzerland

THE DIFFUSION BORONIZING PROCESS OF REACTIVE ATMOSPHERES CONTAINING BORON FLUORIDES

B. Formanek

Silesian Technical University,
ul. Krasinskiego 8b, P.O. Box 221, 40-019 Katowice, Poland

ABSTRACT

In the article present state of the research development and gas technologies of the diffusion boride coatings production has been presented. Thermodynamically analysis of reactions according to the conception and model of boronizing process was presented. Analysis of the reactions: exchange, thermal decomposition and disproportionation in the chamber among boron fluorides and surface of treated steel were presented. The conception of coating of the products with the paste containing boron (the slurry technique) and their diffusion heating in the basis for the industrial technology using the standard furnaces with either protective or vacuum atmosphere. Technological example of boronizing process application was presented for tool steel used for matrix. The developed process of diffusion boronizing and heat treatment of steel parts hardening and are conducted in vacuum furnaces with forced nitrogen circulation.

INTRODUCTION

In the literature date there is some information on some methods of the diffusion boronizing, which differ in the technological conception and determine properties of boride coatings, obtained by them [1,2]. The contact-gas methods of boronizing in the powder mixtures containing the boron carbide is the most widely used in the laboratory and industrial conditions [2,3,4]. Works are being conducted on the application of the atmospheres containing reactive boron compounds to develop economically and technologically effective gaseous technology of the diffusion boronizing [5,6,7]. Covering of products with the special boronizing pastes and then their diffusion heating in the protective or reactive atmospheres seems to be a perspective method [8,9,10]. The research development and characteristics of the diffusion boronizing methods, used in Poland, are presented in other articles [11,12]. An analisysis of the literature shows that the gaseous methods are perspective for the diffusion boronizing processes.

TECHNOLOGICAL CONCEPTION OF THE GASEOUS METHOD OF BORONIZING SCOPE AND EXPERIMENTAL METHODS

A rage of experiments presented in the article comprises:
— development of the conception assumption of boride coating formation by the CVD method;

— thermodynamics analysis of reactions taking place during the diffusion boronizing process;
— selection of technological conditions and parameters of obtaining the coatings;
— determination of the diffusion coating obtained.

The boronizing processes were conducted on the Armco iron samples and carbon and tool steels with the following chemical compositions:

- A - Armco steel;
- B - 0.39% C, 0.64% Mn, 0.25% Si, 0.89% Cr, 0.25% Ni, 0.033% P, 0.031% S;
- C - 0.80% C, 0.22% Mn, 0.26% Si, 0.13% Cr, 0.15% Ni, 0.022% P, 0.023% S;
- D - 0.40% C, 0.49% Mn, 0.85% Si, 1.35% Mo, 1.05% V, 4.92% Cr, 0.32% Ni, 0.025% P, 0.026% S;

On the basis of literature data and our own experience a conception of coating formation, which in the first stage was based on covering the samples surface with the paste containing the boron and then diffusion heating was selected. It was assumed that obtaining an appropriate chemical composition of the boronizing atmosphere, directly at the surface of the treated samples allows to obtain boride coatings with an appropriate structure. It should be taken into consideration that in spite of the advantage of the local boronizing in the pastes, its strong adhesion to the treated object surface often takes place, what should be eliminated.

In the performed thermodynamic analysis of reactions which occur during boronizing process, it was assumed their thermodynamic potential would be defined according to the equation (1)

$$\Delta G_T^\circ = \Delta H_{298}^\circ - T\Delta S_{298}^\circ \tag{1}$$

allowing that for the systems in question the difference among the absolute values for ΔH_T° and ΔS_T° is small, and conditions of the performed processes are difficult to be determined precisely. It has been assumed that the diffusion coatings from the gaseous phase can be obtained in the reactions belonging to the following groups:

— the thermal decomposition

$$MeX_{2n} \leftrightarrow Me + nX_2 \tag{2}$$

— the interaction with the substrate:
- as the result of the exchange reaction:

$$Me_1X + Me_2 \leftrightarrow Me_1 + Me_2X \tag{3}$$

- as result of the formation of the phase with the substrate:

$$Me_1X_{2n} + Me_2 \leftrightarrow [Me_1 + Me_2] + nX_2 \tag{4}$$

— reduction by the hydrogen:

$$MeX_{2n} + nH_2 \leftrightarrow Me + 2nHX \tag{5}$$

— the disproportionation reactions:

$$2MeX_2 \leftrightarrow Me + MeX_4 \tag{6}$$

$$3MeX_2 \leftrightarrow 3Me + 2X_3 \tag{7}$$

The following physical reaction models occurring in the systems were assumed:
B-Fe; BF_3-H_2-Fe; BF_3-H_2-B-Fe; BF_2-H_2-Fe; BF_2-H_2-B-Fe; BF-H_2-Fe; BF-H_2-B-Fe.

For the process simulation of the reactive atmospheres, the contact-gaseous method in the powders mixtures, containing 10-40 wt% KBF_4 and the inert alumina powders, was selected. After the simulation tests the boronizing processes were made in the furnaces for CVD processes.

THERMODYNAMIC ANALYSIS OF REACTIONS DURING THE DIFFUSION BORONIZING PROCESSES

It is evident that in order to initiate the boronizing process it is necessary to create the possible high concentration of the BF or BF_2 boron fluorides in the mixtures of paste components containing the boron. It the performed model tests, in the powders mixtures containing KBF_4 or $NaBF_4$, BF_3 boron fluoride is a direct product of the fluoroborate decomposition and lower fluorides are the result of the BF_3 with the result of the BF_3 with the boron carrier. In the case of the direct inclusion of the

BF$_3$ fluoride into the reaction zone of the chamber the variants of the models: B-BF$_3$-Fe and B-BF$_3$-H$_2$-Fe are simpler.

For the analysis of the positive values of thermodynamic potential of some reactions:

$$BF_3 + 3Fe \rightarrow Fe_2B + FeF_3 \tag{8}$$
$$BF_3 + 3.5Fe \rightarrow Fe_2B + 1.5FeF_2 \tag{9}$$
$$BF_3 + 2Fe \rightarrow FeB + FeF_3 \tag{10}$$
$$BF_3 + 2.5Fe \rightarrow FeB + 1.5FeF_2 \tag{11}$$

it results that the boronizing process is not thermodynamically possible when the BF$_3$ fluoride is the boron carrier in the atmospheres. Even in the case of the hydrogen presence in the reactive atmospheres the probability of the boron coating formation is small, because ΔG_T° thermodynamic potential of the reactions in question within the temperature range of 873-1573K is higher than zero (Fig.1).

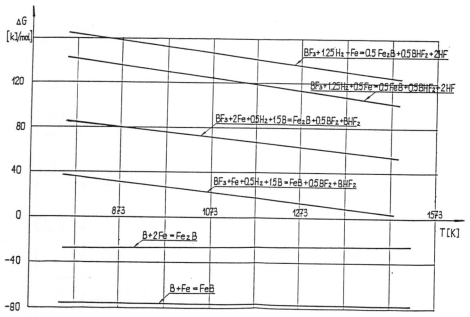

Fig.1 Diagram of the dependence of the thermodynamic potential on the temperature for the reactions taking place in the systems: B-BF$_3$-Fe and B-BF$_3$-H$_2$-Fe.

As it is shown by the calculation, not presented in the article, the partial pressures of particular fluorides change with the increase in the temperature - in turn changing thermodynamic and kinetic of the process [11]. It can be stated that the partial pressure of the BF$_2$ fluorides has a decisive influence on the intensity of the boronizing process. If the changes in the pressure of the boronizing processes are small, it can be assumed that the kinetic conditions of the processes have a great influence on the diffusion boronizing processes. Results of the analysis, presented in Fig.2, with the participation of the boron fluorides and paste allowed to determine technical parameters of the boronizing process. The assumed parameters of the process simulated well and were in accordance with the real conditions of processes occurring in the retorts of typical furnaces for the CVD process.

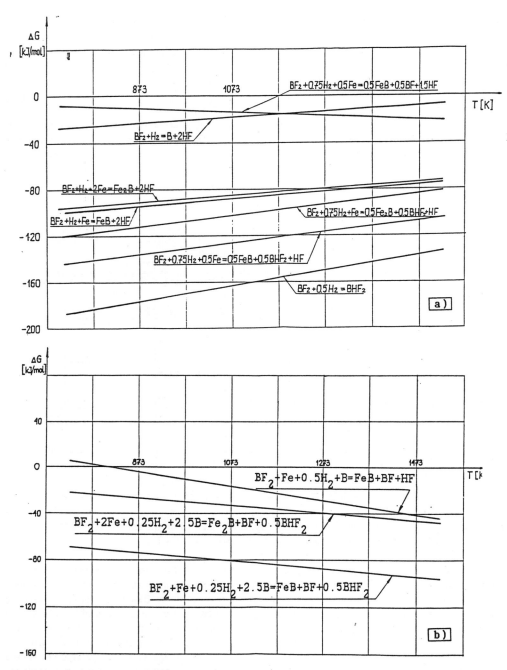

Fig.2 Diagram of the dependence of the thermodynamic potential on the temperature for the reactions taking place in the systems: a). BF_2-H_2 , BF_2-H_2-Fe b). BF_2-H_2-B-Fe.

EXPERIMENTAL VERIFICATION OF THE TECHNOLOGICAL CONCEPTION

It was appropriate to verify the assumed technological conception of the process using the contact-gaseous methods. The reactive atmosphere composition in the container during the boronizing process is difficult to determine and it has not been investigated so far. Good structures of the iron boride coatings were obtained at the 1 mm thickness of the paste containing the boron and chemical bounds. Typical one and two-phase structures of some boride coatings are presented in Fig.3. The obtained structures of the diffusion coatings confirmed validity the technological concept of the developed boronizing process.

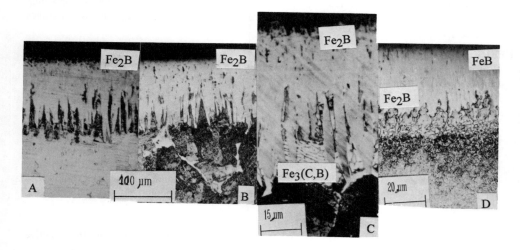

Fig3. Microstructures of boride coatings obtained on the steels with A, B, C, D chemical
compositions.

Taking into consideration the technological process of the boride coating formation in the industrial furnaces for CVD processes the following rules should be obeyed:
— the surfaces of machine parts and tools must be covered uniformly with the paste containing the boron or it compounds;
— the diffusion process should be performed at the low flow of the reactive atmosphere or their constant pressure - e.g. BF_3-H_2-Ar or BF_3H_2-N_2;
— in order to avoid to surface oxidation of the boronized products the retort should be cooled in the protective atmosphere e.g. N_2.

CONCLUSIONS

On the basis of the literature analysis, thermodynamic calculations and experiments the following conclusions can be formulated:

1. The physico-chemical model of the diffusion boronizing process in the reactive atmospheres containing the boron fluorides and the boron in the paste was developed and confirmed experimentally.

2. On the basis of the thermodynamic analysis of the following reactions taking place in the container: the exchange reactions, thermal distribution and disproportionation between the boron fluorides and the surface of the treated steels, it was shown that during the boride coating formation process the transport of the boron to the surface was made by disproportionation

reaction:

$$\frac{n}{2}BF_2 + \frac{nx}{by}Me \rightarrow \frac{n}{2}BF_3 + \frac{n}{by}Me_xB_y \tag{12}$$

$$BF + \frac{2x}{y}Me \rightarrow BF_3 + \frac{2}{y}Me_xB_y \tag{13}$$

3. From the thermodynamic analysis of reactions between BF_3 hydrogen and the surfaces of the treated products, it results that probability of the diffusion boride coatings formation is low. The chemical reactions containing BF_2 and BF give a great probability of the boride coating growth.

The developed method of the boride coating formation in the reactive atmospheres containing metastabile boron fluorides is a technological progress in the diffusion boronizing technologies developed so far.

REFERENCES

1. Worosznin L.G., Lachowicz L.S.: Borirowanie stali, Metallurgia, Moskow, 1978.
2. Formanek B., Swadźba L. : patent RP nr 146460.
3. Mareels S., Wettnick E.: Harter. Techn. Mitt. T.40 (1985), 4, 168.
4. Sommer P.: Harter. Techn. Mitt. T.41 (1986), 4, 194.
5. Hegewald P. and all:Gasborieren, Harter. Techn. Mitt. T.39 (1978), 3, 202.
6. Formanek B. : patent RP nr 146861.
7. Formanek B., Swadźba L., Olejnik J. : Applied vacuum furnaces for the diffusion boronizing and heat treatment steel tools. Proc. of the Conf. "Tool materials", Rydzyna (1985), 233 (in Polish).
8. Formanek B., Swadźba L., Maciejny A. : Diffusion boronizing and aluminizing processes in vacuum furnace. 3-ht Inter. Symp. Surface Modification Technologies, September 1989, Switzerland.
9. VAC HYD Process. patent FRG no 3432114.
10. Formanek B., Swadźba L. : Method of the obtained diffusion coatings on the reactive atmospheres including boron fluorides - patent applied RP.
11. Formanek B., Swadźba L. : Issliedowanije i charakteristika prinimajemych w Polsze tiechnologii difuzionnowo borirowanija. 12 Inter. National Conference "Heat Treatment" Wysokie Tatry, 1989, CSRS.
12. Formanek B., Swadźba L., Maciejny A.: Microstructure and erosion resistance diffusion modified and plasma spaying boride coatings, Proc. of the Conf. "Advances Materials and Technologies" Warszawa, 1992 (in Polish).

Materials Science Forum Vols. 163-165 (1994) pp. 323-328
© 1994 Trans Tech Publications, Switzerland

GAS BORIDING CONDITIONS FOR THE IRON BORIDES LAYERS FORMATION

A. Pertek

Technical University of Poznan, Instytut Technologii Budowy Maszyn,
ul. Piotrowo 3, 61-138 Poznan, Poland

INTRODUCTION

The boriding process enables to produce borides layers on various metallic materials. The borides layers created on the iron and steel have a range of valuable useful properties, such as the high hardness coming up to 2000HV, considerable wear resistance and high against the chemical and gaseous corrosion.

Among the different methods of boriding the most perspective ones are gas methods [1, 2, 3, 4, 5, 6, 7, 8, 9] , which are provided with process control.

The paper will contains study of thermodynamic calculations and experimental investigation, relating to the reactions which lead to the iron borides layers formation as well as the nucleation and growth of these phases.

EXPERIMENTAL AND RESULTS

The nucleation and iron boride growth during the borides layers creation have been examined.

Criteria of gaseous atmosphere and iron borides equilibrium have been established.

The examination of diffusion layers has been made using OM and X-ray method.

Boriding process have been carried on with BCl_3, H_2, Ar[5, 6]. The examinating of gaseous atmosphere and achieved borides of the equilibrium conditions has been done in different BCl_3 content atmospheres and temperature of 900°C.

There were 0,1 mm Fe-Armco foils used as the samples.

The examination methods were as follow:weighing,chemical analysis and X-ray.

Figure 1 shows that samples mass reduction become constant in 60 minutes and does not change in time.The sample mass reduction appears in atmosphere with no hydrogen or small amount of it. In samples with the mass growth,it reaches 16,9%.From chemical examination of boron in samples we learn that boron content is 16,1-16,9% independently of atmosphere content and sample mass. According to Fe-B phase diagram the content 16,23% of boron is for FeB phase.

The X-ray examination results are presented on the figure 2. It shows that in complete boronized samples which reached equilibrium condition to gaseous atmospheres with different mixture ratio BCl_3/H_2 only FeB phase has been found. /Nr5/.

The boride nucleation sequence in iron has been investigated It was found that in the begining short time period first FeB phase starts to nucleate only and next is followed by two borides FeB and Fe_2B. The X-ray diffractions patterns presented on figure 2 show that in temperature of 900°C after 2 second process no borides are found(Nr1),after 4 s process FeB is found(Nr2), after 8s and 5 minutes process FeB and Fe_2B are present(Nr3,4).After 2 hour boriding process FeB only could be found in the whole sample section(Nr5).

Figure 3 shows succesive stadium of iron foils boronized in temperature of 900°C in 5,20 and 120 minutes.

Figure 4 shows microstructures of boronized layers on Fe-Armco.

Boriding kinetics has been investigated in atmospheres with different BCl_3 content.It has been found out that boronized layer depth grows with the growth of BCl_3 content in gas and growth of temperature and time.Two phase borides layers of FeB and Fe_2B are always achieved even if the content of BCl_3 is the lowest(figure 4).

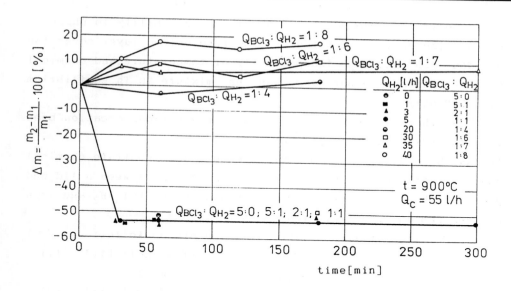

Fig.1.The results of the examination sample mass after boriding

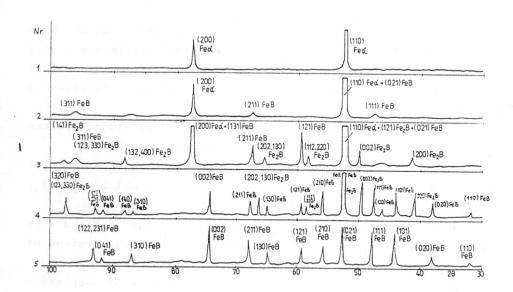

Fig.2.   The   X-ray phase analysis results after   boriding   in
         atmosphere $BCl_3/H_2=1/10$ and in temperature $900^\circ C$ and
         in time: 1 -2s, 2 -4s, 3 -8s, 4 -5min, 5 -2h

Experiments proved that there is no possibility to phase composition control of boronized layer. Boronized layer cover two borides FeB and Fe_2B independently of temperature and chemical content atmosphere. Boriding gaseous mixture is in equilibrium with FeB only.

The thermodynamic analysis of iron borides creation been made according to BCl_3-H_2-Fe system[10].

It has been defined temperature, pressure and substrate chemical content influence the equilibrium conditions between solid and gaseous phases. The eequlibrium condition of BCl_3-H_2 and iron borides FeB and Fe_2B were looked for.

Reaction of iron borides creation are as follow:

$$Fe + B = FeB \qquad \Delta G_T = -79610 + 10,5*T \ (J/mol) \ [11] \qquad (1)$$
$$\Delta G_{1200K} = -67,01 \ (kJ/mol)$$

$$2Fe + B = Fe_2B \qquad \Delta G_T = -87990 + 18,4*T \ (J/mol) \ [11] \qquad (2)$$
$$\Delta G_{1200K} = -65,91 \ (kJ/mol)$$

$$FeB + Fe = Fe_2B \qquad \Delta G_T = -8300 + 7,9*T \ (J/mol) \qquad (3)$$
$$\Delta G_{1200K} = 1,1 \ (kJ/mol)$$

From free enthalpy values examination we can learn that ΔG value for FeB boride is lower than for Fe_2B, what determines higher thermodynamic stability FeB. There is no possibilities of forming Fe_2B from FeB because the free enthalpy values of reactions 3 are positive.

From the data appears that gaseous boriding atmosphere can bee in thermodynamic equilibrium with FeB only.

Many chemical reaction can take place in gaseous mixture of BCl_3 and H_2. In result of which at iron presence HCl, $BHCl_2$, BCl_3, H_2 and $FeCl_2$ can be obtained[10].

For calculating the gaseous phase and borides FeB and Fe_2B equilibrium we make equations including law of conservation of mass, law of mass action and general pressure in system.

Figure 5 shows temperature (a), pressure (b) and substrate chemical content influence on equilibrium condition of gaseous atmosphere and iron borides. Figure shows that even with minimum $FeCl_2$ in substrate concentration (appx. 0,00005%) the equlibrium with boride FeB is obtainable.

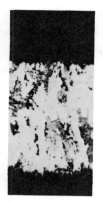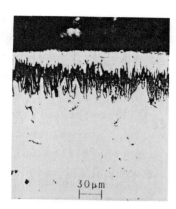

Fig.3.Microstructure of iron foils
boronized in atmosphere
$BCl_3/H_2 = 1/10$ and in temperature
$900°C$ and time:5,20,120 min

Fig.4.Microstructure of
boronized layer
$(FeB+Fe_2B)$ on
Fe-Armco

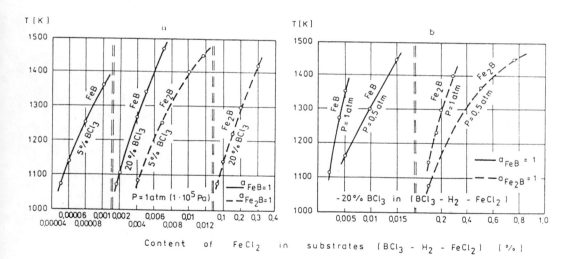

Content of $FeCl_2$ in substrates $(BCl_3 - H_2 - FeCl_2)$ [%]

Fig.5. Influence of temperature (a),pressure (b) and content of
BCl_3 and $FeCl_2$ in substrates on the conditions equlibrium
with FeB and Fe_2B

The equlibrium with Fe_2B boride is obtainable in higher concentration $FeCl_2$ in substrate (appx. 0,005%). In these conditions both borides exist. Increase of BCl_3 in substrates goes with the increase of $FeCl_2$ concentration in substrates.

CONCLUSIONS

From the carried out experiments can be learnt:

1. The phasial content of boronized layer cannot be controled in atmosphere from BCl_3 and H_2. Independently of chemical content of atmosphere the boronized layer consists of two iron borides: FeB and Fe_2B.

2. Gaseous boriding mixture is in equlibrium with FeB only.

3. As a first boride FeB nucleates and just after it FeB and Fe_2B

Experimental results have been confirmed by thermodynamic analysis which says:

1. Gaseous atmosphere achived from BCl_3 and H_2 independently of substrate chemical content, temperature and pressure in the system is in equlibrium with FeB.

2. $FeCl_2$ is probably responsible for no possibilities of boronized layers phase composition control.

REFERENCES

1. Eipeltauer, E. : Metallkundliche Berichte, Berlin, 1951, 12.

2. Toshio Katagiri, : Journ. Jap. Inst. of Metals, 1969, 33, 746-749.

3. Vorosnin, L. G. , Liachovic L. S. : Bororovanie stali. Matalurgia, Moskva, 1967.

4. Cochran, A. A. : Stephenson, J. B. : Metall. Trans. , 1970, 1, 2875-2880.

5. Pertek, A. : Doctoral thesis. Technical University of Poznań, 1980.

6. Pertek, A. , Przyłęcki, Z. : II International Conference "Carbides-Nitrides-Borides", Poznań-Kołobrzeg, 1981, 115-130.

7. Kunst, H. , Schaaber, O. : Härt. Techn. Mitt. , 1967, 22, 275-284.

8. Hegewaldt, F. , Şingheiser, L. : Härt. Techn. Mitt. , 1984, 39, 7-14.

9. Bartsch, K. , Wolf, E. : Zeit. für anorg. allg. chem. , 1979, 457, 31-37.

10. Pertek, A. , Przyłęcki, Z. : Archives of mechanical engineering technology. Polish Akademy of Science, Poznań, 1992, 141-153.

11. Turkdogan, E. T. : Physical Chemistry of High Temperature Technology. Akademic Press, New York, 1980.

Materials Science Forum Vols. 163-165 (1994) pp. 329-334

EROSION-CORROSION RESISTANCE OF BORONISED STEELS

L. Capan and B. Alnipak

Mechanical Engineering Department, Istanbul University, Turkey

ABSTRACT

Boronizing has, in addition to carburizing and nitriding, become the frequently applied diffusion technique in the heat-treatment of ferrous materials.

After borizing, one layer or two may be formed on the surface. These layers are very hard and confers great resistance both to corrosion and to abrasive wear.

Erosion-Corrosion is the first factor which caused damage and abrasion on the material surface. Several studies have been made and put forward the explanations of how the mechanism of corrosion and erosion might interact at the metal surface.

In this study, erosion- corrosion and abrasion resistance of mild steels were studied. Erosion-corrosion resistance of boronized steels are determined and compared with untreated materials. Some significant factors and their influence are also considered.

INTRODUCTION

Erosion-corrosion is the acceleration or increase in rate of deterioration or attach on metal because of relative movement between metal surface and corrosive environment. Result of chemical and electro-cemical reactions, abrasion and metal loss has been observed.There are several factors which effected, the degree of the corrosion rate and abrasionresistance. These factors are summarized as follows; chemical properties of corrosive, temperature, flow velocity of corrosive, metallurgical sturucture of metal surface,turbulance.

Boronizing results in some important improvements in the surface properties of steel. Among the advantages of boronizing, the in various engineering properties such as surface hardness and a high resistance to frictional wear, oxidation and corrosion resistance [1]. Borided layers have high hardness and low surface coefficient of friction, also makes a significant contribution in combating the main wear mechanism and borided parts have an increase fatique life and service performance under corrosive environments [2,3].

Boron has high chemical affinity for oxygen and all borides have a thin oxide film. This plays as a lubricant role on the metal surface[4].

The structure of borided layer depend on the boronizing type, boronizing temperature and time[5]. Boronizing results, in the formation of either a single-phase or double-phase layer of borides. The single-phase consist of Fe2B,which the double phase layer consist of FeB and FeB [6]. A double-phase structure contains sawtooth morphology and diffusion zone shown in Figure 1.

The boron rich FeB is more brittle than FeB layer. Erosion-corrosion of metals has extentsiveley been studied but the role of borided layer and relation with abrasion mechanism has reived considerably less attention.

EXPERIMENTAL STUDIES AND RESULTS

Erosion-corrosion measurements are made by using double cabinet connected with static corrosion test and rotating-disc apparatus as shown in Figure 2. The employed test liquids were tap water, sodium chloride and sulphricacid solutions. Used test materials and their chemical compositions are shown in Table 1. The geometry of a specimen was disc plate of 30by 40by 23 mm3 as shown in Figure 3. Specimens are boronized at 940 C for 5 hours. Addition of Ferro-silicon up to 23% into the borax bath gives the single-phase boride layer of Fe2B and addition of calcine boric acid in to the borax and ferro-silicon system up to 17% coused best results.

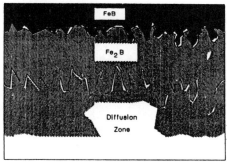

Figure 1: Shematic illustration of Double_phase layer of borides [2].

STEEL TYPE	Chemical Components (W%)			
	C	P	Mn	S
AISI 1020	0.18-0.23	0.04 (Max.)	0.60	0.05 (Max.)
AISI 1040	0.37-0.44	0.04 (Max.)	0.90	0.05 (Max.)
AISI 1050	0.48-0.55	0.90 (Max.)	0.90	0.05 (Max.)

TABLE 1.Chemical components of used steels.

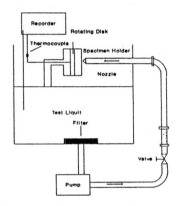

Figure 2: Erosion_Corrosion Test Apparatus

Secontd part of this work was mainly concerned with the search of the properties of boride layer and its underneath. To do this hardness measurements, optical metallography, x-ray micro-analysis were studied. The relation between mass loss and test duration (Figure 4.), effect of velocity on mass loss (Figure 5.), damaged area and velocity (Figure 6.), mean depth of pit and velocity (Figure 7.) are investigaded.

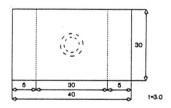

Figure 3. Shape of a test specimen .

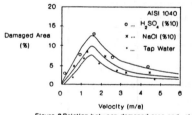

Figure 6:Relation between damaged area and velocity
(t=24 hours)

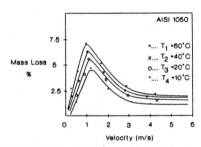

Figure 5: Mass loss velocity relation depend on the
temperature (t=24 hours)

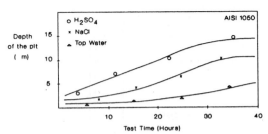

Figure 7: Mean dept of pit versus test time (T=56°C)

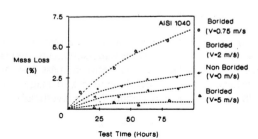

Figure 4 : Mass loss versus test duration

CONCULUSIONS

If we,take into consideration; How is the effect of boronizing on erosion-corrosion resistance of boronized steels at the different corrosive environments and relation with surface layer sturucture, and the mechanism of mass transfer from the surface, the following conclusions could be drawn.

a) Boronised layer layer consist of FeB and Fe2B phases (Figure 1.) The boron rich FeB phase is considered undersirable,because FeB is more brittle than Fe2B layer [3]. The other phase compositions dependt on the type of alloying elements.

   The transition zone structure and thickness of the zone effected by the diffusion rates of the alloying elements [4].

b) The combination of a high surface hardness and a low surface coefficient of the borided layer, which can considerably effect, the erosion-corrosion resistance of ferrouse materials and lubrication effect of thin boron oxide layer of the surface give an increase, the wear resistance of materials [4].

c) The sawtooth morphhology due to prefered diffusion of the alloying elements are formed under tensile and compressive residual stresses. This structure have an increased fatique life and service performance under corrosive enviroments.

d) Hardness of the boride layer can be retained at higher temperatures. Cause of this, the wear resistance of the surface is lower than, for example, that of nitrided cases [9].

e) Erosion-corrosion rate increase with corrosive flow velocity and temperature with the velocity in the low range and decrease with the velocity in the high range. Determination of the critical flow velocity is very important [7]. It has been shown same that the erosion-corrosion rate increase with corrosive concentration with in dilute solutions and decrease with indense solutions [10].

REFERENCES

1) Biddulp,R.H., "Boronizing for Erosion Resistance" Thin Solid Films. 45,(2),341-347,1977.

2) Ficht,W.,"Über Neue, Erkenntnısse Aufdem Gebiet des über flachen Borierens" Harterei-Technische Mitteilungen 29,1974. Heft 2 Juni, 113-119.

3) Mal,K.K.,and Tarkan,S.E, "Diffused Boron ups Hardness,wear Resistance of Metals",Materials Engineering,77,1973,70-71.

4) Bozkurt,N.," Hardening of steels wit Boron diffusion" Doctoral Theis , İ.T.Ü. Phy. Science and Tech. Inst. 1984.

5) Wahl, G.,"Borieren-ein Verfahren zur Erzeugung Harter Oberflachen bei Extremer Verschleissbean Spruchung" Sonderdruck aus VOI-Z117 1975,Nr.17. Seite 785/789,Degussa D076-1-1-188 TD

6) J.Postlethwaite and U.Lotz "Mass Transfer at Erosion-Corrosion Roughened Surfuces" The Canadian Journal of Che.Eng.Vol 64.Feb.1988,75-78

7) H.Nanjo,y:Kurata,D.Asano,N.Sanada and J.Ikenuchi "The Effects of Velocity and Pressure Vibration on Erosion-Corrosion of Mild Steel in Water" Corrosion. Vol 46,No.10,837-842,1990.

8) ASM Handbook,Volume4, Heat Treatment, 1991.

9) Guven,R., ALNIPAK,B.,"The Effect of Nitriding ,n Turbulent Corrosion Created by Particle Deposition in Turbulent Flows" ASM.Heat.Tre.Surf.Eng.Conf.Amsterdam.1991.

10. G.N. Blount,.R.T. Moule and W.J. Tom Linson "Environmental Aspects of Cavitalion Erosion in Simulated Industrial Waters" NACE. Corrosion April.1990. 340-347.

Materials Science Forum Vols. 163-165 (1994) pp. 335-340
© 1994 Trans Tech Publications, Switzerland

AN EFFECTIVE METHOD TO OBTAIN REASONABLY HARD SURFACE ON THE STEEL X210Cr12 (D3): BORIDING

Y. Özmen and A.C. Can

Pamukkale University, Engineering Faculty,
20017 Denizli, Turkey

ABSTRACT

In a variety of industrial area, hard face applications are mostly required. In order to obtain high surface hardness, there are many methods, namely plasma coating, PVD, CVD, thermochemical treatments, etc. Among the methods, boriding can be very effective and economic. This process is simply diffusion of boron into steel.

The most important aim of the present work is to obtain high surface hardness. In this study, X210Cr12 (D3) steel has been borided in four different boriding baths, and cooled in oil and air mediums. As a result, 90-100 μm boride layer, and 1600-2000 HV surface hardness have been obtained on the steel surface.

INTRODUCTION

Surface Engineering discipline, which makes possible the design and manufacture with a combination of bulk and surface properties, has been gained major industrial importance since few decades. Many new surface coating techniques have been developed such as, physical vapour deposition (PVD), chemical vapour deposition (CVD), laser chemical vapour deposition (LCVD), microwave plasma-enhanced chemical vapour deposition (MPECVD), pulsed laser deposition (PLD), electro-deposition (ED), etc. in addition to the traditional surface treatments such as, carbonising, nitriding and boriding, etc. (figure 1) [1-8]. The consequential cause for the increasing importance of this technology is that the destructive effects in most applications concentrate on the exterior of a component. Thereby the surface properties of a component should be varied intentionally from the core.

Each surface treatment has advantages and limitations, which must be considered for a specific application. For the selection of materials and treatments the substrate, the interface and the coating are the most important characteristic factors (figure 2).

The surface engineering treatment techniques together with the traditional surface treatments have a profound influence on a number of engineering properties such as, tribological (friction, wear, etc.), mechanical (hardness, modulus of elasticity, fatigue strength, etc.), chemical (corrosion, diffusion, etc.), thermal, magnetic, electrical, electronic and superconduction, optical, optoelectronic, aesthetic, etc.

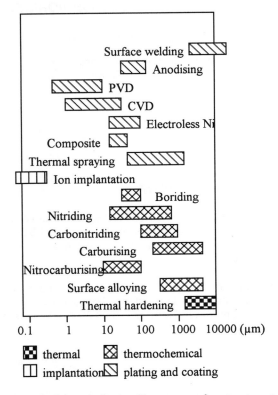

Figure 1. Thicknesses obtained by some surface treatments [1].

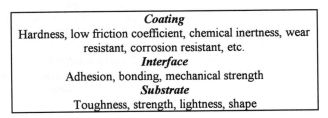

Figure 2. The properties required in a component that characterise the choice of materials and treatments

In this article we tried to investigate how the boriding, one of the thermo-chemical treatments, influences the surface properties of steels, especially of X210Cr12 (D3) steel.

THE STRUCTURE OBTAINED BY BORIDING

Boriding promotes the formation of very hard iron borides, mainly FeB and Fe_2B on the steel surface, respectively. The diffusion zone beneath the boride layer takes place. It is composed of the phases (Fe, M)B and (Fe, M)$_2$B, which are solid solutions derived from FeB and Fe_2B by partial substitution of irons with metallic atoms (M=Cr, Ni, Mn, etc.). Relative abundance of these phases depends on the composition of metallic alloys as well as on the boriding method. They also differ markedly in boron contents, 16.2 wt.% for FeB and 8.82 wt.% for Fe_2B.

The borides grow as columnar morphology in the treating low alloyed steels. This particular morphology improves the adherence of the layer to the matrix. An increase in the content alloying elements lowers the layer thickness and reduces their columnar structure [9-13].

The interatomic bonds in every borides are covalent bonds that based on the use of collective electrons, chiefly belong to iron atoms. Therefore, the borides are very hard and brittle [12].

EXPERIMENTAL DETAILS AND RESULTS

In the present work, cold work tool steel X210Cr12 (D3) has been borided in four different boriding baths at 980±10°C for five hours, and cooled in oil and air mediums (table 1).

Table 1. Boriding bath contents and cooling medium used in boriding process.

Specimen No	Boriding Baths	Cooling Medium
1	30% SiC + 70% $Na_2B_4O_7$	Oil
2	2.5% B_4C + 10%KBF_4 + 87.5% SiC	Oil
3	40% B_4C + 60% $Na_2B_4O_7$	Oil
4	10% B_2O_3 + 20% SiC + 70% $Na_2B_4O_7$	Oil
5	10% B_2O_3 + 20% SiC + 70% $Na_2B_4O_7$	Air

All of the boriding medium has become like paste. In order to prevent the vaporisation, boriding baths have been covered by graphite powders. After cooling, specimens have been sectioned and prepared for the metallurgical and micro hardness tests.

As a result of boriding 90-100µm boride layer and 1600-2000 HV(0.025) micro hardness have been obtained on the surface of X210Cr12 steel. Micro hardness curves and micro structures are seen in figure 3 and 4, respectively.

DISCUSSION

As can be known, X210Cr12 contains ~2.1%C and ~12%Cr. Numerous contributions, not all in agreement with each other, can be seen in the literature about the effects of carbon and chromium. In the boride layer of steels, there is a systematic increase in hardness and quantity of (Fe, Cr)B and hardness of (Fe, Cr)$_2$B phases with increasing chromium content, as can be seen in figure 5 and 6. Chromium, an element with an atomic number lower than iron, tends to insert itself preferentially and systematically in the (Fe, Cr)B phase and to spread from the matrix towards the surface.

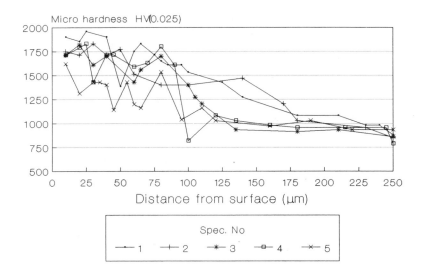

Figure 3. Micro hardness of borided X210Cr12 steel .

Figure 4. Micro structure of borided X210Cr12 steel(200x).

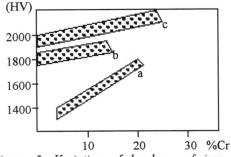

Figure 5. Variation of hardness of iron borides versus wt.%Cr; a:(Fe,Cr)$_2$B phase, b:Fe$_2$B phase, c:FeB phase [9].

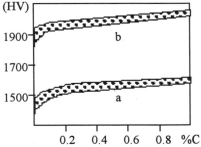

Figure 6. Variation of hardness of iron borides versus wt.%C; a:Fe$_2$B phase, b:FeB phase [9].

In most of literature for different conditions, the hardnesses obtained as a result of boriding have been given in table 2.

Table 2. Micro hardness values obtained by different researchers
as result of boriding of some steels

Contents of steels %C	%Cr	Micro hardness (HV)	Treatment	Reference
2.1	12	1600-2000	Liquid boriding	Present study
0.01	1.5	1500-1750	Powder boriding	11
2.1	12	1600-1800	Powder boriding	12
0.2	5.65	1500-1850	Powder boriding	14
0.45	-	1400-1800	Powder boriding	15
0.4	13	2000-2500	Powder boriding	16
0.45	-	1300-1900	Gas boriding	17

In agreement with the findings of some researchers, with increase in the carbon content in steels the thickness of boride layer decreases and the interface between Fe_2B and the matrix is more regular. The carbon tends to accumulate below the boride layer, forming a polyphase zone rich in carbides and borocarbides of types Fe_3C, $Cr_{23}C_6$ and Fe_7C_3 [9, 14, 15].

CONCLUSION

-As a result of boriding 1600-2000 HV(0.025) micro hardness, and 90-100µm boride layer have been obtained.

-All of the boriding medium have become like paste. It should be tried to developed to become liquid.

REFERENCES

1) Bell, T.: Metals and Materials, August 1991, 478
2) Bedford, G.M.: Metals and Materials, November 1990, 702.
3) Staines, A.M.: Metals and Materials, December 1985, 739.
4) Sankaran, V.: Advances in Materials Technology:Monitor, 1992, 24/25, 1.
5) Sheppard, K.G., et al: Processing of Advanced Materials, 1991, 1, 27.
6) Saenger, K.L.: Processing of Advanced Materials, 1993, 2, 1.
7) Shanov, V., et al: Processing of Advanced Materials, 1993, 3, 41.
8) McCune, R.C., et al: Surface and Coating Technology, 1992, 53, 189.
9) Badini, C., et al: Surface and Coating Technology, 1987, 30, 157.
10) Carbucicchio, G., et al: J. of Mat. Sci., 1982, 17, 3123.
11) Rosso, M., et al: The Future of Powder Metallurgy, P/M'86, Düsseldorf, 7-11 July 1986, Proceeding, Verlag Schmid, 381.
12) Özmen, Y.: M.Sc. Thesis, Dokuz Eylül University, İzmir, 1990.
13) Matuschka, G.V.: Borieren, Beratungsstelle für Stahlverwendung, Merkblatt, 1979, 446.
14) Carbucicchio, G., et al: Proceedings, Metals Soc. (Book 310), London, 3.44.
15) Épik, A.P., et al: Protective Coatings on Metals, Consultant Bureau, London, 1966, 2, 133.
16) Kotlyarenko, L.A., et al:Protective Coatings on Metals, Consultant Bureau, London, 1966,3,83.
17) Takeuchi, E., et al: Wear, 1979, 55, 121.

Materials Science Forum Vols. 163-165 (1994) pp. 341-346
© 1994 Trans Tech Publications, Switzerland

SUCCESSFUL BORONIZING OF NICKEL-BASED ALLOYS

H.J. Hunger and G. Trute

Elektroschmelzwerk Kempten GmbH, Germany

ABSTRACT

Surface hardening of steels with the aid of commercial EKabor boronizing agent is a thermochemical boron diffusion process that has been practised successfully for many years.

However, when the process is used for nickel-based alloys, the resultant layer is of sub-standard quality. Experiments and thermodynamic calculations of the reaction processes involved show that two competing processes occur during boronizing, namely boronizing and siliconizing.

The temperature at which treatment is carried out and the nickel content of the base material determine which of these processes is more likely to occur. The formation of nickel silicides is due to the composition of the boronizing powder. The authors show that perfect boride layers can be produced on nickel-based alloys if the boronizing powder does not contain silicon.

SUCCESSFUL BORONIZING OF NICKEL-BASED ALLOYS

Boronizing is a thermochemical surface-hardening process in which boron atoms diffuse into a metal surface to form metal borides. The hard boride layer thus formed has good resistance to abrasive and adhesive wear.

As yet, the only boronizing process to have established itself is the powder-pack method. Heating causes volatile boron compounds from the boronizing powder to diffuse into and react with the surface of the workpiece.

The following reactions occur when ferrous materials are boronized:

$$8 \ BF_3 \ + \ B_4C \ \ ---> \ 12 \ BF_2 \ + \ C$$
$$12 \ BF_2 \ + \ 8 \ Fe \ \ ---> \ 4 \ Fe_2B \ + \ 8 \ BF_3$$
$$12 \ BF_2 \ + \ 4 \ Fe \ \ ---> \ \ 4 \ FeB \ + \ 8 \ BF_3$$
$$2 \ BF_3 \ + \ 4 \ Fe \ \ ---> \ 2 \ Fe_2B \ + \ 3 \ F_2$$
$$2 \ BF_3 \ + \ 2 \ Fe \ \ ---> \ \ 2 \ FeB \ + \ 3 \ F_2$$

Analogeous reactions occur when nickel is boronized:

$$12 \ BF_2 \ + \ 8 \ Ni \ \ ---> \ 4 \ Ni_2B \ + \ 8 \ BF_3$$
$$12 \ BF_2 \ + \ 4 \ Ni \ \ ---> \ \ 4 \ NiB \ + \ 8 \ BF_3$$
$$2 \ BF_3 \ + \ 4 \ Ni \ \ ---> \ 2 \ Ni_2B \ + \ 3 \ F_2$$
$$2 \ BF_3 \ + \ 2 \ Ni \ \ ---> \ \ 2 \ NiB \ + \ 3 \ F_2$$

Boronized ferrous and nickel-based alloys are illustrated in Figs. 1 - 2.

Whereas the boride layer formed on steels is perfect, it is substandard on nickel-based alloys.

Electron microprobe analyses yielded the first clues to the cause of this phenomenon. As can be seen in Figures 3 - 6, extensive silicon enrichment has occured in the surface region. The results show that nickel silicides of the kind Ni_3Si and $NiSi$ are formed there during heating.

The composition of the boronizing powder and the thermodynamic conditions are responsible for this enrichment.

Conventional boronizing powders contain SiC by way of "inert diluent". However, the SiC apparently is not inert. The silicon compounds react with the fluoride in the powders to form gaseous silicon tetrafluoride. Possible reactions are as follows:

$$Si \ + \ 4 \ BF_3 \ \ ---> \ \ SiF_4 \ + \ 4 \ BF_2$$
$$3 \ SiO_2 \ + \ 4 \ BF_3 \ \ ---> \ 3 \ SiF_4 \ + \ 2 \ B_2O_3$$
$$SiC \ + \ 4 \ BF_3 \ \ ---> \ \ SiF_4 \ + \ 4 \ BF_2 + C$$

Silicon tetrafluoride enters into the following reactions at the surface of nickel-based alloys:

$$SiF_4 \ + \ Ni \ \ ---> \ \ NiSi \ + \ 2 \ F_2$$
$$SiF_4 \ + \ 3 \ Ni \ \ ---> \ Ni_3Si \ + \ 2 \ F_2$$

Hence, two competing reactions occur when nickel-based alloys are boronized, namely boronizing and siliconizing. Wether boride layers or boride-silicide mixed layers are formed depends on the thermodynamic conditions.

Experimental results obtained to date are shown in Table 1.

Siliconization of Ni containing materials during boronizing

Thermal treatment EKabor 1 / 850°C / 6h

Material	Ni %	Siliconization	
Nickel	100	yes	pure silicide
Hastelloy B 2.4800	65	yes	silicide-boride mixed layer
Incoloy 825	44	yes	" "
Alloy 20	32	yes	" "
X 5 NiCr 26 15 1.4944	26	yes	silicide-boride double layer
X 15 CrNiSi 25 20 1.4841	20	no	pure boride
X 5 CrNiMo 18 10 1.4401	10	no	pure boride

Thermodynamic calculations show that siliconization does not occur in ferrous alloys under normal boronizing conditions.

In nickel-based alloys, the probability of siliconization increases with rise in nickel content. If the nickel is pure, only siliconization occurs under normal boronizing conditions.

Since boronizing of nickel-based alloys constitutes a market with great potential, the question arises as to how siliconization can be avoided. We have solved this problem by developing a boronizing powder that contains no SiC. The boronizing results obtained with the aid of this powder are shown in figure 4.

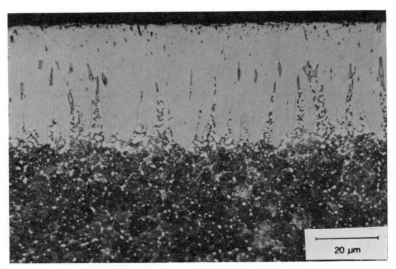

Fig. 1   Boride-layer on Steel 100 Cr 6

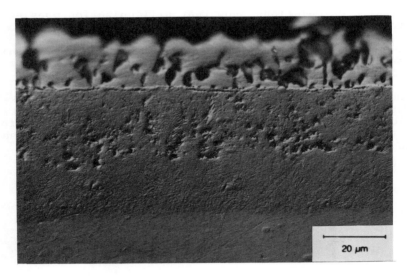

Fig. 2   Boride-silicide-layer on Hastelloy B

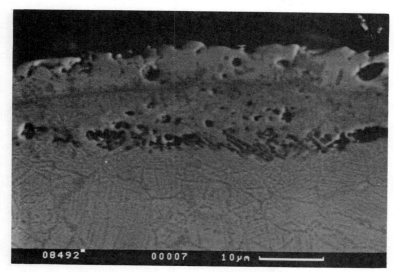

Fig. 3   Electron-microprobe-analysis of
          boride-silicide-layer on Hastelloy B
Fig. a   Secundaryelectrons

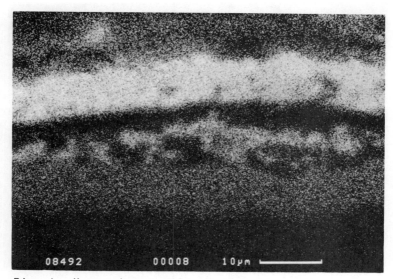

Fig. b X-ray-image silicon

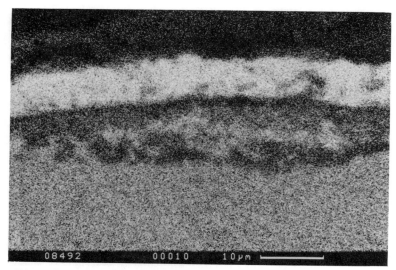

Fig. 3c   X-ray-image nickel

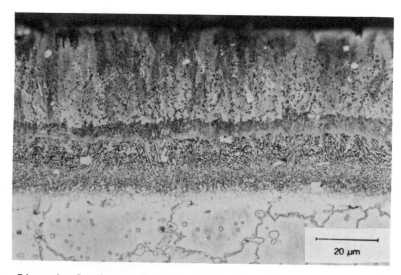

Fig. 4   Boride-layer on Hastelloy B
boronized with new type of
boronizing powder

VII.   INDUCTION AND FLAME HARDENING

Materials Science Forum Vols. 163-165 (1994) pp. 349-354
© 1994 Trans Tech Publications, Switzerland

SURFACE HARDENING OF STEELS AND CAST IRONS BY A DC PLASMA TORCH

H. Liao, Y. Wang and C. Coddet

LERMPS, Institut Polytechnique de Sevenans,
BP 449, F-90010 Belfort Cédex, France

ABSTRACT

The thermal transfer between a DC plasma and a substrate was studied for the understanding of the phenomena and was applied to the heating of steels and cast iron surfaces in order to induce phase transformations and subsequent hardening.

Different parameters: plasma power, speed of the torch relative to the substrate, nature and flow rates of the plasma gases, distance between the torch and the substrate, and cooling modes were considered. Materials selection included the steels 35CD4 (AFNOR), 20CD4 (AFNOR), 100C6 (AFNOR), C55 (E.N) and cast irons (grey iron). Hardening depths up to 1,5 mm and hardening widths up to 5 mm in one pass were obtained. Procedure, results and future prospects are described.

INTRODUCTION

The surface heat treatment of metals and alloys is an important technique in industry to obtain hard, wear and fatigue resistant surfaces. Salt bath, flame, induction, electron beam, laser beam and plasma jet are the possible heat sources to realize this type of treatment. [1],[2],[3]. Among these, the plasma jet has some particular advantages: the power density is rather high (about 80 kW/cm^2) and the parts may be easily cooled after the heating; moreover, it is not necessary to use a vacuum chamber as with the electron beam.

The present work was carried out in order to determine the feasability of the surface heat treatment of steels and cast irons with a DC plasma torch. The morphology of the quenched zone, the microhardness profiles and the thermal exchange properties were studied according to the following parameters:
a) plasma power, b) transferred arc power, c) speed of the torch relative to the substrate, d) nature and flow rate of the plasma gases,
e) distance between the torch and the substrate.[4].

EXPERIMENTAL

In general, a plasma torch consists of a cathode and a tubular anode, powered by a DC current generator. A secondary current generator may be connected between one electrode and the substrate to establish a transferred arc. (figure 1)

This work was carried out with a Plasma-Technik "F4" torch, a 80 kW current source for the plasma jet and a 9 kW secondary current source for the transferred arc. The parameters ranges considered in this work are shown in the table 1.

The chemical compositions of the substrates considered in this work are shown in the table 2. Of course, there must be, for the treated materials, an austenitic transformation during the heating up, and a transformation from the austenitic state to the martensitic one during the cooling.

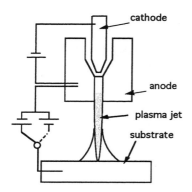

Figure 1: The plasma torch and the DC current sources

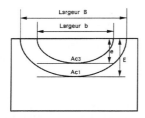

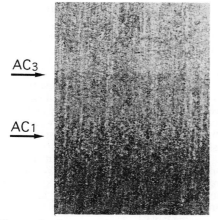

Figure 2: Aspect, width and depth of the quenched zone (1 cm= 150 μm)

Table 1: Parameters ranges

plasma power kW	10 - 55
torch velocity mm/s	10 - 50
stand off distance mm	30 - 60
plasma gas Ar l/min	30 - 50
auxiliary gas H_2 l/min	1- 6

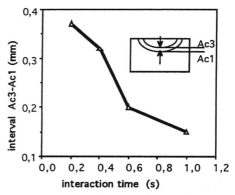

Figure 3: Evolution of the domain between Ac1 and Ac3 versus the interaction time.

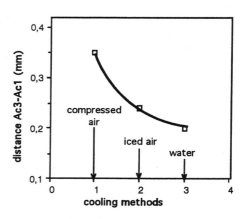

Figure 4: Interval domain between Ac1 and Ac3 versus the cooling rate (mode).

Table 2: compositions in weight per cent of the substrates (balance: Fe)

	C	Mn	Si	Ni	Cr	Mo	Cu	S	P	Ti
20CD4	0,15	0,86	0,28	0,14	0,84	0,20	0,18	0,010	0,014	
100C6	1,00	0,30	0,27	0,21	1,71	0,04	0,14	0,030	0,013	0,02
35CD4	0,36	0,77	0,28	0,16	0,96	0,28	0,10	0,010	0,019	
C55	0,52	0,60	0,28	0,05	0,04	0,05	0,52	0,017	0,020	
Cast Iron	3,25	0,75	1,92		0,02			0,07	0,25	

After being heated, the surface must be cooled rapidly to obtain the martensitic phase. If thick enough, the substrate can cool the heated zone by conductivity, it is the "selfquenching" process. Compressed air, iced air (temperature of about 2 °C) and water were also used in order to increase the cooling rate.

After treatment, the specimens were cut into small sections to observe the microstructure and to measure the microhardness.

RESULTS AND DISCUSSION

First of all, on a cross section of a specimen correctly treated, one observes two semi elliptic zones (figure 2) limited by two isotherms, respectively Ac3 (the end of the austenitic transformation) and Ac1 (the begining of the austenitic transformation). For a given steel, the distance between Ac3 and Ac1 is a function of the interaction time between the plasma jet and the substrate, of the cooling speed [5] and the type of arc (energy input).

The systematic study of the phenomenon on transverse sections of the specimens showed that:
- the shorter the interaction time, the larger the interval between Ac3 and Ac1 (figure 3).
- the higher the cooling speed of the substrate, the smaller the Ac3-Ac1 interval (figure 4).
- the use of a superimposed transferred arc decreases this interval (figure 5), thus showing a larger heat input into the substrate.

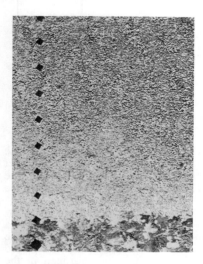

Figure 5: Aspect of the zone quenched with a superimposed transferred arc. (C55 steel) (1 cm = 100 μm)

The figure 6 illustrates the whole structure evolution from the surface to the inner for a 35CD4 specimen.

It can be observed successively:
- a martensite zone with large grains + residual austenite
- a martensite zone with small grains
- a mixed zone containing martensite + pearlite + residual austenite
- a ferrite+ pearlite zone (initial structure)

decarburization zone, martensite, residual austenite

Martensite, residual austenite

Martensite, little pearlite

Martensite, pearlite,

Ferrite, pearlite (initial structure)

Figure 6: Evolution of the quenching structure for a 35CD4 steel versus depth. (1 cm = 200 μm)

For a cast iron, a typical microstructure is shown in the figure 7; from the surface to the center, one observes, apart from the graphite lamellae :
- martensite + residual austenite
- martensite + residual austenite + bainite
- bainite + pearlite
- pearlite

The quenching depth of the cast iron is observed to be smaller than that of steels under the same treatment conditions because the graphite lamellae act as an obstacle to the heat transfer.[6]

Martensite,
residual
austenite,
graphite

Martensite,
residual
austenite,
bainite.
graphite

Bainite,
pearlite
graphite

Pearlite,
graphite

Figure 7: Evolution of the quenching structure for a grey cast iron versus depth. (1 cm = 70 μm)

Microhardness profiles were plotted following two directions on the transverse sections of the quenching zone:
- a direction perpendicular to the surface to determine accurately the quenching depth [7].
- a direction parallel to the surface to study the quenching homogeneity.

A typical microhardness profile such as the one on the figure 8 shows:

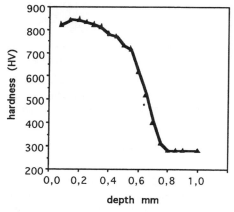

Figure 8: Microhardness profile of a quenched 35CD4 steel.

- a small surface zone where sometimes the hardness decreases probably due to an overheating or a light decarburization.
- a plateau where the hardness is rather high and depends on the treatment conditions.
- a transition zone where the hardness decreases abruptly; in the case of specimens treated with a superimposed transferred arc this last zone is very small and even not visible under an optical microscope (see figure 5 for the structure and 9 for the hardness profile).

According to these microhardness profiles, the influence of each parameter on the quenching quality for a given material can be evaluated. The figures from 10 to 13 show the effects of the electrical power (power density), of the scanning speed of the plasma torch, of the stand off distance and hydrogen flow rate on the hardness and quenching depth for different materials.

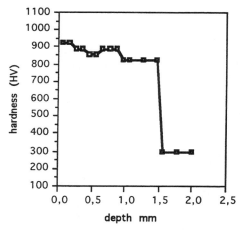

Figure 9: Microhardness profile of a C55 steel quenched with a superimposed transferred arc.

The volume of the quenched zone increases with the increase of the plasma power density (figure 10) and of the interaction time (figure 11) as well as with the decrease of the stand off distance (figure 12).

The plasma gas composition and flow rate also play an important role on the plasma power and energy transfer to the substrate. The figure 13 shows, for example, the role played by a small proportion of hydrogen in the plasma. This behaviour may be explained by the thermal conductivity characteristics and mass enthalpies of the different gases [8].

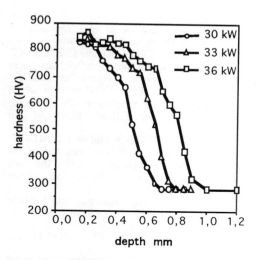

Figure 10: Evolution of the hardness profiles versus the plasma power density. (35CD4 steel)

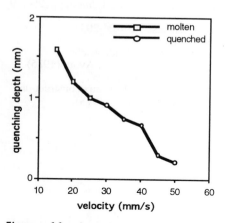

Figure 11: Quenching depth versus the interaction time (torch velocity). (I=600A, Ar=50 l/min, H2=4.5 l/min, 35CD4 steel)

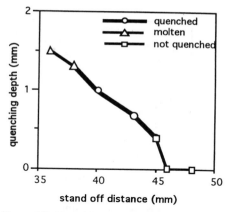

Figure 12: Quenching depth versus the stand off distance. (I=600A, Ar=50 l/min, H2=4,5 l/min, 35CD4 steel)

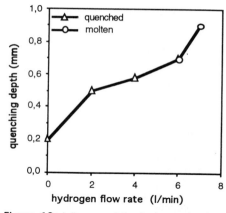

Figure 13: Influence of the hydrogen level on the quenching depth. (Ar 50=l/min, power=30 kW, distance=40 mm, torch velocity=30 mm/s 35CD4 steel))

Moreover, the cooling conditions also play a very important role. The selfquenching mechanism appears not very efficient unless the treated volume is very small as compared to the substrate volume. If the specimen is not very large, the substrate must be cooled with additional methods. Different tests were carried out (figure 14), the results of which show that from tested media, water is the best for increasing the hardness and quenching depth.

The substrate initial temperature can also influence the cooling speed, especially when the treatment is performed via continuous scanning. In this case, the supplementary cooling decreases the body temperature and increases the cooling speed and the hardness of the quenching zone[9].

By using a superimposed transferred arc during the treatment, the quenching depth was increased up to 1,5 mm, and the hardness is more homogeneous; the tolerance on the parameters is also larger.

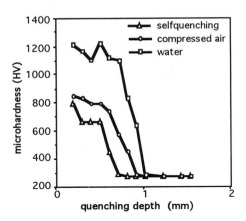

Figure 14: Influence of the cooling modes to quenching depth. (C55 steel)

CONCLUSION

This study showed that a plasma jet combined or not with a superimposed transferred arc can be used in the field of superficial heat treatment of steels and cast irons.

The quenching depth increases with the plasma power density, however with a limit which must not be overtaken, otherwise a melting of the surface will occur. The use of a transferred arc increases the quenching depth and supplementary cooling increases the hardness and quenching depth. The maximum depth was obtained in the case of the steel C55; for cast irons, the depth is a little smaller because the graphite lamellae act as a thermal barrier in the substrate.

In conclusion, this technique allows the surface heat treatment of materials with good performances in relatively simple conditions; no special preparation is required, the work can take place in the open air and complex shapes may be treated, locally if necessary.

REFERENCES

1) GUILLAIS J. C. and LEROUX C.
"Procédés électriques dans les traitements et revêtements de surface".
Ch. 6, pp 331-362, Ed. Dopée, Avon, (1985)
2) ZIMMERMANN M.
"Perspectives des traitements thermiques par faisceaux d'électrons"
Traitement thermique, N°187, pp15-25, (1984)

3) VANNES A. B.
"Lasers et industries de transformation"
Ed. C.A.S.T-INSA de LYON (1986)
4) WANG Y., ELKEDIM O. and CODDET C.
"Traitement thermique superficiel des aciers et des fontes à la torche à plasma d'arc soufflé"
"L'électricité dans l'élaboration et la transformation des matériaux" 7e Colloque Université Industrie. ESEM, Orléans, 4 juin 1992
5) FARIAS D., DENIS S. and POURPRIX Y.
"Relations between thermal calculation and métallurgical transformations during the laser surface hardening treatement in plain carbon steel and S.G.cast iron"
2nd International Seminar on Surface Engineering with high energy beams LISBON, Portugal, (1989)
6) WANG Y., LIAO H., ELKEDIM O. and CODDET C.
"Traitement combiné de trempe superficielle à la torche à plasma et de dépôt cermet Cr3C2-NiCr en vue de résoudre un problème d'usure.
ATTT 92, Internationaux de France du traitement thermique. PYC Edition, pp125-139, 1992
7) Norme française NF A04-203
8) LAROCHE G.
"Les plasmas dans l'industrie"
Ch.7, pp22-53, Ed. Dopée, Avon, (1985)
9) BAUS R. and CHAPEA W.
"Application du soudage aux constructions"
Ed. EYROLLES, Paris, (1978).

Materials Science Forum Vols. 163-165 (1994) pp. 355-366
© 1994 Trans Tech Publications, Switzerland

ON THE EFFECT OF AN ADDITIONAL ROLLING ON THE FATIGUE BEHAVIOUR OF SURFACE INDUCTION HARDENED COMPONENTS

P.K. Braisch

Institut für Werkstoffkunde, TH Darmstadt, Germany

A B S T R A C T

Surface strengthening is the appropriate way to increase the strength of components which present a linear or notlinear load profile in his different cross-sections. Many advantages in this domain presents surface induction hardening (IH).

The fatigue resistance of IH-components is determined by the basic strength of the material and increases with the strengthened depth, up to a critical depth, hyperbolical. A governing influence plays the residual stress state.

The question arises, if the fatigue resistance of under the viewpoint of fatigue behaviour optimal surface induction hardened components can further be increased by a cold plastic deformation of the induction hardened surface layer, for instance by an additional rolling.

The paper presents the results of a recent investigation concerning the mentioned question.

Appropriate surface induction hardened components achieve a rotating or alternating bending fatigue resistance of typical 900 MPa. This amount is valid in very closed limits as well for alloyed and unalloyed structural steels as for notched and unnotched specimens.

As the results of the concerned investigation show, is the effect of an additional rolling detrimental on the alternating bending fatigue resistance of appropriate surface induction hardened specimens.

This detrimental effect correlates with the change of the residual stress state on the crack initiation site.

Introduction

Surface strengthening is an appropriate way to increase the
strength of a component in his different cross-sections with a
linear or notlinear load profile. Many advantages presents surface
induction hardening (IH).

The fatigue resistance of IH-components is determined by the core
strength and increases hiperbolical with the strengthened depth,
up to a critical depth. At this depth, surface induction hardened
components achieve a rotating or alternating bending fatigue resistance of typical 900 MPa (see references no. 2 to 9 in /1/).

The question may be arise, if the fatigue resistance of under the
viewpoint of fatigue behaviour optimal surface induction hardened
components can be further increased, for instance by an additional
rolling.

Fundamentals

Firstly a theoretical consideration, similar as it has been done
for long time in other domains of surface strengthening /2/. The
profile of the local effective fatigue resistance R_{bf} in the cross
section of a surface strengthened component is interpreted as the
combined effect of the local intrinsic fatigue resistance R_{bf}^{+} and
the local residual stresses σ^{rs} (Fig. 1). The contribution of σ^{rs}
is considered by a proportional coefficient (m) /3/. At alterna-
ting load both the tension and the compression phase must be taken
into account (Fig. 1 shows the situation for the tension phase).

The profile of the effective fatigue resistance itself can be
defined by three basic figures (Fig. 2): the local fatigue
resistance on the surface $R_{bf,s}$, the local fatigue resistance on the
core border $R_{bf,c}$ and the strengthened depth SD respectively the to
the radius of the component related hardened depth r_f.

From the comparison of this profile with the load profile, both in
function of the related surface distance, results in an unequi-
vocal geometrical manner, how to choice the material and how to
prescribe the adequate SD (in Fig. 2 the load profile is
represented simplistic by a broken line; generally it is a
function of the notch factor K_t).

For the case of IH it is proved by extensive experimental investigations that $R_{bf,s}$ lies in the range of 880 - 920 MPa : for notched
and unnotched specimens, independently of the state of surface or
type of steel, but hardened under the consideration of the specific metallurgical requirements /4/.

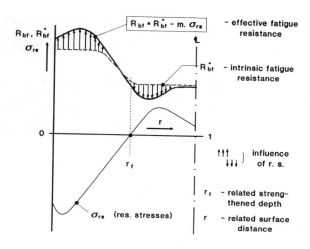

Fig. 1: Combined effect of local intrinsic fatigue resistance and residual stresses (at tensile load)

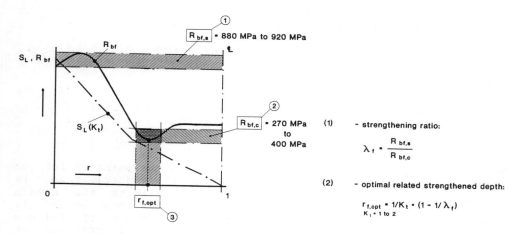

Fig. 2: Profile of the effective fatigue resistance; (1), (2) and (3) - basis rates

$R_{bf,c}$ depends from the choiced type of steel and his initial structure, which can be normalized, hardened and tempered or, at microalloyed steels, corresponding to a controlled cooling from forging
temperature.

The maximum attainable SD is a function of both the chemical
composition and the initial structure of the choiced material
(Fig. 3). The specific geometrical conditions of the part,
influencing the heat flux and the development of residual
stresses, determine additional limits for the attainable max. SD.

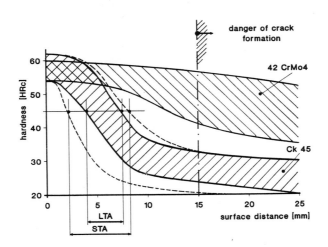

Fig. 3: Hardenability at long time austenitization (LTA)
-------- and at short time austenitization (STA)

The optimal related strengthened depth from a viewpoint of fatigue
behaviour, $r_{f,opt}$, can be deduced from eqs. (1) and (2). $r_{f,opt}$ is
in the same time a critical rate, because it delimits the domain
of failure due to crack initiation on the core border (:
subcritical strengthened components) from the domain of failure
due to crack initiation in the hardened layer (: critical or
overcritical strengthened domain) /1/. In function of the actual
geometrical situation (absolute value of the dia., notched or
unnotched specimen) and the hardenability of the considered steel,
it can be deduced, if in the considered case $r_{f,opt}$ respectively
the critical strengthened depth can be achieved or not (for
orientation use following figures: for low alloyed, hardened and
tempered steels λ = 2.2 respectively for unalloyed, normalized
steels λ = 3.2; from eq. (2) it then results that $r_{f,opt}$ (for K_t =
1) = 0.54 resp. 0.68 and $r_{f,opt}$ (for K_t = 2) = 0.27 resp. 0.34).

Additional surface strengthening

From Fig. 4 it becomes evident that the question of an additional
surface strengthening of IH-components can be posed rational only
in the case that all possibilities to increase the fatigue resistance mentioned above and symbolized by arrow (1), are exhausted.
So the adequate form of the question is: is there a possibility to
increase the fatigue resistance by raising $R_{bf,s}$ in excess of the
limits of 880 to 920 MPa, achievable by IH oneself (arrow (2)),
namely up to a second optimal related strengthened depth $r''_{f,opt}$.

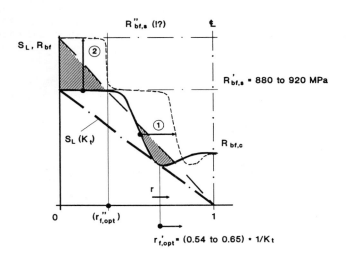

Fig. 4: Conditions for increasing oscilating bending
------- endurance beyond the limits achievable by IH

So the investigation has to start from the level achievable by IH
oneself of each of the two components of $R_{bf,s}$, the local intrinsic fatigue resistance $R*_{bf,s}$ (in the order of 1,000 to 1,300 MPa
/5/) and the local residual stresses σ_s^{rs} (in the order of 800 to
1,400 MPa; see references /5/ and /6/ quoted in /1/). Considering
the case of bending tests and taking into account the compression
phase, it is evident from a theoretical viewpoint and from
experiments /6/, that an increasing of $R_{bf,s}$ can be achieved only
via a significant increasing of $R*_{bf,s}$ beyond the mentioned limits.

In the following we present some results of a recent investigation
on the effect of an additional rolling on the fatigue behaviour of
IH-specimens.

Experimentals

Notched specimens (K_t = 2.01) of DIN grade steel 42 CrMo 4 (V), hardened and tempered (Rm = 950 - 1,000 MPa), then overcritical induction hardened but not tempered (in the notch area SD = 4.2 mm > $r_{f,opt}$ = 0.26*12.5 mm = 3.2 mm; in the shaft area SD = 7.0 mm > $r_{f,opt}$= 0.54*12.5 mm = 6.8 mm; for details of calculation see formulas in Fig. 2 and dimensions of the specimen in Fig. 5). These specimens, called in the following 'as-IH', were compared in alternating bending tests with as-IH-specimens rolled under conditions shown in Fig. 5, called in the following 'additional rolled' (details of the rolling effect see Figs. 6 to 10).

The used testing machine was an horizontal pulsator with mechani-cal resonant drive with electro-mechanical load controll (25 to 30 Hz; Carl Schenk, Darmstadt, 1952). The load was calibrated by means of strain gauges.

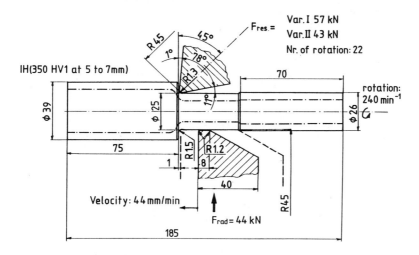

Fig. 5: Dimensions of specimen and rolling conditions

Results

The major result of the fatigue tests is the fact that the fracture plane of the additional rolled specimens lies at a distance of about 8 mm out of the notch bottom (Figs. 11 and 12). This behaviour contrast significantly to the failure of equivalent specimens in the as-IH-state, at which the fracture starts from the bottom of the notch (see ref. no. 2 in /1/).

The crack initiation site itself lies on the surface (Fig. 13). Based on Figs. 9 and 10, which show the results of X-ray investigations and reveal the change of the residual stress state due to rolling, it can be assumed that the depth of the layer influenced by the rolling operation, using the conditions of variant I, is about 0.3 to 0.5 mm.

Fig. 14 shows the fatigue behaviour in alternating bending tests of the additional rolled specimens (5) in comparison with that of specimens in the as-IH-state (4). The latter correlate satisfying with the fatigue behaviour in rotating bending test of specimens investigated in earlier works (1) to (3) (see quoted references in /1/).

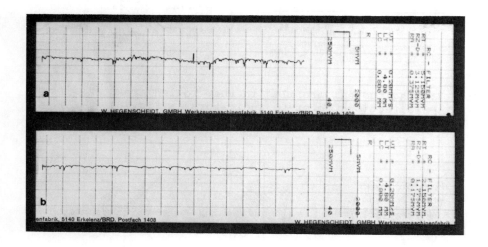

Fig. 6: Surface roughness in the shaft area in the as-IH-
------ state (a) and after an additional rolling (b)

From the distribution of number of cycles up to failure of the additional rolled specimens in comparison of that of as-IH specimens, it can be concluded that the endurance is significant lowered by the rolling operation. The amount of 824 MPa, given in Fig. 14 and corresponding to the actual load on the crack initiation site, leads to the conclusion that the endurance in the additional rolled shaft area lies in a range of 750 to 800 MPa. This result means that the effect of the additional rolling opera-

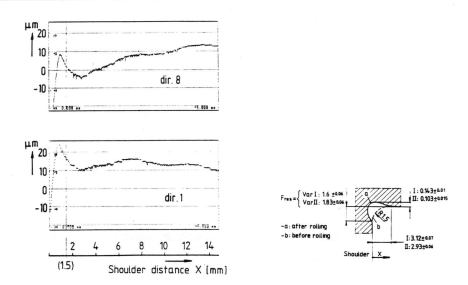

Fig. 7: Longitudinal profile in the rolled area
------- (directions refer to indicators in Fig. 8)

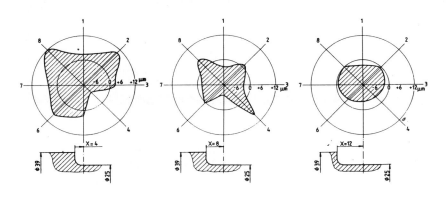

- residual stresses (ax) and half value breadth on the surface at shoulder distance "x"

distance x [mm]	2	3	6	7	8	9	20
r. s., dir. 1 [MPa]	–	–	–	–	-1140	–	-1230
r. s., dir. 8 [MPa]	-1125	-890	-1020	-1040	-1050	-915	-900
HVB, dir. 8 [°]	–	4.4	4.9	4.9	4.8	4.5	4.3

Fig. 8: Circumferencial profiles and surface residual stresses
------- at different shoulder distances in the rolled area

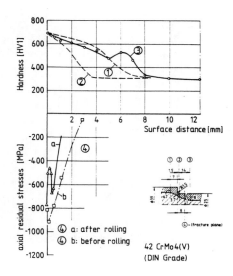

Fig. 9: Hardness profiles and residual stress profiles
------- near the fracture plane

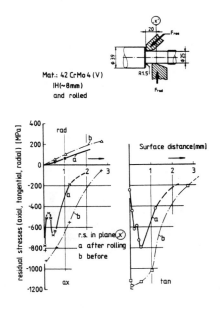

Fig. 10: Change of the residual stress state in the shaft
-------- area due to rolling

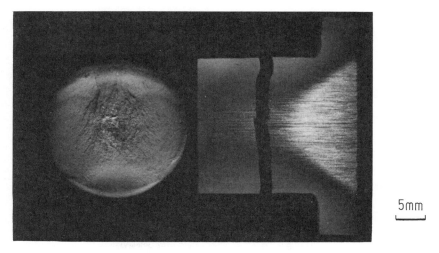

Fig. 11: Fracture plane and longitudinal section of the
-------- notched area

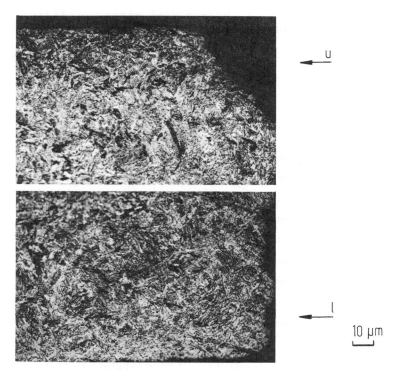

Fig. 12: Microstructure of the upper and lower edge
-------- of the fracture plane

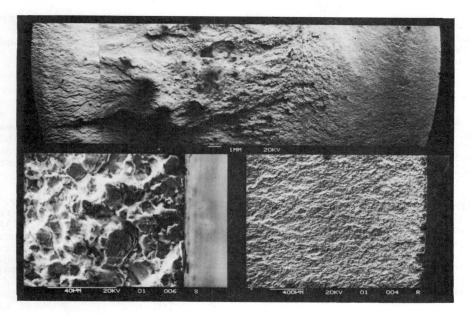

Fig. 13: SEM-pictures of the fracture plane

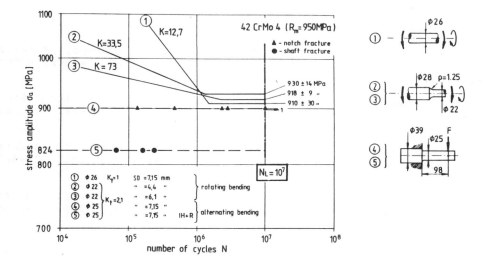

Fig. 14: Comparison of S-N-curves of specimens in the
-------- states 'as-IH' and 'additional rolled'

tion is equivalent to a diminution of fatigue resistance in the order of 10 to 20 %. Because a statistical confirmation of such a result is out of interest, the fatigue tests with rolled specimens were restrained to a reduced number of samples.

Discussion and summary

An appropriate analysis of the problem shows that an increase of the endurance of under the viewpoint of fatigue behaviour optimal surface induction hardened components beyond the limits attainable by IH only, using an additional surface strengthening operation, would impose the raise of the local intrinsic fatigue resistance in the surface layer of a certain thickness.

Using rolling as an additional strengthening operation, this goal can obviously not be achieved. On the contrary, based on X-ray measurements it seems that the effect is a considerable diminution of the local intrinsic fatigue resistance up to a depth of about 0.3 to 0.5 mm. It can be assumed that this diminution results from changes on a microstructural level in the meaning of LCF and would may be revealed by TEM. Light microscopy investigations by 1,000-fold magnification reveal no visible structural changes.

The global result of the discussed investigation shows that the effect of an additional rolling of optimal surface induction hardened components is detrimental and reduces the alternating bending fatigue resistance in the order of 10 to 20 %.

References

/1/ Braisch, P.: 1st ASM Heat Treatm. Conf. in Europe, Material Science Forum, Vols 102 -104 (1992) Pt. 1, pp. 319-334. Trans Tech Publications, CH.

/2/ Woodvine, J. G. R.: Carnegie Scholarships Memoirs, Iron Steel Inst. 13 (1924), pp. 197-237; quoted by H.J. Spies et al. in HTM 46 (1991) 5, pp. 288 - 293.

/3/ Macherauch, E., Kloos, K. H.: Advances in Surface Treatments, Ed. A. Niku-Lari, Vol. 4 (Residual Stresses), pp. 529-561, Pergamon Press.

/4/ Braisch, P. et al.: Proc. of the 7th Intern. Conf. on the Strength of Metals and Alloys, Montreal, Canada, 12-16 August, 1985, Vol. 2, pp. 983-88.

/5/ Kloos, K. H., Braisch, P.: to be published.

/6/ Braisch, P., Kloos, K. H.: Sur&ace Engineering, Ed. P.Mayr, DGM-Inform.-Gesellschaft, Oberursel (Germany), 1993, pp. 77 - 82.

Materials Science Forum Vols. 163-165 (1994) pp. 367-374
© 1994 Trans Tech Publications, Switzerland

QUALITY CONTROL OF INDUCTION HARDENED CAST IRON CYLINDER HEAD VALVE SEATS

Ch. Ch. Lal

Chief Metallurgist
Consolidated Diesel Company, Whitakers, NC 27891, USA

ABSTRACT

This presentation at the ASM Conference dealt with the quality control during induction hardening of cast iron cylinder head valve seats for diesel engines. The details of the heat treat equipment, hardening operation, and statistical quality control procedures were presented. Various aspects of selecting coil design, frequency, power duration of heating, and coil proximity were discussed. The procedure of sectioning valve seats and application of metallograph, scanning electron microscope, and microhardness testing machine to evaluate microstructure, surface texture, and case depth were described. The utilization of statistics in conducting machine capability study and controlling the case depths of intake and exhaust valve seats were discussed. The results of a joint project with TOCCO were presented in which the problem of excessive circumferential variation in case depth was solved through modification of inductor design and optimization of induction hardening process parameters. Experiences with eddy current equipment and magnetic particle inspection machine in sorting out defective parts and disposition of discrepant material involving surface cracks, non-conforming microstructure and unacceptable case depths were shared. The photomicrographs of induction hardened case showing excessive retained austenite, microcracks, and incipient melting conditions were exhibited. An overall quality control plan was presented which helped to ensure that hardness, microstructure, and case depths of intake and exhaust valve seats consistently conform to required specifications.

During induction hardening of cast iron cylinder head valve seats, heat is generated within the part by transfer of electromagnetic energy from an induction coil or inductor to the localized valve seat area. An alternating electric current flows from the RF transformer to the water-cooled copper coil establishing a highly concentrated, rapidly alternating magnetic field within the coil. The magnetic field thus established induces an electric potential in the part to be heated. Since the part presents a closed circuit, the induced voltage causes the flow of current. The electrical resistance of the part to the flow of the induced current causes localized heating of the cylinder head in the valve seat area. Due to the large mass of the cylinder head, hardening takes place by a self-quenching mechanism when the power is turned off. A desired heating pattern and case depth can be achieved by varying the coil design, operating frequency, alternating current-power input, heating time, and the proximity between inductor and the valve seat. In order to achieve case depths of 1.5 - 2.0 mm for a 45 degree valve seat about 30 mm in diameter, a typical process using TOCCO equipment consists of frequency: 350 kHz, power: 10 kW, heating time: 8 seconds, and inductor back off (proximity): 1.25 mm. Figures 1 and 2 depict intake and exhaust valve seats on the combustion face of a cylinder head and cross-sections showing induction hardened patterns for metallurgical evaluation.

Consolidated Diesel Company uses cylinder heads of standard gray cast iron with a hardness range of BHN 180 - 235. A typical microstructure in the valve seat area contains predominantly ASTM Type 'A' graphite flakes, size 4 to 5, in a primarily pearlitic matrix with less than 5% free ferrite and 1 - 2% uniformly distributed carbides. Ideal microstructures before and after hardening are shown in Figures 3 and 4. The amount of free ferrite in excess of 5% in the base material of valve seat is undesirable since it may not transform completely during austenitizing cycle and result in poor microstructure and low wear resistance. Additon of copper and tin equivalent of 0.7% in the cast iron alleviated an excessive ferrite problem without any significant adverse effect on machinability. Induction hardened valve seat inspection procedure along with various photomicrographs of the hardened case showing martensitic matrix of properly hardened material, ghost pearlite resulting from partially transformed microstructure, various amounts of retained austenite, incipient melting condition resulting from excessive heating, and micro-cracking initiating at the tips of the graphite flakes have been previously published by the author (Ref 1).

At Consolidated Diesel Company the valve seats of 4-B and 6-B series diesel engines with horse power ratings of 80 - 300 are induction hardened to enhance wear resistance and fatigue properties. Highly alloyed valve seat inserts are used for 6-C series engines with horsepower up to 400. There are beneficial compressive stresses at the hardened sufrace and the need of subsequent machining is elimnated due to the minimal distortion of the material. The induction hardening process provides improved resistance to valve seat recession at lower cost. Exhaust valve seats are more prone to recession due to rapid reduction rates of mechanical and fatigue properties of cast irons at elevated temperatures encountered during engine operation. Recession involves metal to metal contact of valve face and seat promoting adhesive and abrasive wear resulting from particle removal of low strength cast iron material (Ref 2). Therefore, a martensitic structure associated with considerably higher strength and better wear resistance should reduce metal pick up minimizing recession. Patches of free ferrite at the valve seat promote metal pick up accelerating recession. For optimum engine performance, the matrix microstructure to a depth of 0.5 mm from the valve seat should be predominantly martensitic with less than 20% retained austenite and no free ferrite, exhibiting matrix hardness in excess of Rc 50 (converted from Knoop microhardness). The case depth specification is 0.89 mm minimum with upper process control limit of 1.8 mm to minimize excessive retained austenite and incipient melting conditions. The process is monitored using Moving Average and Moving Range charts plotted by "Datamyte" computer (Figure 7). In the past, although the parts conformed to print specification, there was significant circumferential case depth variation specially for exhaust valve seats. The case depth was shallowest at the zero degree position with an average variation of .5 mm around the seats. An energy monitor was incorporated in the equipment to improve the quality of induction hardening process and reduce the amounts of undesirable phases in the microstructure.

A joint project with TOCCO was initiated to enhance the uniformity of induction hardened pattern through inductor design modification. Various factors such as position of inductor loop, coupling of inductor to valve seat, heating face width, coil split position, concentricity of loop to seat, and the use of ferrotron flux intensifier were critically evaluated. A final optimum inductor design (Figure 8) resulted in improving the worst case Cpk value from 0.32 to 1.64 at zero position of the valve seats.

ACKNOWLEDGEMENT

The author greatly appreciated Bill West, George Pfaffmann, and Mike Hammond of TOCCO; and Peter Jackson, Richard McNamee, Eddie Davis, Bruce Pruitt, Brenda Odom, and Katrina Thorne of CDC for their valuable technical discussions and assistance.

REFERENCES

1) Charanjit, Charlie Lal, ASM Heat Treatment and Surface Engineering Conference, Materials Science Forum Vols 102-104 (1992) pp. 365-372.

2) Robert E. Bisaro, Metal Progress, October 1973

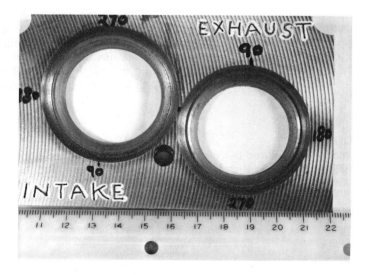

Figure 1: Photograph showing intake and exhaust valve seats on the combustion face of a cylinder head.

Figure 2: Cross-sections through the valve seat showing induction hardened patterns for metallurgical evaluation.

Figure 3: Ideal microstructure of the base cast iron showing graphite flakes in a pearlitic matrix.   400X

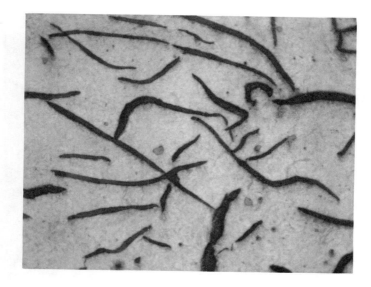

Figure 4: Microstructure of the hardened case showing graphite flakes in a fully martensitic matrix.   400 X

Figure 5: Photomicrograph showing undesirable microstructure of the base material exhibiting excessive ferrite along the graphite flakes in the valve seat area.   200 X

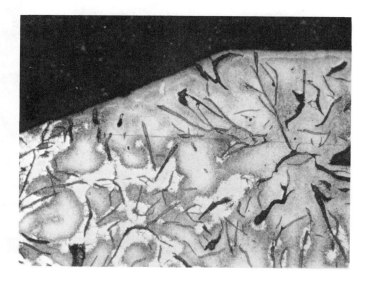

Figure 6:   Photomicrograph showing poor microstructure with untransformed free ferrite in the induction hardened valve seat area resulting from prior structure shown in Figure 5.   100 X

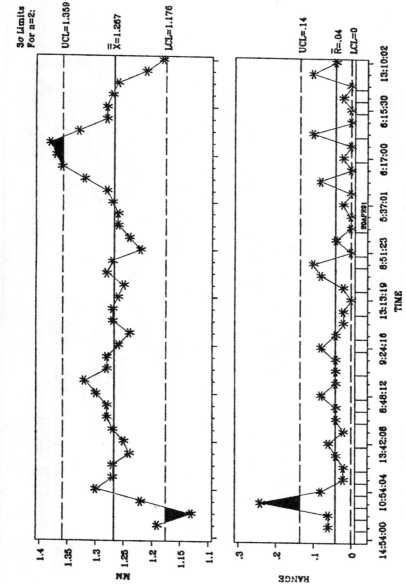

Fig. 7: Moving Average Moving Range Chart

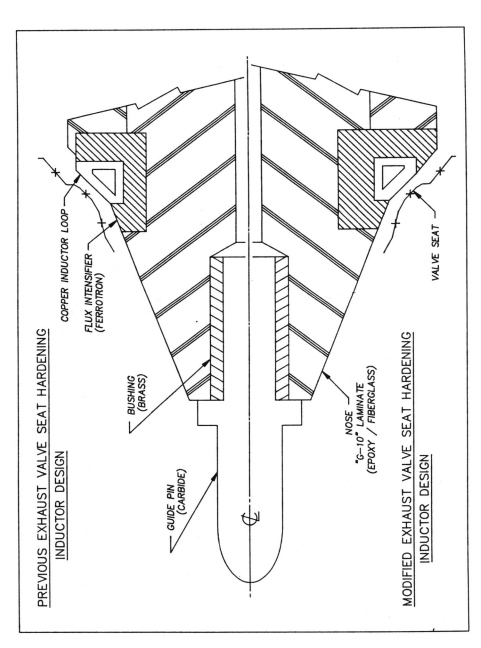

PREVIOUS EXHAUST VALVE SEAT HARDENING INDUCTOR DESIGN

COPPER INDUCTOR LOOP

FLUX INTENSIFIER (FERROTRON)

BUSHING (BRASS)

GUIDE PIN (CARBIDE)

VALVE SEAT

NOSE "G—10" LAMINATE (EPOXY / FIBERGLASS)

MODIFIED EXHAUST VALVE SEAT HARDENING INDUCTOR DESIGN

FIGURE 8 :   INDUCTOR DESIGN MODIFICATION USING FERROTRON TO MINIMIZE CIRCUMFERENTIAL VARIATION IN CASE DEPTH OF CYLINDER HEAD EXHAUST VALVE SEATS.

VIII. LASER-BEAM TREATMENT

Materials Science Forum Vols. 163-165 (1994) pp. 377-404

INDUSTRIAL APPLICATIONS OF SURFACE TREATMENTS WITH HIGH POWER LASERS

H.W. Bergmann [1], D. Müller [1], T. Endres [1], R. Damascheck [1], J. Domes [1] and A.S. Bransden [2]

[1] Universität Erlangen-Nürnberg, Lehrstuhl Werkstoffwissenschaften 2, Metalle, Erlangen-Nürnberg, Germany

[2] GEAT-Industrielaser GmbH, Nürnberg, Germany

ABSTRACT

Surface treatment with high power lasers has been a subject of concentrated research activities within the last years. These activities have resulted in a scientific knowledge of the processes and in a development of the industrial implementation. This presentation outlines the specific advantages of the four most important laser surface treatment processes (transformation hardening, remelting, alloying, cladding) with respect to typical components. Depending on the intended application, laser surface treatments can either improve wear behaviour, enhance fatigue properties or increase corrosion resistance. In some cases all three properties can be improved.

1. INTRODUCTION

1.1 CHARACTERIZATION OF SURFACES

Surfaces and surface layers can be described in terms of, chemical composition, crystallography and mechanical stresses present in the surface. **Fig. 1** indicates that topography, roughness and rippleness as well as surface defects like pores, cracks or grease layers are the major geometrical aspects apart from layer thickness and layer sequence. The chemical behaviour is determined by the bulk composition, the concentration profiles as well as the wettability and activity. The microstructure can be characterized by the type, size and distribution of phases and defects. For many structural applications, not only external but also internal stresses are important.

1.2 SURFACE TREATMENTS

Surface treatments are as old as materials technology itself. The motivation for surface treatments in general is the generation of components with either an improved durability or with enhanced tribological properties or with better corrosion behaviour. Throughout history engineers have modified the surface by means of mechanical, thermal and chemical treatments to help components to withstand surrounding conditions. However, *surface engineering* as a scientific subject in its own right has been in existance for only 20 years, and been brought into existance through the development of new physical and chemical technologies. Electro heat, PVD and CVD as well as plasma- and beam technologies have been the driving forces in *surface engineering* as an interdisciplinary technology. More specialized applications and the need for high efficiency materials has necessitated the development of property profiles which are not simply achievable by alloy development and by bulk treatments, but rather require the custom tailoring of surface properties.

Laser surface treatments (LST) offer a wide range of possibilities to achieve desired surface properties. Whether LST is the best or cheapest way to complete the required task can only be decided depending on the individual component, its size, number and base material.

Surface treatments are subdivided into those which make modifications via microstructural

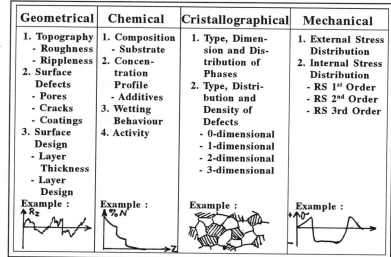

Geometrical	Chemical	Cristallographical	Mechanical

1. Topography
  - Roughness
  - Rippleness
2. Surface
  Defects
  - Pores
  - Cracks
  - Coatings
3. Surface
  Design
  - Layer
    Thickness
  - Layer
    Design
	1. Composition
	  - Substrate
	2. Concen-
	  tration
	  Profile
	  - Additives
	3. Wetting
	  Behaviour
	4. Activity
		1. Type, Dimen-
		  sion and Dis-
		  tribution of
		  Phases
		2. Type, Distri-
		  bution and
		  Density of
		  Defects
		  - 0-dimensional
		  - 1-dimensional
		  - 2-dimensional
		  - 3-dimensional
			1. External Stress
			  Distribution
			2. Internal Stress
			  Distribution
			  - RS 1st Order
			  - RS 2nd Order
			  - RS 3rd Order

Example :	Example :	Example :	Example :

Fig. 1: *Characterization of surfaces*

changes (and hence the resultant properties) of a surface layer, those which alter the shape and behaviour of the surface itself and finally cladding techniques where the layer can consist of a completely different material. For all three treatments, thermal, chemical and mechanical processes (as well as mixed ones such as thermochemical processes) are known to alter the base material leading to a profile of properties that fit better the requirements in service. In **Fig. 2** examples of established processes for LST are shown.

In the generation of surface layers with properties different to those of the substrate, the generation of an adequate layer thickness is important. For example, case hardening of shafts is only recommended if the case depth is less than a tenth of the shaft diameter. If this is not the case conventional techniques should be used. Dies that have to be regrinded (after they have worn about a hundred microns) either have to be rehardened or the original case depth

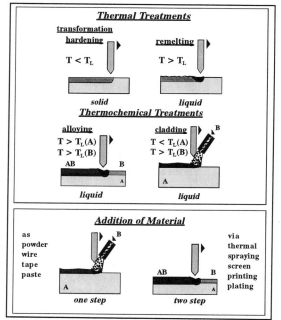

Fig. 2: *Processes of laser surface treatments*

has to be sufficiently thick to allow regrinding without a second hardening process. In addition to the selection of the layer thickness, the compatibility of layer and substrate has to be considered. Hard surface layers require a sufficient support from either the substrate or from a subsequent layer to avoid layer break through (egg-skin effect). Chemical compatibility is essential in order to avoid the generation of brittle phases which could lead to cracking or splitting off. This is the case especially for liquid processes but also for some diffusion processes.

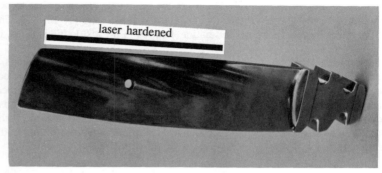

Fig. 3:   *Laser hardened steam turbine blade*

Often the optimization of a single property is of most interest. There are, however, examples where all previously mentioned aspects have to be improved simultaneously in order to prolong the life time of the component. A typical example is shown in **Fig. 3**. A turbine blade of a steam turbine must withstand erosion, fatigue and corrosion. Laser hardening the leading edge of the blade results in the improvement of the resistance to all three phenomena [1, 2]. Improved properties, however, are not the only consideration in the implementation of LST. Complex and expensive equipment can only be justified if it allows a reduction in labour costs (by a high degree of automation) or if reduced waste production is possible (due to process control). Suitable applications are those where local treatments of almost finished components with complex geometry can be carried out by a laser and where only a negligible amount of post-treatment is necessary. Hence laser surface treatments are in competition with numerous advanced and established processes, and thus can only be successful if, in addition to the improved properties, the necessity of local treatments, high flexibility, compatibility in manufacturing lines and cost effective production as well as on-line process control can be achieved.

1.3   SELECTION OF SUITABLE LASER SOURCES

As known in the literature the correct selection of an adequate laser source (type of laser, maximum output power, etc.) as well as a suitable handling system for beam and/or workpiece manipulation is essential [3, 4]. Wavelength, mode of operation (cw, pulsed), mean power and beam properties (cross-section, intensity distribution and focusability) are some of the major aspects that have to be considered. At the present time, LST of metallic components is mainly done by high power CO_2-lasers [5, 6, 7]. Successful processing with Nd:YAG-lasers has been reported for small area treatments within a tooling centre [8]. Here, additional fastening and alignment events can be avoided. If only

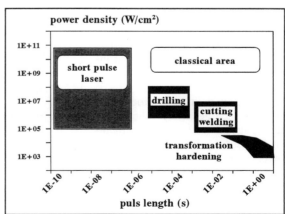

Fig. 4:   *Laser materials processing in order of intensity and interaction time*

the surface properties themselves and not the behaviour of a layer has to be changed, other lasers irradiating in the visible or the UV range are suitable tools [9, 10, 11]. Changing the wettability of polymers by Excimer laser irradiation, as used in the textile industry, is a current example [12].

The required depth in which a surface layer has to be modified depends on power density and interaction time. **Fig. 4** shows a well known graph in which the regime of power density and interaction time for different laser processes are indicated [13]. If thermal processes in the solid state are considered the heat penetration depth estimated from the one-dimensional heat conduction equation provides a good estimation of the depth a material is altered by the irradiation. If liquid processes have to be taken into account, the situation is more difficult (as discussed in subsequent sections).

2.   PROCESS SYSTEMATIC OF LASER SURFACE TREATMENTS

The principle of laser surface treatment is the modification of a surface as a result of the interaction between a beam of coherent light and the surface within a specified atmosphere (vacuum, protective or processing gases) (see **Fig. 5**). The light generated in a resonator is directed onto the surface of a sample via an optical transmission system (mirror systems or fiber-optics). Starting from a given mean optical output power, the required power density and intensity distribution throughout the beam is modified by beam focussing and/or beam shaping optics such as lenses, mirrors, scanner units or beam integrators. Via the movement between the beam and the workpiece, a track pattern can be successively generated at the surface of a component. The interaction time is than determined by the cross-section of the beam and the feed rate.

Depending on the type of process and the workpiece geometry either translation stages, portal systems or robots can be used by which such relative

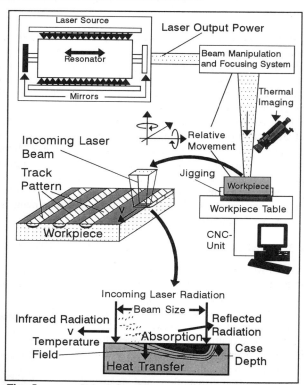

Fig. 5:   *Principle of laser surface treatment*

movement is achieved. The selection of a suitable handling system for the beam and/or workpiece depends primarily on the precision required, the processing speed and the masses that have to be handled. The time for fastening and alignment of the workpiece as well as the investment costs are other important considerations.

As currently accepted the various processes of LST can be subdivided into thermal processes (e.g. transformation hardening and remelting) or thermochemical processes (e.g. alloying or cladding).

When the surface composition is altered, single processes are known where the material is added simultaneously during the melting process (one-step processes) and others are known that make use of pre-coating or preplacement techniques (two-step processes). For the processes mentioned, different types of beam shaping units are suitable. For solid state processes square or rectangular beams are recommended. Liquid state processes like remelting often require line forming optics. For cladding operations optics are needed that generate a sufficiently large melt pool which

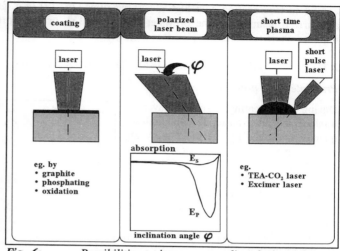

Fig. 6: *Possibilities to increase coupling during LST*

allows the feeding of additional material directly into the melt bath.

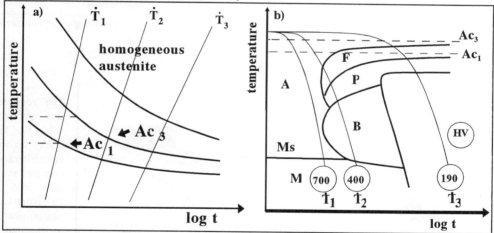

Fig. 7: a) TTA-diagram of a hypoeutectic steel
b) TTT-diagram refering to a)

2.1 SOLID-STATE PROCESSES

The electrical to optical conversion efficiency for the generation of laser light is relatively low and is highest for CO_2-lasers (typically 12 % efficiency). Unfortunately, the absorption of infrared light by metals at room temperature is also low. Therefore it is essential to enhance the coupling efficiency especially for solid state processes. The situation is less critical for liquid-state processes as other absorption mechanisms (such as increased penetration depth, plasma absorption and multiple reflection etc.) are involved [14, 15]. For solid-state treatments, enhanced coupling can be achieved by surface coating with high absorbing films [16, 17] (e.g. graphite coatings), by irradiating a polarized beam under the Brewster angle [16, 18], or by improving the coupling via a super-imposed short time plasma [19] (see **Fig. 6**).

2.1.1 LASER TRANSFORMATION HARDENING

Principles and processing windows

During laser heating a surface layer is transformed into austenite, and subsequently quenched by the bulk or by an external media. Both processes are described classical with TTA- and TTT-diagrams. In laser hardening heating rates of 1000 K/s and more are present. Regarding this, one has to recognize the shift of the transformation temperatures (and especially the temperature necessary for the formation of homogeneous austenite) to higher values due to the TTA-diagram (see **Fig. 7a**). The microstructural changes that occur when the steel is quenched from the austenitic state to room temperature can be seen in the TTT-diagram given in **Fig. 7b**. In laser hardening generally the cooling rates obtained from heat conduction into the substrate are high enough for martensitic transformation.

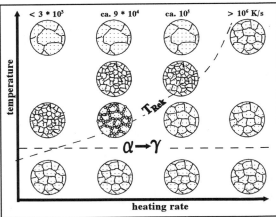

Fig. 8: *Kinetics of phase transformation during laser heating*

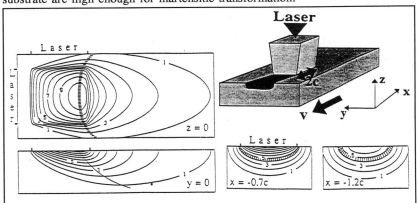

Fig. 9: *Stationary temperature field during laser hardening*

The relatively short period at which the layer is above austenitization temperature can lead to incomplete austenitization as the time for carbon diffusion may be insufficient [20]. Therefore the carbon distribution in the substrate has an essential influence on the generated spatial sequence of microstructures and the resulting hardening profiles [21]. Hence, the thermal cycle is different for volume elements at different depths below the surface, a complex sequence of microstrutures is generated from the surface to the bulk. Accordingly the phase transformation during heating causes a changing of the grain size of the steel. Depending on the heating rates and maximum temperatures effects of recrystallization and overaging can cause a finer microstructure as well as a coarser one. In **Fig. 8** kinetics of phase transformation during fast heating in the solid state can bee seen.

One possibility to improve the quality and efficiency of laser hardening is to use a beam of rectangular cross-section and a rectangular [22] or chair like intensity distribution [23] across the beam. This allows to keep a large area of the sample sufficiently long above transformation temperature (Ac$_3$) without melting the surface, see **Fig. 9**. A second improvement is achieved if the maximum temperature generated by laser irradiation [20, 24, 25] and the intensity distribution [26] are controlled. The principle of a typical control unit is depicted in **Fig. 10**. In this set-up a pyrometer is used to measure the surface temperature and a control unit actuates the output power of the laser so that constant surface temperature is obtained. Superimposing a TEA-CO$_2$-laser allows an efficient and controlled hardening in a very flexible way. However, a surface film of about one to 3 μm thickness is molten and has to be removed after processing [27].

Alloys of iron and carbon can solidify in the stable Fe-C-system (cast iron) or in the metastable Fe-Fe$_3$C-phase diagram (steels). Changing from plain carbon steel to other types of steels different hardening profiles are found [28], their principle features can be seen in **Fig. 11**. Depending on the composition and pretreatment different processing windows in terms of controll temperature and feed rate can be observed [29]. For some of the more common steels **Table 1** indicates typical case depths and processing parameters.

material	DIN-Nr.	max.T$_{contr}$ (°C)	v (m/min)	hardness (HV0,3)	case depth (mm)	liquid N$_2$ quenching
C45W	1.1730	1350	0,3	750-800	1,2	no
C60W	1.1740	1250	0,2	800-900	1,3	no
25CrMo4	1.7218	1300	0,3	550-600	1,35	no
34CrMo4	1.7220	1300	0,3	600-700	1,4	no
42CrMo4	1.7225	1250	0,3	700-750	1,4	no
50CrMo4	1.7228	1250	0,3	800-850	1,4	no
50CrV4	1.8159	1275	0,3	800-850	1,45	no
21CrMoV51	1.7709	1300	0,3	480-500	0,8	no
50NiCr13	1.2721	1250	0,2	800-900	1,6	no
56NiCrMoV7	1.2714	1300	0,2	850-950	1,4	no
90MnCrV8	1.2842	1250	0,2	900-1000	1,2	yes
100Cr6	1.3505	1250	0,2	900-1000	1,15	yes
X38CrMoV51	1.2343	1250	0,2	650-750	1,25	no
X155CrVMo121	1.2379	1300	0,2	750-850	1,0	yes
X210Cr12	1.2080	1250	0,5	850-950	1,2	yes
X210CrW12	1.2436	1250	0,5	800-900	1,3	yes
GGG60	0.7060	1150	0,1	850-950	1,1	no

Table 1:   *Results from laser hardening of different steels (DIN)*

Internal stresses

When a component is heated it will deform due to thermal expansion. If the temperature is not homogeneous throughout the component this will lead to distortion as not only elastic but also plastic deformation is involved [30]. The fact that the flow stress of a material and the thermal expansion behaviour are temperature dependend themselves leads finally to a distortion of the sample and the occurance of internal stresses within the workpiece. **Fig. 12a** shows the classical hysteresis in heating and cooling of a not transformable material during heating and cooling of a surface layer. Plastic deformation due to stress release and thermal expansion lead to a distribution of the residual stresses with tensile stresses at the surface and compressive stresses in the substrate. A martensitic transformation in the heated surface layer, however, causes compressive stresses at the surface and tensile stresses in the component (see **Fig. 12b**). From the two figures it is obvious that surface hardening is intrinsically coupled with a change in the internal stress distribution and it is the skill of the engineer to modulate the process or to change the geometry of the component in such a way that the desired stresses (generally compressive stresses) are placed where they are needed while undesired ones (in general tensile stresses) are located in areas where they can be neglected. A typical stress profile across a laser hardened track and the profile from the surface into the bulk are given in **Fig. 13a and b** [31, 32]. In the hardened track the

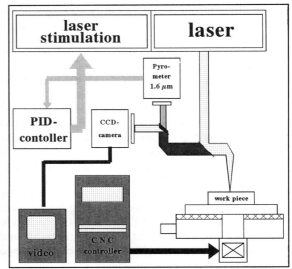

Fig. 10:   *Arrangement for temperature controlled laser hardening*

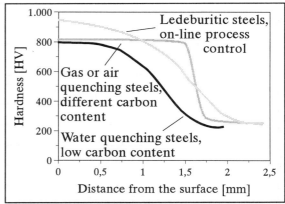

Fig. 11:   *Hardness profiles for different kinds of steel*

martensitic transformation causes an expansion when the lattice is transformed. This effect causes compressive stresses in the hardened area. Near the track tensile stresses are present in the heat effected zone which then drop to the level of the bulk with increasing distance. A similar behaviour of the residual stresses can be observed in the depth profile. The martensitic transformation leads also to compressive stresses in the transformed layer and tensile stresses in the heat affected zone. Depending on the degree of austenitization and subsequent martensitic transformation laser processing can generate different internal stress situations. **Fig. 14** shows the influence of the control temperature, feed rate and composition for some selected steels. It is obviously that the value of the compressive stresses is not influenced strongly by the process parameters control tempera-

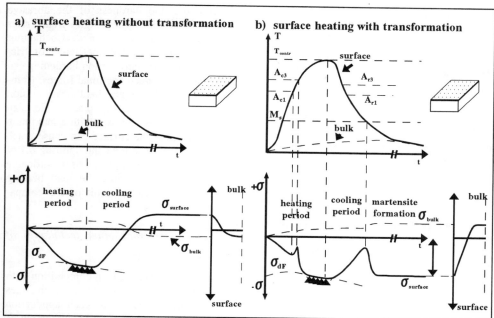

Fig. 12: *Volume hysteresis and residual stresses during case hardening of a*
a) non transformable and b) transformable material

ture and feed rate as well as by the carbon concentration of the investigated steels. The tensile stresses near the track, however, seem to depend on the control temperature and the feed rate. High temperatures and low feed rates result in low tensile stresses. The reason for this behaviour could possibly be found in the temperature gradient formed by the different process parameters.

If a martensitically transformed microstructure is heated a second time, the annealing causes a typical deviation from linear expansion for a given steel. Hence, on the second thermal cycle austenitization and subsequent martensite formation will be absent at a certain distance from the irradiated area and therefore tensile stress formation in the heat effected zone will result [33]. Overlapping tracks, tracks that meet each other but also separated ones could therefore have tensile stresses in between them or in the annealed zone [31, 34]. Looking at **Fig. 15** one can realize that a laser track positioned in different distances from another track forms a stress field that is independend from the distance to the first track. It is possible, however, to overcome this disadvantages if the reheating occurs prior to the transformation. For example hardening of a fast rotating cylinder with an axial feed rate will generate completely compressive stresses over a large area if a sufficiently high rotation speed is applied [35] (see **Fig. 16**). This treatment may then be used to improve the duration limit of notched samples or to improve the fatigue behaviour of axes and shafts which locally differ in their diameter. However, sufficiently high output powers have to be applied so that a glowing ring can be formed around the area to be treated.

Wear Behaviour

Local hardening increases not only the hardness but also the wear resistance. Abrasive wear is characterized by a principle wear profile. Low wear rates are present when the surface hardness

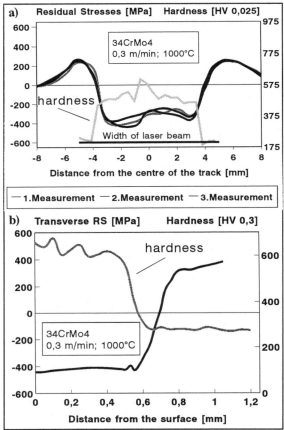

Fig. 13:   *Typical profiles of residual stresses after laser hardening*
a) across the laser hardened track
b) depth profile

is above the hardness of the abrasive materials otherwise high wear rates occur [36]. The adhesive wear behaviour is influenced by the coefficient of friction. Here, transformation hardening can be used to reduce frictional forces. Components with high load transmission forces sometimes fail due to Hertzian stresses. These stresses with a maximum value below the surface can cause crack generation by fatigue and splitting off of particles (pittings) [37]. Laser hardening can be used to shift the load bearing capacity above the limit of the comparative stresses.

Comparison with other techniques of case hardening

As laser hardening is in competition with accepted classical hardening techniques one has to decide for each application specifically whether laser hardening is the correct tool. The evaluation depends on the size of the work piece, its geometry and the number of pieces that have to be treated. Situations favourable for laser applications are given in **Fig. 17**. Comparing with other techniques laser hardening has advantages due to the possibility of local treatments with well defined energy input into the component. Therefore only zones that need to be hardened are treated and a low distortion can be obtained. Furthermore a CNC controlled handling system increases the flexibility of the process so that often changing geometries can easily be hardened. A general comparison of advantages and disadvantages for various treatments are indicated in **Table 2** [39]. Surface hardness and the possible improvement in the duration limit are schematically shown in **Fig. 18** for different surface treatments. This illustration only shows the areas of classical surface treatment processes. If laser hardening must be classified into the area of flame or induction hardened heat treatable steels cannot be determined today. Extensive fatigue tests must confirm that laser hardening offers better fatigue properties than induction hardening. Concerning this figure we also have, however, to take into account, that together with the improvement in properties the costs of the processes will also increase [39].

Duplex treatments

In general the disadvantage of Duplex treatments are their high costs, as two steps of production are involved. However, a combination of laser hardening with a thermochemical pre- or post-

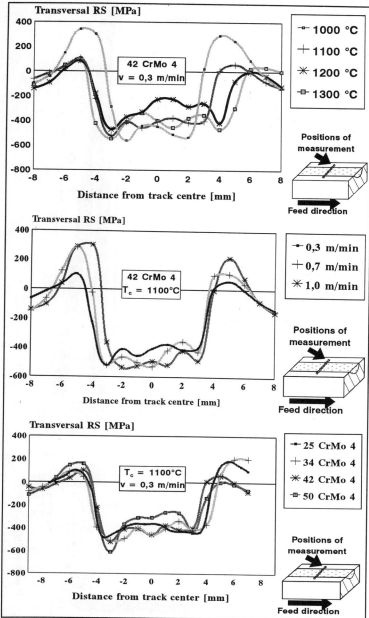

Fig. 14:   *Influence of control temperature, feed rate and carbon concentration on the residual stresses after laser hardening*

treatment is reasonable if long diffusion times can be avoided [36, 37]. This is indicated for components where at the surface a certain tribological behaviour has to be realized simultaneously with a sufficiently large case depth to ensure an enhanced duration limit or to avoid fatigue wear due to Hertzian stresses.

An example of hardness profiles obtained by combinations of laser hardening and nitriding is shown in **Fig. 19**. Performing laser hardening before the nitriding treatment one has to recognize that the hardened track will be heat treated during the following step and therefore the nitriding temperature has to be adapted on the steel substrate. If this is done correctly a supporting layer with increased hardness will remain below the nitrided layer. Laser hardening after the nitriding process results in a martensitic transformed layer similar to a laser hardened one. Due to the increased Nitrogen concentration in the diffusion layer higher hardness values can be obtained at the surface. However, retained austenite has to be expected too for high amounts of Nitrogen. If a nitride layer is present at the surface low control temperatures and high feed rates should be applied, otherwise a destroying of the layer has to be

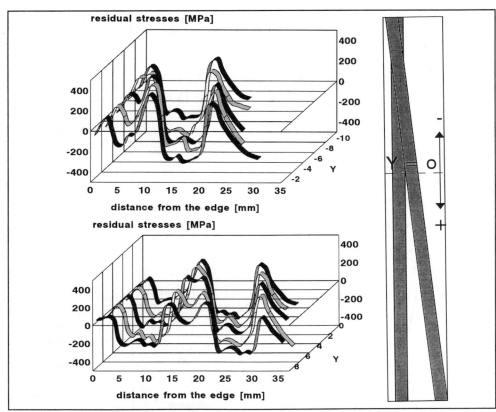

Abb. 15:   *Residual stresses in overlapping laser hardened tracks of a 0,45 % C steel*
                    (1100°C, 0.3 m/min)

expected due to recombination and
diffusion of Nitrogen.

Other solid state treatments

Transformation hardening, resulting in improved mechanical properties as has been shown in the
previous section for the laser hardening of steel, is unfortunately
unique in metallurgy. Other base
metals like some Copper and Titanium alloys show also a martensitic
transformation but without a remarkable hardness increase. Precipitation hardening after a thermal
homogenization is important for a
much larger array of metals, but

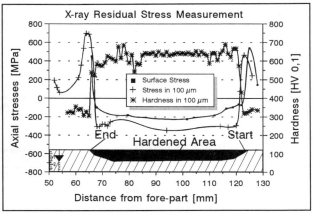

Fig. 16:   *Residual stresses from overlapping free laser*
                    hardening of shafts

this hardening mechanism is either not applicable for surface treatments or can be obtained cheaper by other methods. This restricts solid state treatments with lasers to steel hardening or softening either via recrystallization, recovering or overaging [42].

More recently shock hardening is discussed for better fatigue properties [43], but this technique is still in the early stages of development. Finally one has to mention laser surface treatments in the solid or liquid state which are applied to alter the magnetic domain structure of transformator sheets [44, 45].

2.2   LIQUID STATE PROCESSES

2.2.1. REMELTING

Ledeburitic surfaces on ductile and grey cast iron generated via chill casting or remelting show good properties regarding wear resistance with high load bearing capacity [46-48]. For the production of crack- and pore-free layers by laser remelting it is necessary to specify the moulds, the carbon aquivalent and the formation and size of graphite similar to the standards for the TIG remelting process [49, 50]. Often a pre- and/or postheattreatment is performed to avoid martensite formation in the heat affected zone [51]. Fine lamellar graphite is easily dissolved in the melt pool. The dissolution of spheroidal graphite, however, requires larger contact time with the melt. The interaction time is limited by the movement of the spheroids due to the metal hydrostatic pressure and differences in the density [52]. Accordingly one has to recognize carbon losses due to the burning of graphite at the surface.

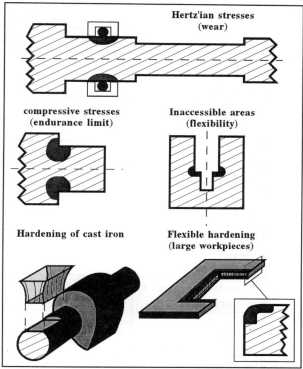

Fig. 17:   *Geometries suitable for laser hardening*

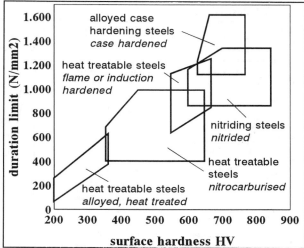

Fig. 18:   *Possible improvement of the duration limit by different surface treatments*

	Surface Hardening Process			
	Flame	Induction	Laser	E-beam
Spatial resolution	-	-	+	+
Accessibility	0	-	0	0
Intensity modulation	-	0	0	+
Low technical effort	+	+	0	-
Low investment costs	+	+	-	-
Flexibility	0	-	+	0
Low distortion	-	-	+	+
Self quenching	-	-	+	+
Quality of result	-	0	+	+
Surface oxidation	-	0	0	+
Treatment of large components	+	+	0	-
Single pieces	+	-	0	+
Complex geometry	0	-	+	+
Small pieces	-	0	+	+
Large production	-	+	+	+
Industrial value	0	+	+	+

Table 2: *Laser hardening compared with competitive processes (+: good, -: poor, 0: adequate)*

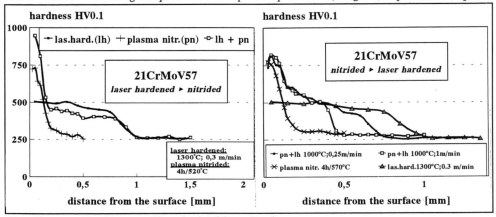

Fig. 19: *Hardness profiles obtained by a Duplex treatment*

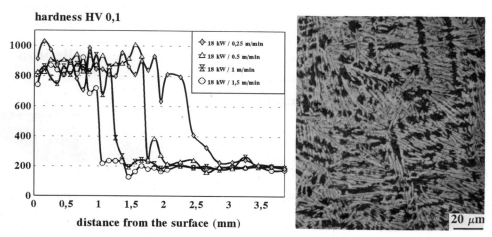

Fig. 20:   *Microstructure and hardness profiles obtained from laser remelting of cast iron*

Fig. 20 shows a remelted microstructure existing of transformed γ-dendrites and ledeburite eutectic together with typical hardness profiles [53]. With increasing output power in combination with line focussing optics laser remelting becomes an economic alternative to TIG remelting [54-56]. The reason is that the expenditure for a handling system only slightly increases while the cost per kW decreases considerable with increasing output power. Many and extensive publications refer to the realizable mechanical and tribological properties of laser remelted cast iron it can therefore be refered to the current literature [57, 58].

More recently the importance of residual stresses in laser remelted layers is realized (see **Fig. 21 and 22**). Remelting without preheating results in high amounts of retained austenite and all qualities of cast iron are forming cracks in the ledeburitic layer. Measurements of the residual stresses show high tensile values beneath the remelted layer. Due to the high amounts of retained austenite no residual stresses could be measured by the $\sin^2\psi$-method using the α-Fe (211)-peak. If a fin is remelted over the whole width with one track the tendency towards the formation of cracks decreases, the problem of retained austenite, however, remains. Using preheated samples the laser treatment causes compressive stresses across the remelted track for the materials GG-25, GGG-40 and GGG-60. The residual stresses along the laser track are zero for GGG-40 if 600°C preheating temperature are applied, otherwise tensile stresses can be observed. Remelting a GGG-60 results in low compressive stresses if a preheattreatment of 500°C is applied and tensile stresses at 600°C in contrast to the above. In technical processes usually a preheating at 500°C is applied followed either by subsequent quenching or post annealing. Therefore the internal stresses remaining after remelting of preheated GGG-40 with and without subsequent postheating was determined. In both cases longitudinal and transversal compressive stresses between 50 and 150 N/mm² were found. A similar behaviour appears for the other cast iron qualities. Generally compressive stresses increase with increasing post heat treating temperature. In all cases preheating of 500°C was sufficient to avoid tensile stresses. The results could be varified not only for flat and fin samples but also for real components. Remelting of cams for example resulted in compressive stresses in the surface between 50 and 150 N/mm² which hold throughout the melted zone. In the heat affected zone the sign of stresses changes [54].

If cast iron is remelted in a CO_2 atmosphere flat and smooth beads are found. Technically this can be used for smoothing of large milled components which have otherwise to be smoothened by

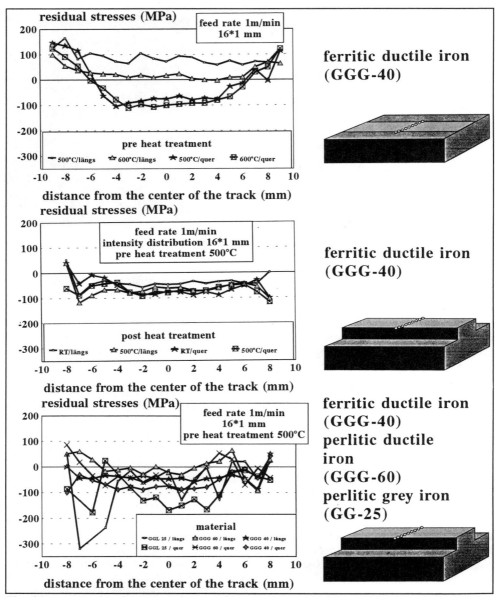

Fig. 21:   *Residual stresses after remelting of cast iron*

manual grinding and polishing. **Fig. 23** shows a deep drawing tool where the original roughness was reduced by remelting the surface using suitable track patterns. The reduced roughness and sufficient contour tolerances allow to reduce tremendously the amount of manual post treatments so that improved wear resistance and longer lifetime of the die could be combined with reduced costs.

To a certain amount there is also a need of flexible remelting smaller cast iron parts with complex geometry. This could become a market for fibre transmitted Nd:YAG laser beam remelting.

2.2.2 ALLOYING

With laser treatment it is possible to change the chemical composition of the surface by adding other materials to the melt pool. Some of these processes where only the corrosion behaviour has to be improved use cladding techniques and try to avoid the intermixing with the substrate [59, 60]. For such treatments which are based on Cr, Mo or Ti coatings the reader is refered to review articles in the literature [61, 62]. More often, however, it is the aim to increase the surface hardness. The achievable increase in hardness due to solid solution hardening is relatively low. In addition a course grain structure occurs in the resolidified bead if single phase materials are treated [63]. Changes in the transformation hardening by laser alloying often result in an inhomogeneous hardness distribution and tends to cracking. Therefore the majority of laser surface treatments that

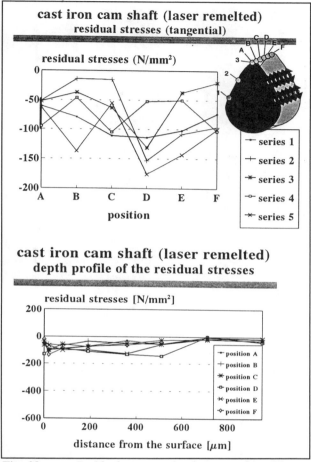

Fig. 22   *Residual stresses after laser remelting of cams*

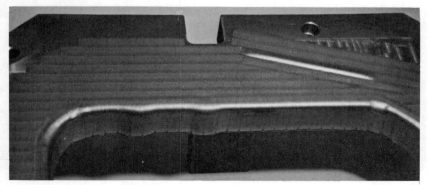

Fig. 23:   *Reducing the roughness of milled surfaces by laser remelting*

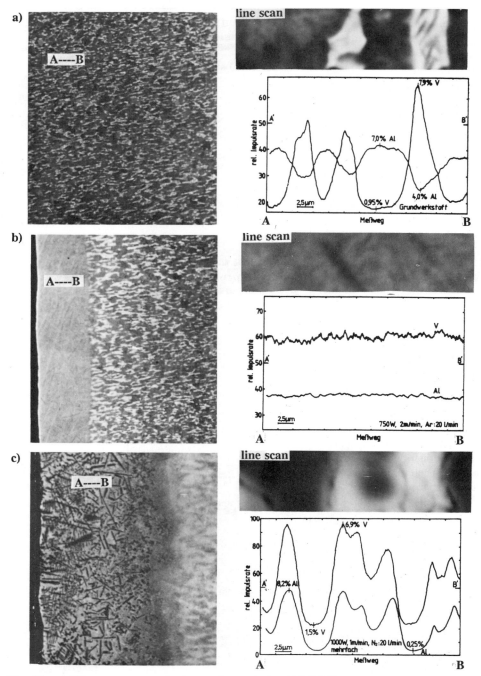

Fig. 24: *Microstructure and microprobe analysis of TiAl6V4*
a) starting structure   b) after surface remelting   c) after gas alloying with N_2

incorporate liquid processes are based on dual phase hardening.

The most prominent alloy systems suitable for dual phase hardening are the alloying of Al-base materials with Si or Fe and Ni [64-66], the alloying of Ti-based alloys with N, C [67-69] or transition metals (Ni, Co, Mo) [70] and the alloying of graphite [71,72] and carbides [73-75] into steel surfaces. Some applications of the three examples mentioned shall be discussed in more detail.

Gas alloying of Ti

The challenges of Ti-alloys are their low specific weight, high strength, good corrosion resistance and in case of some of the intermetallic alloys good high temperature properties. However all Ti-alloys suffer from low wear resistance and the tendency to fretting [76]. Remelting Ti-alloys in a reactive atmosphere (eg. Ar/N_2 or Ar/CH_4 mixtures) it is possible to generate crack free surfaces consisting of a hard phase material (TiN, TiC) embedded in a matrix consisting of Ti solid solution crystals [77, 78]. The quality of the layer depends on gas alloying in a correct processing window and may need additional preheating. From microprobe analysis it is possible to separate some of the complex metallurgi-

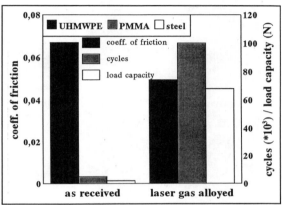

Fig. 25:   *Properties after laser gas alloying of TiAl6V4*

cal processes involved. In **Fig. 24** a typical α-β TiAl6V4-alloy is shown that has been remelted under Ar atmosphere. In the remelted layer the differences in the Al and V concentration of the substrate are leveled and a martensitic α' microstructure is observed. Remelting under N-atmosphere leads to the formation of TiN dendrites and the alloying elements Al and V are enriched in the interdendritic zone. With this increasing TiN content the layer hardness increases but unfortunately also the internal stresses and the tendency to crack formation. Up to 700 HV stress relieve annealing after gas alloying can help to avoid crack formation [79]. The change in some of the surface properties is shown in **Fig. 25** in comparison to untreated material. In this figure the wear properties obtained from a pin on disc and a ball on disc test respectively of laser gas alloyed TiAl6V4 against the plastics UHMWPE and PMMA (often used for surgical implants) as well as against steel are illustrated. One can see that the good corrosion behaviour of the Ti-alloy can be combined with the hardness and the wear resistance of the nitrided layer. On the other hand different surface roughness and friction coefficients occur after polishing compared to the base material similar as for other dual phase hardened layers. For a possible application all modifications have to be taken into account and often specific tests are required before one can decide for a given component, whether gas alloying is the desired treatment.

As mentioned above also intermetallic compounds like TiAl can be nitrided. A typical component is a valve, shown in **Fig. 26**, gas nitriding of the valve seats requires a preheating of 750°C to ensure a crack-free hardening with a sufficiently high hardness plateau and an alloyed depth of 0.6 mm. Preheating, which is also common for laser cladding of steel valves, generates a compressive stress situation in the surface after cooling and saves expensive laser energy.

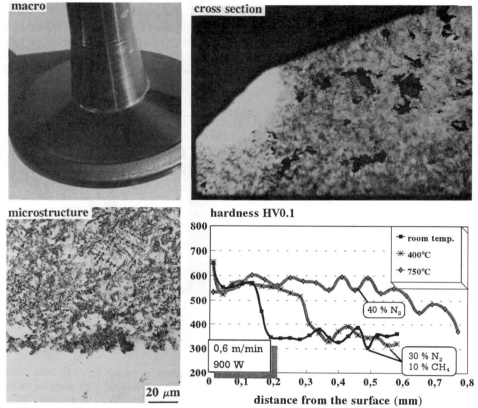

Fig. 26:   *Laser alloyed valve of TiAl*

Alloying of graphite into mild steel

Hollow camshafts manufactured by forging a mild steel with an inside pressure has often been discussed as an alternative to massive camshafts. However mild steels suffer from their low wear resistance that make the application mentioned impossible without a suitable surface treatment. It is possible for such components to generate a hard surface layer by melting graphite into the steel surface [71]. The control of the process parameters amount of graphite and volume of the melt bath (via output power and feed rate) enables to produce a carburized layer with a well defined carbon concentration. The carbon concentration can be increased up to ledeburitic compositions. This enables to produce mild steel camshafts with ledeburitic layers which are known to have good properties for this application. **Fig. 27** shows the variety achievable in microstructure and hardness by this process. One should mention that the process leads to commercially attractive alloying rates and the relatively low rippleness needs only a little final grinding operation.

Surface alloying of eutectic Al-Si-alloys with Silicon

Wear resistant Aluminium alloys are used for motor components like pistons and cylinder liners. Good wear resistance of the cylinder wall is achieved by hyper-eutectic compositions (17 % Si) or via Nicasil coatings of the eutectic alloy. High tooling costs of hyper-eutectic alloys and the

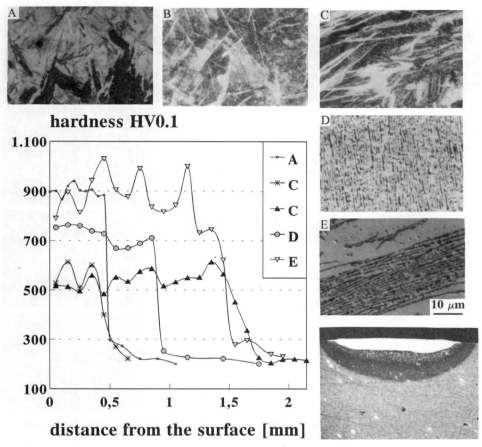

Fig. 27:   *Laser carburizing of a mild steel (DIN St52-3)*

environmental problems involved with galvanic coatings have, however, initiated an increasing interest in surface alloying of Al-Si-alloys. Remelting of such material results in a fine resolidified microstructure with a moderate hardness increase [80, 81]. Coarse Si-precipitations, however, which are desired from the tribological point of view, can only be achieved if a sufficiently high amount of Si is melted into the surface [82], often together with nucleids of the Si-phase [83]. Typical microstructures, surface topographies and resultant hardness profiles are given in **Fig. 28**. Other developments are leading into the direction of replacing Si-additions by iron, where better wear resistance is reported for higher loads [84]. Apart from this a number of other transition metals have been tried to melt into the surface [65, 85-87].

2.2.3 CLADDING

Among all laser processes, cladding offers the widest variety of possibilities to alter a component at its surface. Coating with a completely different material enables surface properties considerably different to those of the bulk [88-91]. Laser cladding enables a low dilution of the clad layer with the bulk material combined with a metallurgical bonding between the layer and the substrate

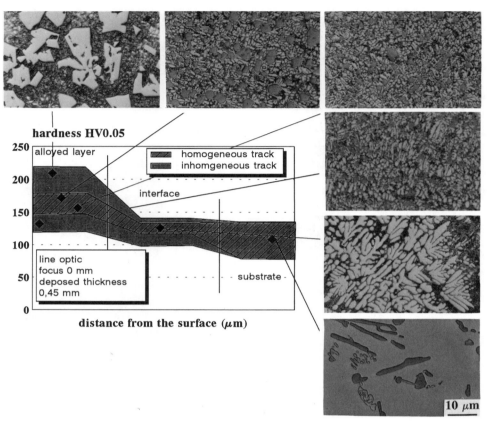

Fig. 28:   Microstructure and hardness profiles of a laser alloyed Al-Si cylinder liner

compared with other cladding techniques as thermal spraying methods [92]. The process can be performed using the addition material in form of powders [93, 94], wires [95, 96], ribbons [97] or pastes [98]. Moreover, thicknesses of several mm can be achieved by monolayer coatings and multilayer arrangements can even further enlarge the thickness and help to completely suppress dilution of the surface by the bulk material. Saying this, one has to stress that laser cladding is also the most costly process, because the price for the coating material in form of powder, wire of foil can often outweight the laser treatment costs. In addition the higher energy input could lead to a higher distortion and internal stress level. Therefore, it is even more essential to control the process. Many investigations in the last years have dealt with pyrometric temperature measurement and output power control to avoid a local superheating of the workpiece and to avoid an increasing intermixing due to deeper melt depths, which occur when the component as a whole is heated up during laser processing. Other investigations were related to plasma spectroscopic process control and monitoring of the melt bath with a pilot beam in order to keep the melt bath stable. As it is almost impossible to report all these experiments in detail the reader is refered to some recent overview articles [98, 99].

Laser cladding of cast iron can lead to the formation of cracks and pores in the layer so that conditions, e.g. preheating have to be applied [100]. As an example of a laser cladded part **Fig.**

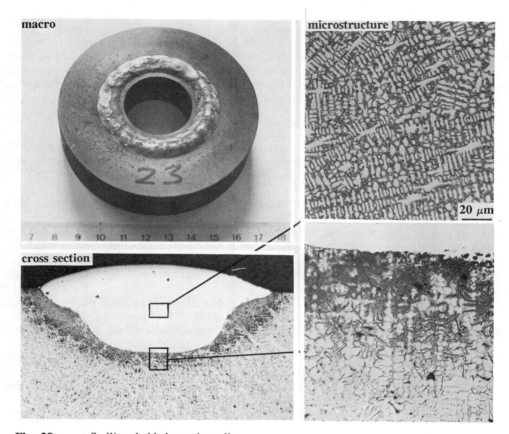

Fig. 29:   Stellite cladded cast iron die

29 shows a stellite coated cast iron die, used for glass bottle production. Together with the microstructure and the hardness profile through the cladded layer with such coatings it is possible to improve the wear and oxidation behaviour. Again a sufficient preheating is required to avoid surface cracking. Further examples for a successful laser cladding of components are combustion engine valves [101, 102], forging tools [103, 104], piston ring seats [95] and turbine blades.

CONCLUSIONS AND FINAL REMARKS

Laser surface treatments have been substantially developed over the last decade. This development has come about through incremental improvements based on the contributions of many disciplines. This technology has been brought about by the development of suitable laser sources, optical systems and process control strategies. Moreover, the benefits of a new technology have to be demonstrated over a long period of intensive testing (like every other conventional surface technology). Further refinements to these classic surface processes will offer economical solutions to the customer. With a systematic development of the basic mechanisms of laser surface treatments, it is possible to offer a range of solutions for a potential customer. Depending on the required property profile of a component it has to be decided whether a laser surface treatment is a cost-effective and efficienct treatment for an extented lifetime.

ACKNOWLEDGEMENTS

The authors thank the following institutions for financial support of different aspects of research involved in this paper: the BMFT, the DFG and the Bavarian Ministry of Economy. Among others the following companies contributed to the investigations: Aesculap AG, Tuttlingen, Audi AG, Ingolstadt, Mercedes Benz AG, Stuttgart, Messer Griesheim Steigerwald Strahltechnik, Puchheim, Thyssen HOT, Nürnberg, GEAT-Industrielaser, Nürnberg and Heinz Kleiber, Erlangen.

REFERENCES

[1] B. Brenner, G. Wiedemann, B. Winderlich, S. Schädlich, A. Luft, D. Stephan, W. Reitzenstein, H.-T. Reiter, W. Storch: Laser Hardening - an Effective Method for Life Time Prolongation of Erosion Loaded Turbine Blades on an Industrial Scale, ECLAT'92, DGM, Göttingen, 1992, p. 199-204

[2] K. Messer, H.W. Bergmann, D. Müller: Laserhärten von Turbinenschaufeln mit Hilfe eines CO_2-Lasers, internal study, University of Erlangen, FLE, 1993

[3] T. Endres, H.W. Bergmann, D. Müller, H. Reichelt: Laserstrahlschweissen mit einem 18 kW Hochleistungs-CO_2-Laser und einem 5-Achsen-Handhabungssystem, LASER'91, Springer-Verlag, 1992, p. 453-458

[4] M. Gonschior: CAD/CAM-Systeme für die Lasermaterialbearbeitung, LASER'91, Springer-Verlag, 1992, p. 569-572

[5] H.W. Bergmann, R. Kupfer, D. Müller: Laser Hardfacing, Steel & Metals Mag., 28(1990)4/5, p. 275-283

[6] H. Paul, L. Morgenthal: Technologien der Laseroberflächenveredelung und Anwendung an Bauteilen, ECLAT´90, p. 311-319

[7] B.L. Mordike: Metallurgical Aspects of Laser Surface Treatments, ECLAT'92, DGM, Göttingen, 1992, p. 171-180

[8] M. Wiedmaier, E. Meiners, T. Rudlaff, F. Dausinger, H. Hügel: Integration of Materials Processing with YAG-Lasers in a Turning Center, ECLAT'92, DGM, p. 559-564

[9] E. Schubert: Modifikation von metallischen Oberflächen mit Excimerlasern: Grundlagen und Anwendungen, Dissertation, Erlangen, 1991

[10] R. Kupfer: Laserstrahlbohren mit Kupferdampflasern, Dissertation, Erlangen, 1992

[11] W. Pompe, B. Schultrich, H.-J. Scheibe, P. Siemroth, H.-J. Weiß: Laser Ablation and Arc Evaporation, ECLAT'92, DGM, Göttingen, 1992, p. 445-450

[12] J. Breuer, S. Metev: Einfluß der UV-Laserbestrahlung auf funktionelle Gruppen in Polymerwerkstoffen, in: M. Geiger, F. Hollmann: Strahl-Stoff-Wechselwirkung bei der Laserstrahlbearbeitung, Meisenbach, Bamberg, 1993, p.55-62

[13] G. Herziger et.al.: Werkstoffbearbeitung mit Laserstrahlung - Teil 1: Grundlagen und Probleme, Feinwerktechnik & Messtechnik, 92(1984)1, p. 156-163

[14] E. Beyer: Einfluß des laserinduzierten Plasmas beim Schweißen mit CO_2-Lasern, Dissertation, Darmstadt, 1984

[15] M. Beck, F. Dausinger: Modelling of Laser Deep Penetration Welding Process, in: European Scientific Laser Workshop, Sprechsaal-Verlag, Coburg, p. 201-216

[16] T. Rudlaff: Arbeiten zur Optimierung des Umwandlungshärtens mit Laserstrahlen, Dissertation, Stuttgart, B.G. Teubner, 1993

[17] K. Wissenbach: Umwandlungshärten mit CO_2-Laserstrahlung, Dissertation, Darmstadt, 1985

[18] F. Dausinger, T. Rudlaff: Steigerung der Effizienz des Laserstrahlhärtens, ECLAT´88, DVS 113, p. 88-91

[19] R. Jaschek, R. Taube, K. Schutte, A. Lang, H.W. Bergmann: Materials Processing Using a Combination of a TEA-CO_2-Laser and a cw-CO_2-Laser, ECLAT'92, DGM, Göttingen,

1992, p. 673-679

[20]   S.Z. Lee, E. Geissler, H.W. Bergmann: On-line Computer Controlled Laser Hardening, LIM-5, p. 301-312

[21]   H.U. Fritsch, H.W. Bergmann: Experimentelle Untersuchungen und mathematische Simulation des Einflusses der Kohlenstoffverteilung auf den Prozeß des Laserhärtens (Teil 1), Härterei Technische Mitt., 46(1991)3, p. 145-154

[22]   R.L. Pierce: The Segmented Aperture Integrator in Material Processing, J. of Laser Applications, Spring (1990), p. 18-22

[23]   D. Burger: Beitrag zur Optimierung des Laserhärtens, Dissertation, Stuttgart, 1988

[24]   E. Geissler: Dissertation, Erlangen, 1993

[25]   A. Drenker, K. Wissenbach, E. Beyer: Adaptive Temperaturregelung beim Umwandlungs-härten mit CO_2-Laserstrahlung, Laser und Optoelektronik, 23(1991)5, p. 48-53

[26]   H. Willerscheid: Prozessüberwachung und Konzepte zur Prozessoptimierung des Laser-strahlhärtens, Dissertation, Aachen, 1990

[27]   R. Jaschek: unpublished results

[28]   J. Bach, R. Damaschek, E. Geissler, H.W. Bergmann: Laser transformation hardening of different steel, ECLAT'90, Sprechsaal, Coburg, p. 265-282

[29]   H.W. Bergmann, E. Geissler: Laser hardening of steels, ECLAT'90, Sprechsaal, p. 321-331

[30]   G. Faninger: Thermisch bedingte Eigenspannungen, Härterei Technische Mitt., 31(1976)1/2, p. 48-51

[31]   J. Domes, D. Müller, H.W. Bergmann: Residual Stresses in Temperature Controlled Laser Hardened Steels, 3rd Europ. Conf. on Residual Stresses, Frankfurt 1992, DGM, 1993

[32]   M. Boufoussi, S. Denis, J.Ch. Chevrier, A. Simon, A. Bignonnet, J. Merlin: Prediction of Thermal, Phase Transformation and Stress Evolutions during Laser Hardening of Steel Pieces, ECLAT'92, DGM, Oberursel, p. 635-640

[33]   J. Domes: priv. communications

[34]   K.-D. Schwager, B. Scholtes, E. Macherauch, B.L. Mordike: Residual Stresses and Microstructures in the Surface Layers of Different Laser Treated Steels, ECLAT'92, DGM, p. 629-634

[35]   R. Damaschek, H.W. Bergmann, D. Müller: Laserstrahlhärten rotationssymetrischer Bauteile, Härterei-Kolloqium 1993, Wiesbaden, in preparation

[36]   W. Wahl: Abrasive Verschleißschäden und ihre Verminderung, VDI-Berichte, Nr. 243, p. 171-187

[37]   E. Broszeit, O. Zwirlein, J. Adelmann: Werkstoffanstrengung im Hertzschen Kontakt - Einfluß von Reibung und Eigenspannungen, Z. Werkstofftechnik, 13(1982), p. 423-429

[38]   H.W. Bergmann: Laser Beam Hardening - State of the Art, Int. Conf. on Surface Eng., Bremen, 1993

[39]   DIN 3990

[40]   H.W. Bergmann, D. Müller, M. Amon, J. Domes: Kombination des Laserstrahlhärtens mit einer Kurzzeitnitrierbehandlung, Härterei Techn. Mitt., in print

[41]   R. Zenker, U. Zenker: Laser Beam Hardening of a Nitrocarburized Steel Containing 0.5% C and 1 % Cr, Surface Eng., 5(1989)1, S. 45-54

[42]   H.W. Bergmann: Short Term Annealing by Laser Treatment, SPIE vol. 801, p. 296-301

[43]   H.W. Bergmann, R. Jaschek: Untersuchung zum Absorptionsverhalten von Kurzpulslasern bei flächiger Bestrahlung von dünnen Oberflächenschichten, in: M. Geiger, F. Hollmann: Strahl-Stoff-Wechselwirkung bei der Laserstrahlbearbeitung, Meisenbach, Bamberg, 1993, p. 21-26

[44]   B. Weidenfeller, W. Riehemann: Einfluß verschiedener Laseroberflächenbehandlungen von Elektroblechen auf ihre magneteischen Eigenschaften, M. Geiger, F. Hollmann: Strahl-

Stoff-Wechselwirkung bei der Laserstrahlbearbeitung, Meisenbach-Verlag, Bamberg, 1993, p.193-198

[45]   A. Gillner, K. Wissenbach, F. Bölling, M. Hastenrath, H.D. Riehn: Verringerung des Wirbelstromverlustes von kornorientiertem Elektroblech durch Laserbehandlung, LASER'87, Springer-Verlag, Berlin, 1987, p. 422-427

[46]   P.L. Peppler: Chilled Cast Iron Engine Valvetrain Components, SAE Technicla Paper Series, 880667, 1988

[47]   G. Werner, J. Ziese: Anwendung physikalischer Oberflächenmeßmethoden an gegossenen Nockenwellen mit dem Ziel, durch definiertes Nitrieren und Oxidieren die Lebensdauer zu erhöhen, Härterei Technische Mitt., 39(1984)4, p. 156-164

[48]   K. Heck: Local Electric Arc Melting Process for the Generation of Wear Resistant White Iron Layers on Grey Cast Iron Camshafts, World Electrical Congress, Moskau, 1977, p. 1-12

[49]   F.H. Reinke: TIG Hardening of Cast Iron Camshafts, Heat Treatment of Metals, (1981)1, p. 17-23

[50]   H.W. Bergmann: Properties of Laser Surface Melted Cast Iron, Opto Elektronik Mag., 3(1987)3, p. 298-302

[51]   H.W. Bergmann: Laser Surface Melting of Iron Base Alloys, in: C.W. Draper, P. Mazzoldi: Laser Surface Treatment of Metals, Martinus Nijhoff Pub., 1986, p. 351-368

[52]   R. Dekumbis, P. Magnin, G. Barbezat: Graphite Dissolution and Porosity Formation during Remelting of Nodular Cast Iron, ECLAT'86, DGM, Oberursel, p. 195-204

[53]   A. Luft, W. Reitzenstein, N. Zarubova, J. Cermak: Untersuchungen zum Laserstrahlumschmelzhärten von Gußeisen, Schweißen & Schneiden, 43(1991)3, p. 137-141

[54]   D. Müller, J. Domes, H.W. Bergmann: Eigenspannungen und Gefügeausbildung nach dem Randschichtumschmelzen von Nockenwellen mit linienfokussierter CO_2-Laserstrahlung, Härterei Techn. Mitt., 47(1992)2, p. 123-130

[55]   A. Zwick, A. Gasser, E.W. Kreutz, K. Wissenbach: Surface Remelting of Cast Iron Camshafts by CO2 Laser Radiation, ECLAT'90, p. 389-398

[56]   S. Mordike: Grundlagen und Anwendung der Laseroberflächenbehandlung von Metallen, Dissertation, Clausthal-Zellerfeld, 1991

[57]   H.W. Bergmann: Current Status of Laser Surface Melting of Cast Iron, Surface Eng., 1(1985), p. 137-155

[58]   C.H. Chen, C.J. Altstetter, J.M. Rigsbee: Laser Processing of Cast Iron for Enhanced Erosion Resistance, Metallurgical Trans. A, 15A(1984)4, p. 719-728

[59]   E. Vandehaar, P.A. Molian, M. Baldwin: Laser Cladding of Thermal Barrier Coatings, Surface Eng., 4(1988)2, p. 159-172

[60]   V.M. Weerasinghe, W.M. Steen, D.R.F. West: Laser Deposited Austenitic Stainless Steel Clad Layers, Surface Eng. 3(1987)2, p. 147-153

[61]   J.H.P.C. Megaw, A.S. Bransden, D.N.H. Trafford, T. Bell: Surface Cladding by Multikilowatt Laser, 3rd Int. Coll. on Welding & Melting by Electron & Laser Beams, Lyon, 1983

[62]   K.P. Cooper: Surface Treating by Laser Melt/Particle Injection, SPIE vol. 957, p. 42-53

[63]   H.W. Bergmann, D. Müller: Erstarrung, Handbuch "Oberflächenbearbeitung mit CO_2-Lasern", VDI-Verlag, in preparation

[64]   P. Gundel, A.S. Kalinitchenko, H.W. Bergmann: Laser Surface Alloying of AlSi12 and AlCr2.5Zr2.5 with Transition Metals an Si, Opto Elektronik Mag., 4(1988), p. 510ff

[65]   E. Hornbogen, A.G. Crooks: Metallurgical Aspects of Laser Alloying of Aluminium, ECLAT'90, Sprechsaal, Coburg, p. 613-624

[66]   E. Blank, O. Hunziker, S. Mariaux: Load Bearing Capability of Laser Surface Coatings for Aluminium-Silicon Alloys, ECLAT'92, DGM, Oberursel, p. 193-198

[67] S.Z. Lee: Thermochemische Behandlung von Titan mit dem CO_2-Laser, Dissertation, Erlangen, 1992

[68] J. Folkes, W.M. Steen, D.R.F. West: Laser Surface Alloying of Titanium Substrates with Carbon and Nitrogen, Surface Eng., 1(1985)1, p. 23-29

[69] A. Gasser, E.W. Kreutz, K. Wissenbach: Gaslegieren von TiAl6V4 mit CO_2-Laserstrahlung, ECLAT'88, DVS 113, p. 157-150

[70] A. Weisheit, B.L. Mordike: Laser Surface Alloying of Titanium with Nickel and Cobalt, ECLAT'92, DGM, Oberursel, p. 229-234

[71] D. Müller, H.W. Bergmann, T. Heider, T. Endres: Laser Carburizing of a Low Carbon Steel, ECLAT'92, DGM, Oberursel, p. 293-297

[72] J. Shen, F. Dausinger, B. Grünenwald, S. Nowotny: Möglichkeiten zur Optimierung der Randschichteigenschaften eines Einsatzstahls mit CO_2-Lasern, Laser und Optoelektronik, 23(1991)6, p. 41-49

[73] F. Behr, Ch. Arndt, E. Haberling: Production of Surface Layers on Tool Steels by Alloying and Addition of Hard Particles Using Laser Energy, ECLAT'90, Sprechsaal, Coburg, p. 411-416

[74] K. Wissenbach, A. Gasser, A. Gillner, E.W. Kreutz, W. Amende: Einschmelzlegieren mit Hochleistungslasern, Laser und Optoelektronik, (1985)4, p. 376-384

[75] B.L. Mordike, H.W. Bergmann: Surface Alloying of Iron Alloys by Laser Beam Melting, in: B.H. Kear, B.C. Giessen, M. Cohen (eds.): Rapidly Solidified Amorphous and Crystalline Alloys, Elsevier, 1981, p. 463-483

[76] T. Bell, Z.L. Zhang, J. Lanagan, A.M. Staines: Plasma Nitriding Treatments for Enhanced Wear and Corrosion Resistance, in: K.N. Stafford (ed.): Coatings and Surface Treatments for Corrosion and Wear Resistance, 1985, Chichester, Ellis Horwood, p. 164-177

[77] H.W. Bergmann: Thermochemische Behandlung von Titan und Titanlegierungen durch Laserumschmelzen und Gaslegieren, Z. Werkstofftechnik, 16(1985), p. 392-405

[78] D. Müller, H.W. Bergmann, H. Sibum, L. Bakowsky, E. Kappelsberger: Lasergaslegieren von Ti-Werkstoffen, Dünnschichttechnologien'90 - Band I, VDI-Verlag, p. 222-236

[79] U. Fink, H.W. Bergmann: Laser Surface Treatments of Implants, ECLAT'90, Sprechsaal, Coburg, p. 451-460

[80] B.L. Mordike, H.W. Bergmann, U. Luft: Laserumschmelzen von Aluminiumlegierungen, Opto Elektronik Mag., 3(1987), p. 435ff

[81] T.-C. Zou, Ch. Binroth, G. Sepold: Eigenschaften laserumgeschmolzener Aluminiumlegierungen, ECLAT'90, Sprechsaal, Coburg, p. 671-680

[82] A.S. Bransden, T. Endres: High Power Laser Processing of Aluminium Alloys, ECLAT'92, DGM, Oberursel, p. 117-123

[83] T. Endres, Dissertation, Erlangen, 1993

[84] E. Blank, Laserkolloquium, Erlangen 1990

[85] E.W. Kreutz, N. Pirch, M. Rozsnoki: Solidification in Laser Surface Alloying of Al an AlSi10Mg with Ni and Cr, ECLAT'92, DGM, Oberursel, p. 269-281

[86] H. Haddenhorst, E. Hornbogen: Laser Fusion of Ceramic Partcles on Aluminium Cast Alloys, ECLAT'92, DGM, Oberursel, p. 245-251

[87] H. Lindner, H.W. Bergmann, T. Endres: Remelting and Alloying of Al-Si Alloys, ECLAT'90, Sprechsaal, Coburg, p. 661-670

[88] A. Schüßler, K.-H. Zum Gahr: Oscillating Sliding Wear of TiC and TiN Laser Hardfacings, Mat.-wiss. u. Werkstofftechnik, 22(1991), p. 10-14

[89] J. Singh, J. Mazumder: In-situ Fotmation of Ni-Cr-Al-R.E. Alloy by Laser Cladding with Mixed Powder Feed, LIM-3, 1986, p. 169-179

[90] St. Nowotny, J. Shen, F. Dausinger: Verschleißschutz durch Auftragschweißen von Stellit 21 mit CO_2-Laser, Laser und Optoelektronik, 23(1991)6

[91]   W. Amende, G. Nowak: Hard Phase Particles in Laser Processed Cobalt Rich Claddings, ECLAT´90, p. 417-428

[92]   P.J.E. Monson, W.M. Steen: Comparison of Laser Hardfacing with Conventional Processes, Surface Eng., 6(1990)3, p. 185-193

[93]   T. Takeda, W.M. Steen, D.R.F. West: Laser Cladding with Multi Elemental Powder Feed, LASER´85, Springer-Verlag, Berlin, p. 394-398

[94]   K.P. Cooper, J.D. Ayers: Laser Melt-Partcle Injection Processing, Surface Eng., 1(1985)4, p. 263-272

[95]   Ch Binroth, R. Becker, G. Sepold: Zusatzdraht erweitert die Einsatzgebiete der Lasermaterialbearbeitung, ECLAT´90, p. 717-720

[96]   D. Burchards, A. Hinse, B.L. Mordike: Laserdrahtbeschichten, Z. Mat.wiss. und Werkstofftechn., 20(1989), p. 405-409

[97]   V. Fux, A. Luft, D. Pollack, St. Nowotny, W. Schwarz: Laser Cladding with Amorphous Hardfacing Alloys, ECLAT'92, DGM, Göttingen, 1992, p. 387-392

[98]   E. Lugscheider, H. Bolender, H. Krappitz: Laser Cladding of Paste Bound Hardfacing Alloys, Surface Eng., 7(1991)4, p. 341-344

[98]   W. König, D. Scheller: Kombination von Sensorsystemen zur Überwachung umschmelzender Verfahren der Laserstrahlbearbeitung, in: M. Geiger, F. Hollmann: Strahl-Stoff-Wechselwirkung bei der Laserstrahlbearbeitung, Meisenbach, Bamberg, 1993, p. 89-94

[99]   G. Backes, A. Gasser, E.W. Kreutz, K. Trojan, K. Wissenbach: Prozeßdiagnose beim Beschichten mit CO_2-Laserstrahlung, LASER'91, Springer-Verlag, Berlin, 1992, p. 249-254

[100] W.J. Tomlinson, A.S. Bransden: Laser Surface Alloying Grey Iron with Cr, Ni, Co, and Co-Cr Coatings, Surface Eng., 6(1990)4, p. 281-286

[101] K. Mori, S. Takagi, S. Ohishi, I. Torii, S. Yamada: Application of Laser Cladding for Engine Valve, Joint Sy,p. on "Creation of High Function Layers on Materials", Osaka, 1989, p. 149-152

[102] F. Küpper, A. Gasser, K. Wissenbach, E.W. Kreutz: Cladding of Valves with CO_2 Laser Radiation, ECLAT´90, Sprechsaal, Coburg, p. 461-468

[103] W. König, L. Rozsnoki, P. Kirner: Laser Beam Surface Treatment - Is Wear no longer the Bug Bear of Old, ECLAT'92, DGM, Göttingen, 1992, p. 217-222

[104] P. Feinle, G. Nowak, W. Amende: Auflegieren von Randschichten mit CO_2-Laserstrahlen unter Einbeziehung von Kupfer- und Aluminiumlegierungen, LASER'87, Springer-Verlag, Berlin, p. 452-458

Materials Science Forum Vols. 163-165 (1994) pp. 405-410
© *1994 Trans Tech Publications, Switzerland*

MECHANICAL PROPERTIES AND MICROSTRUCTURE OF LASER TREATED Al ALLOYS

J. Noordhuis and J.Th.M. De Hosson

Research Group Materials Science, Department of Applied Physics,
University of Groningen, Nijenborgh 4, NL-9747 AG Groningen, The Netherlands

ABSTRACT

This paper concentrates on the mechanical properties and microstructural features of an Al-Cu-Mg alloy, Al 2024-T3, that was exposed to laser treatments at various scan velocities. As far as the mechanical property is concerned a striking observation is a minimum in the hardness value at a laser scan velocity of 1/2 cm/s. Usually an increasing hardness with increasing laser scan velocities is reported in the literature. This remarkable property could be explained based on the microstructural features observed by transmission electron microscopy.

After subsequent shot peening, in all cases the formation of precipitates was observed, independent of the laser scan velocities originally applied. This phenomenon of precipitation, induced by shot peening afterwards is most striking at a high concentration of alloying elements in solid solution. Since this is the case in samples treated at a low scan velocity (i.e. an increased homogenization time) as well as in samples treated at a high scan velocity samples (high quench rates), the parabola shape is maintained even after shot peening.

1. INTRODUCTION.

At the first ASM Heat Treatment and Surface Engineering conference we reported [1] on our novel approach to strengthen Al-Si alloys by a laser treatment followed by an additional shot peening treatment. It turned out that a laser treatment of a cast aluminium-silicium alloy leads to a very fine dispersion of the silicon phase in the Al-matrix. In addition, it was shown that the laser treatment amplifies substantially the effectiveness of a subsequent shot peening treatment. In particular the maximum attainable hardness and compressive stress turned out to increase upon increasing quench rate, i.e. upon increasing the laser scan velocity. The appearance of shot peening induced precipitates forms the most striking observation in that study.

This paper concentrates on the mechanical properties and microstructural features of an Al-Cu-Mg alloy, Al 2024-T3, that was exposed to laser treatments at various scan velocities followed by shot peening treatments. Aluminium alloys containing Cu, Mg and small weight percentages of other alloying elements have shown to develop high strength after appropriate heat treatments. Nevertheless improvement of the mechanical properties along conventional processing routes has its limits, and alternative techniques like splat cooling and laser melting are increasingly being explored. Conventional hardening treatments of aluminium-base alloys usually exploits the beneficial effects of a high concentration of precipitates. In contrast, hardening by laser treatments is achieved primarily by the small grain sizes that are formed as a result of the rapid solidification process. The flow stress varies as $d^{-\alpha}$ where α depends on the cell size and cell wall-type. Since the cell size is inversely related to the solidification rate, commonly an increasing hardness is observed with increasing laser scan velocity.

2. EXPERIMENTS.

Samples of commercial Al 2024-T3 (4.4% Cu, 1.5% Mg, 0.6% Mn and Bal. Al.; solution treated, cold rolled and naturally aged) were sandblasted to obtain a rough well absorbing surface. After ultrasonically cleaning the samples were irradiated using a transverse flow Spectra Physics 820 CW-CO_2 laser under a protective argon atmosphere. At the surface the power of the beam was 1300 W. The focus point of the ZnSe lens with focal length of 127 mm lay 5 mm above the surface, resulting in a spot diameter of 0.75 mm. Single tracks were made at laser scan velocities ranging between 1/8 to 25 cm/s. On the samples for analysis by X-ray diffraction several adjacent tracks were laid down.

It should be emphasized that of all the elements in Al 2024, magnesium has by far the lowest boiling point (1390 K), and it is therefore to be expected that a substantial fraction of magnesium will actually evaporate.

Shot peening was carried out in a conventional blast cleaning apparatus, applying glass beads with an average diameter of 720 μm. The air pressure was 2.6 bar. Vickers hardness measurements were carried out just below the surface of a taper sectioned sample, with a 100 gram weight. TEM samples were prepared using a mechanical dimpler, followed by ion milling in the cold stage of a Gatan ion mill, in such a way that the electron transparent area is located in the centre of the laser track at a depth of approximately 30 μm. The hardness measurements provide information of this area as well.

X-ray analysis applying Cu-radiation, revealed the stress state after laser melting and subsequent shot peening, utilising the <422> reflection. Line profile analyses on the <111> reflection were carried out to obtain information about the dislocation density. For further experimental details we refer to [2].

3. RESULTS.

After laser melting a cellular structure develops during solidification. Cracks are occasionally observed in samples treated at low as well as high laser scan velocity, indicating a reduced ductility at both ends of the scan velocity range. The measured cell sizes are depicted in Fig. 1. The intermetallic compounds mainly consist of the $CuAl_2$ (θ) phase, and the $CuMgAl_2$ (S) phase. These phases were identified with by X-ray diffraction.

Because of the rapid hardening right after the laser treatment, all hardness measurements presented in the following are performed on samples aged for at least two weeks. The surface hardness as a function of laser scan velocity, is displayed in Fig. 2. This graph exhibits a parabolic shape, with a minimum value at a scan velocity of ·1/2 cm/s. After subsequent shot peening the shape of the hardness curve remains approximately the same, although it shifts to higher values.

TEM specimens taken from samples exposed to different laser scan velocity samples before shot peening show marked differences in microstructural features. The samples treated at a lower scan velocity, 1/8 cm/s to 1/2 cm/s, are mainly precipitate free although in some areas coarser precipitates could be observed. However, it is uncertain whether these are partly dissolved cell boundaries or real precipitates.

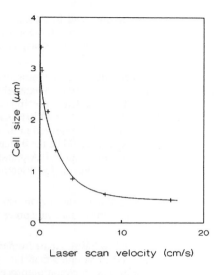

Fig 1 Cell size as a function of laser scan velocity.

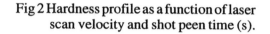

Fig 2 Hardness profile as a function of laser scan velocity and shot peen time (s).

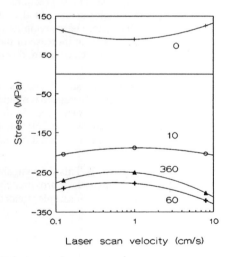

Fig 3 Residual stress as a function of laser scan velocity and shot peen time.

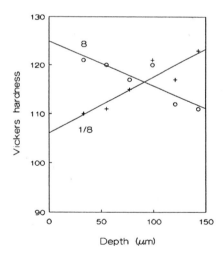

Fig 4 Hardness profile as a function of depth of two samples treated at a low and a high laser scan velocity.

The most striking feature is the formation of helical dislocations as being observed in all samples treated at low laser scan velocity. Most of these dislocations exhibit a uniform equilibrium shape with respect to pitch and radius. At the lowest scan velocity however, non uniform shapes are occasionally observed.

In samples treated at laser scan velocities between 1 cm/s to 4 cm/s the precipitation of the tetragonal θ' platelets, with an average size of 150 nm, is observed. Since these platelets are oriented parallel to the {100} faces of the α Al, we may conclude that these precipitates are not the so-called S' precipitates (Al$_2$CuMg). The S' precipitates grow locally on {120} planes, giving {110}, {100} or {130} as an overall growth plane [3][4]. Beside the platelets a more globular precipitate is also observed in these samples, mainly nucleating on the edges of the θ' plates and on dislocations. This could be the θ phase as well as the S' phase. Since in alloys with a copper:magnesium ratio less than 8:1 Al$_2$CuMg precipitates can start to play a role [5], the existence of the S and S' phase cannot be excluded. Electron diffraction patterns have not yet revealed the nature of these precipitates.

Samples treated at laser scan velocities of 8 cm/s to 25 cm/s show only a cellular structure, in which it is seen that the cell boundaries act as obstacles for dislocations motion. GP-zones, that could precede the formation of the θ' plates, are not observed.

After shot peening, in all samples a higher dislocation density was observed by TEM. Apart from the increased dislocation density however, now precipitates could be detected in samples that did not develop observable precipitates before shot peening. A typical example of this precipitation process induced by shot peening in a sample treated at a low laser scan velocity of 1/8 cm/s. The concentration of precipitates seems also to be higher than before peening in a samples treated in the scan velocity range of 1 cm/s to 4 cm/s. This might indicate an increased solid solution in these samples, but the smaller average size, 100 nm, will also result in a higher concentration. Again, most of these precipitates are θ'-platelets, aligned with the cubic axes. A second type of precipitate, smaller in size but also of the shape of a platelet, is identified as the S' phase, initially growing on [120] planes. These precipitates have a much smaller width-to-thickness ratio and will therefore contribute less to an increase of the flow stress than the θ'-plates (at the same volume fraction). The very broad size distribution, down to only 2 nm, makes this contribution more uncertain.

X-ray stress measurements reveal a similar parabola shaped curve for the residual stress state, as is displayed in Fig. 3. Before shot peening the stress state has a tensile character with an average magnitude of 100 MPa. After shot peening the stress is inverted to a compressive one with a magnitude in the order of 250 MPa, while maintaining the parabola shape. The magnitude of the stress after 360 s shot peening is lower than after 60 s. This is not due to an inaccuracy of the measurement. In fact it turned out to be reproducible by peening the 60 s sample for an extra 300 s period.

Line profile analysis was applied to obtain information on the dislocation density [6]. This shows again a minimum as a function of laser scan velocity. The increase in dislocation density is approximately the same for all different samples except for the decrease after 360 s of shot peening of samples treated at a high scan velocity.

4. DISCUSSION AND ANALYSIS.

To discuss the experimental results, the laser scan velocity range is divided into two regimes: the low scan velocity regime (1/8 cm/s to 1/2 cm/s) and the high scan velocity regime (1/2 cm/s to 25 cm/s).

low scan velocity regime

Since the cooling rates are slower at the low laser scan velocity, compared to the 2 cm/s, the latter of which precipitation of the θ' plates is observed, at first sight one would expect to see an increasing nucleation an growth of precipitates in these samples. In contrast, below 1 cm/s no significant precipitation is observed. The explanation for this observation lies in the decreasing vacancy concentration with decreasing laser scan velocity. Apparently at the lower laser scan velocity the vacancies are annealed out, a.o. by absorption on dislocations that causes the formation of helical dislocations, and before the temperature reaches values that would favour the nucleation and growth of observable precipitates. Therefore the rise in hardness values in the low velocity regime can not be caused by precipitation hardening but due to two other contributions:

First, the dissolution of the cell-walls will play an important role. Since there exists only a small temperature interval during which homogenization takes place [5], the cooling rate must be low enough to allow sufficient time for diffusion. The copper (and magnesium) brought into solid solution will contribute to an enhanced hardness but for reasons described above, copper will not form any precipitate. The dissolution may continue as long as not all cell walls have been fully dissolved, provided that the cooling rate is low enough. Obviously this situation has not been reached in practice, and it explains why the hardness is still increasing with decreasing laser scan velocity. In some additional experiments performed on an alloy with 2.32 w% Cu, it was observed that only near the edges of the low scan velocity laser tracks the cell walls were completely dissolved. From this experiment we conclude that to a first approximation only half of the copper is in solid solution in Al2024 treated with a scan velocity of 1/8 cm/s. A calculation of the diffusion distance in a 1/8 cm/s track, during cooling in the interval between 660 °C and 450 °C, applying an analytical model [7] yields a distance of approximately 0.1 μm. A calculation for magnesium yields a similar value. These value, that should only be regarded as a first approximation, are too small for significant homogenization. Nevertheless, it shows that homogenization may play a role only at the lowest scan velocities.

An indication that supports homogenization as being the responsible mechanism for the hardness increase in the low scan velocity regime, lies in the observation that in a laser track of 1/8 cm/s scan velocity the hardness increases with depth, as displayed in Fig. 4. This is a result of the increased homogenization time with increasing depths. In laser tracks where this mechanism does not operate, e.g. a 8 cm/s track, hardness decrease is detected with increasing depth. This decrease is also observed in other materials [2] where indeed homogenization is not a likely process to occur.

Beside homogenization, another contribution to the increased hardness in the low scan velocity regime concerns the formation of helical dislocations. Since glide of a helical dislocation is restricted to its cylindrical surface on which it is wound, these dislocation are essentially immobile. The formation of helical dislocations is observed in all samples treated at low laser scan velocity. The large quantity of vacancies is present because of the substantial cooling rates from high temperatures. The reason that the formation is possible only in samples treated at the low laser scan velocity is that only here the cooling rate at intermediate temperatures is low enough to allow for adequate vacancy diffusion. Once the temperature is too low for significant vacancy diffusion, the dislocations become immobile over the distance that has been affected by climb or by the interaction with the vacancy loop. The equilibrium shape of a helical dislocation is one of a constant pitch and loop radius [8]. Since a non-equilibrium shape is sometimes observed, we conclude that in the sample treated at the lowest scan velocity the anneal time is still not long enough for all the dislocations to reach this equilibrium situation. This in turn can be caused by the continuous formation of new helical segments, which means that not all vacancies are used, or by an increased locked segment through relaxation to the equilibrium shape. These explanations are also in line with the fact that the hardness curve is still rising at the 1/8 cm/s point. Nevertheless we do not expect that this mechanism will dominate over the role of homogenization as described first.

high scan velocity regime

The increased hardness in the high velocity regime, reflects the general observed trend, and is caused by two reasons. First, the cell size decreases upon increasing the laser scan velocity and causes a hardening. Furthermore, the observed precipitates will contribute to the hardness values, and when this contribution gets large the contribution arising from the small cell size may become negligible. The presence of these precipitates is observed in samples with scan velocities of 1 cm/s to 4 cm/s.

The absence of GP zones in samples that have not yet developed precipitates, in the range between 8 cm/s and 25 cm/s, might be explained by the fact that the amount of Cu in solid solution is less compared to conventionally heat treated Al 2024. In Al with 2 w% Cu, GP-I zones do not appear, and the precipitation sequence starts with GP-II [5]. Furthermore it has been reported that in rapidly quenched samples the early stages in the precipitation sequence are accelerated and that GP-II zone formation is suppressed in favour of the θ' phase. If both effects are present in our samples the θ' phase is indeed the first precipitate to be observed.

After shot peening an increase in hardness that is independent of the laser scan velocity is observed. This is not what one would expect since the microstructure is quite different at different scan velocities. The precipitation induced by shot peening, that was observed earlier in a laser melted Al-Si alloy [2], explains why the observed values are the same for the low and high scan velocities: After shot peening precipitates are present independent of the scan velocity regimes, whereas the amount of alloying elements in solid solution is expected to be much less than before shot peening [1][2].

The minimum value of the hardness of samples treated at a scan velocity of 1/2 cm/s is still present after shot peening, and is also visible in the curves of the residual stresses as well as of the dislocation density vs laser scan velocity. Apparently the amount alloying elements brought into solid solution during laser melting has a minimum for this scan velocity. At higher scan velocities more copper and magnesium is retained during the rapid solidification process, and for the lower ones the homogenization time during cooling is longer.

5. CONCLUSIONS.

The hardness curve observed in this study, with a minimum at a laser scan velocity of 1/2 cm/s, is quite different from the monotonically increasing curves, usually detected in many other laser treated materials. The reason for this observation is the minimum in the solid solution concentration. By varying the quench rate it is possible to achieve different stages in the precipitation sequence.

The mechanisms by which the hardening occurs are the same as the ones in conventionally hardened specimens, except for the contributions originating from the small cell sizes and from the formation of helical dislocations. The latter ones however do not seem to contribute significantly.

Shotpeening leads to a further hardness increase, which can be attributed to the creation of (additional) θ' and S' precipitates. An increased dislocation density contributes to the increased hardness as well.

references

[1]   De Hosson, J.Th.M., Noordhuis, J.:Mat. Sci. Forum 1992 (Ed. E. Mittemeijer), Vols. 102-104, 393.

[2]   Noordhuis, J., De Hosson, J.Th.M., Acta Metall et Mat. 1992, 40, 3317. and Noordhuis, J., De Hosson, J.Th.M., Acta Metall et Mat. 1993, 41, in press.

[3]   Wilson, R.N., Partridge, P.G., Acta Metall.,1965, 13, 1321.

[4]   Stolz, R.E., Pelloux, R.M., Met. Trans.,1976, 7A, 1295.

[5]   Mondolfo, L.F., Aluminum alloys structure and properties, 264 (1976).

[6]   Williamson, G.K., Smallman R.E., Philos. Mag., 1956, 1, 34.

[7]   Ashby, M.F., Easterling, K.E., Acta Metall., 1984, 32, 1935.

[8]   Hirth, J.P., Lothe, J., Theory of dislocations, McGraw Hill, New York, 1982, 623.

Materials Science Forum Vols. 163-165 (1994) pp. 411-416
© *1994 Trans Tech Publications, Switzerland*

SURFACE ALLOYING OF Ti6Al4V ALLOY BY PULSED LASER

A. Zambon [1], E. Ramous [1], M. Bianco [2] and C. Rivela [2]

[1] DIMEG - Università di Padova, Italy

[2] Istituto RTM, Vico Canavese (Torino), Italy

ABSTRACT

The feasibility of gas surface alloying of Titanium alloys by laser has been already studied. However relevant difficulties in the aforesaid process result in surface chemical heterogeneity and roughness.
We tried to overcome these problems by optimizing the nitriding and carburizing treatments with a pulsed Nd-YAG laser.
Treated samples were examined by metallographic techniques, XRD and by hardness and roughness tests.
Results showed that carburized layers exhibit better properties compared with nitrided ones, both as hardness profiles and surface roughness. Preliminary wear tests confirmed the better behaviour of carburized layers.

INTRODUCTION

It is well-known that the utilization of titanium, which exhibits both good mechanical properties and low specific gravity, is hindered by its high friction coefficient, with a consequent low wear resistance.
Its is well established the possibility to improve titanium surface properties with suitable treatments such as salt bath or gas and plasma nitriding. In any case the layer thickness is very limited, and should be increased, specially for wear resistant applications under high specific loads, at least up to some tenths of millimeter. This could be achieved enhancing Nitrogen absorption by liquid phase alloying: the formation of nitrides would take place upon cooling of the surface layer consisting of TiN, helped by the fact that its estimated solidification temperature is about 3290 °C [1], compared to that of titanium (1660 °C).
The process of liquid phase alloying is made possible by high power laser beams which permit to melt in very short interaction times surface layers 1 to 2 mm thick, without altering the microstructure of the inner part of the workpiece. The possibility to perform titanium nitriding by means of laser irradiation under a N_2 atmosphere or even in air has already been described [2], [3], [4].

Another possible surface alloying, aimed at obtaining wear resistant surfaces, is to promote Carbon absorption from a surrounding atmosphere, in order to obtain Titanium carbides rich surface layers, again by means of liquid phase alloying.

The formation of Titanium carbides during laser irradiation has already been investigated by Walker et al. [2].

Unfortunately the laser sources which have so far been used, that is mainly cw CO_2 lasers, are very expensive due to the high power densities needed to promote surface melting.

Far cheaper laser sources are available, such as pulsed Nd-YAG or excimer lasers.

In the present work the feasibility of titanium laser nitriding and laser carburizing by means of Nd-YAG laser is studied and a comparison of the respective microstructures, microhardness profiles and tribological behaviour in pin on disk tests is proposed.

METHODS AND MATERIALS

The material used in all the tests was the well-known alloy Ti6Al4V. The tests were carried out with a Nd-YAG Lumonics JK series 700 laser unit whose main characteristics were:

Mean Power	0 - 250	W
Pulse Energy	0.1 - 35	J
Pulse duration	0.5 - 20	ms
Frequency	0.2 - 500	Hz

A secondary He-Ne laser permitted to accurately spot the 1.06 μm laser beam onto the target, while the resonating cavity permitted to lase a rectangular pulse. Various parameters permit to define the laser processing:

S_s [mm] = spot size (diameter)
v [mm/s] = traverse speed, that is the speed of a worktable which displaces the specimen
t_s [s] = pulse duration
f [Hz] = frequency
P_i [W] = single pulse power

Other parameters are related to the above said, that is:

E_i [J] = P_i*t_s = pulse energy
P_o [W] = P_i*t_s*f = mean power
P_s [W/mm2] = P_i/A = specific power
ON/OFF ratio= t_s*f = pulse duration / period
$d_i = f*S_s/v$ = pulse density

In particular d_i defines the number of laser pulses that hit one definite point on the surface during the treatment.

The tests arrangement consisted of a pulsed laser beam which impinged perpendicularly on the specimen surface, which was traversing at constant speed beneath it.

The alloying gas was supplied by means of a nozzle in close proximity of the surface to be lased, under an angle of 45° in the same direction of the traverse speed, so that the gas could also shield the just treated surface.

RESULTS and DISCUSSION

Laser Nitriding

Scope of the tests was the identification of the proper operative parameters, in order to obtain a surface as smooth as possible. The different tests immediately showed the influence of the frequency and of the traverse speed on surface roughness. The higher is the pulse density, that is the product of the pulse frequency by the time necessary to traverse the spot diameter, the better is the surface aspect, up to a pulse density d=200 (value obtained with high

frequencies, due to the fact that the worktable could not traverse at speed lower than 1 mm/s).
In optimizing surface roughness the parameter "ON/OFF", that is the fraction of the pulse period during which the laser beam is actually striking the specimen surface, which with the power density per pulse and the pulse frequency defines the energy transfer in a definite point of the specimen surface, has been determined to be 0.06.
Different combinations of lasing frequency, power, traverse speed and pulse density have been considered.
The influence of such parameters on surface roughness, on nitrided layer thickness and on the hardness profiles was appreciable and can be roughly summarized as follows:
the absence of surface cracks seems to be related to the frequency: good values are over 100 Hz; surface roughness seems to be related to the density. In specimen treated with the same density, frequency and mean power, surface roughness seems to decrease with pulse duration, that is increasing single pulse power;
nitrided layer thickness is related, being the ON/OFF ratio constant, to the product P_i times d_i: values of 6000W were related to layer thickness of say $10\mu m$, 30000W to about 15 μm, 480000W to $160\mu m$;
surface hardness is related to specific power.
The aforesaid technological characteristics are obviously related to the microstructure of the samples. Laser processing is extremely rapid and takes fractions of a second to be completed, so that the resulting structure is a non-equilibrium one.

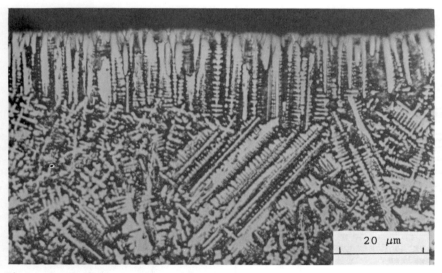

Figure 1. Dendrite configuration in nitrided sample.

A compact layer of columnar dendrites originating from the specimen surface is visible (figure 1), which consist, according to the Ti-N phase diagram, of TiN, with a Nitrogen amount over 30 at%, which solidifies at 2950-3290°C. During this first stage of the solidification process some latent heat of fusion is released, which promotes nitrogen diffusion towards the bulk. In the second stage of solidification, dendrites formation beneath the surface layer takes place, corresponding to the peritectic reaction (at about 2350 °C) between alpha Ti and delta TiN. Moreover, according to [3], this dendrite layer consists of non stoichiometric TiN_x, with x ranging between 0.65 and 0.85.

Underneath this zone, as the nitrogen content further decreases, a second peritectic transformation takes place, between beta Ti and alpha Ti. Upon cooling the beta phase transforms in martensite.

Such micrographic configuration results in hardness profiles ($HVN_{0.3}$) exhibiting surface hardness values in the range 700-1000 kgf/mm2, rapidly decreasing to the bulk hardness. The 550 kgf/mm2 hardened zone ranges, for the most significant samples, in the interval 100-160 μm (figure 2a).

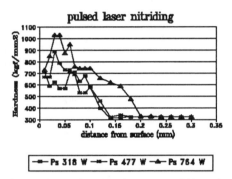

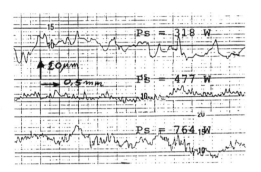

Figure 2a. Microhardness profiles in nitrided specimen

Figure 2b. Corresponding surface roughness profiles

Laser Carburizing

Microhardness profiles in carburized specimen (figure 3a) generally showed higher values than those obtained with N_2 atmosphere;

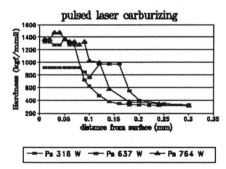

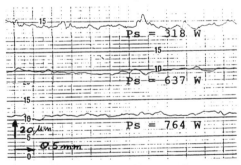

Figure 3a. Microhardness profiles in carburized specimen

Figure 3b. Corresponding surface roughness profiles

in similar working conditions this result not only concerns the value of the surface microhardness (figure 4a), but also the hardened layer thickness is higher. Moreover surface roughness results lower (figure 4b).

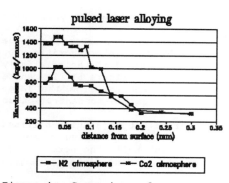

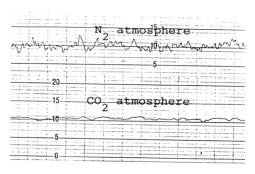

Figure 4a. Comparison of
microhardness profiles of nitrided
and carburized layers

Figure 4b. Corresponding surface
roughness profiles

The high surface hardness values, together with the good surface roughness, let
us foresee a good tribological behaviour.
It can be hypothesized that such favourable combination of elevated hardness and
low roughness may be caused by the additional energy coming from the exothermic
reaction of titanium oxidation. In effects $CO2$ dissociates, at elevated
temperatures in CO and O_2 according to the:
$2 CO_2 \longrightarrow 2CO + O_2$.
Diffraction analysis performed on laser carburized specimen confirmed such
hypothesis, showing, together with the Gamma TiC peaks, also delta TiO ones [5].

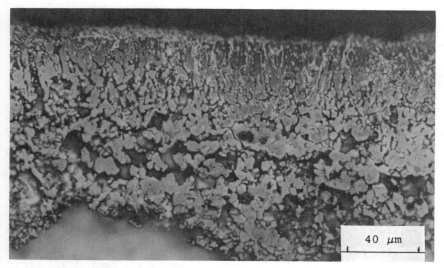

Figure 5. Carburized layer morphology, with dendritic TiC and former
interdendritic TiO.

The metallographic observations have been carried out to point out, in
particular, the TiC phase. The reagent consisted of Sodium Molibdate 3g, HCl

5ml, Ammonium Difluoride 2g, in 100ml H_2O, which colors in brownish yellow the TiC phase and in greenish blue the titanium matrix.

In the laser treated sample, with the CO_2 atmosphere, two phases can be identified: a dendritic phase, which consists of TiC, and an interdendritic one, which, according to the diffractometric measurements, consists of the relatively low-melting TiO, which is strongly corroded by the reagent (figure 5). It can be reasonably supposed that the different brown tones proceeding from the laser treated surface towards the bulk, are due to different carbon concentration so that non stoichiometric TiC phases are produced.

Comparative wear tests without lubrication were carried out on annular nitrided and carburized tracks obtained with the optimized conditions, using a pin on disk configuration, the counter being a 10 mm AISI 52100 ball. The normal loads were 10N and 20 N, the relative speed 0.15 m/s and the sliding distance ranging between 1 and 6 km.

The results are summarized in the plots of figure 6a,b from which the better wear behaviour of the carburized layer is clearly visible. It can be observed that, in the 20N test the nitrided layer behaviour becomes even worse than that of the base Ti6Al4V alloy, probably due to surface delamination.

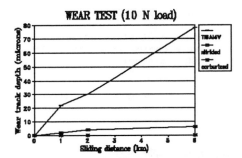

Figure 6a. Comparative wear tests plot.

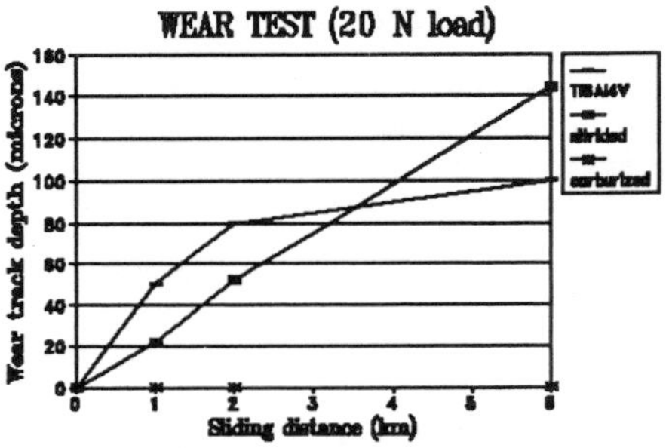

Figure 6b. Comparative wear tests plot.

CONCLUSIONS

Laser surface nitriding and carburizing of the Ti6Al4V alloy has been performed by means of a pulsed Nd-YAG laser, using Nitrogen and Carbon Dioxide gas flows.

The carburized specimen exhibited higher hardness and lower surface roughness compared to that of nitrided specimen. Diffraction analysis as well microscopy observations of the latter specimen alloyed layers permitted to identify a compound structure consisting of dendritic TiC and interdendritic TiO.

Wear tests results showed a better behaviour of the carburized samples compared to the nitrided ones.

References

1) ASM Handbook, Alloy Phase Diagram, vol. 3, pag 2.229
2) Walker A., Folkes J., Steen W.M. and West D.R.F.: Surface Eng. 1985, 1, 23-29
3) Bell T., Bergmann H.W., Lanogan J., Morton P.H. and Staines A.M.: Surface Eng. 1986, 2, 133-143
4) Cantello M., Bianco M.,Lorenzi M., Giordano L., Ramous E.: Proc. XXIII AIM Nat. Cong. (Italy) sept. 1990, 427-433
5) Bianco M., Rivela C., Zambon A. Costantini G.: Proc. XIV National Cong. Heat Treatments May 1993, 73-83

Materials Science Forum Vols. 163-165 (1994) pp. 417-422

EFFECT OF PROCESS PARAMETERS ON THE MICROSTRUCTURE, GEOMETRY AND MICROHARDNESS OF LASER-CLAD COATING MATERIALS

K. Komvopoulos

Department of Mechanical Engineering, University of California, Berkeley, CA 94720, USA

ABSTRACT

The effects of independent process parameters on the microstructural characteristics, geometry and microhardness of Fe-Cr-W-C coatings deposited on low-carbon steel by laser cladding were investigated experimentally. The rapidly solidified microstructures possessed fine dendritic and feathery type morphologies. The small characteristic arm spacing of the dendrites in the harder microstructures indicated that the cooling rates must have been in the range of 10^3 to 10^6 K/s. Refined and homogeneous microstructures exhibiting microhardnesses as high as 900 kg_f/mm^2, thin coatings with a metallurgical bond, and small heat-affected zones (HAZ) were obtained by adjusting the laser power, process speed, and powder feed rate of cladding material. The primary dendrites and the interdendritic matrix were rich in Fe and Cr, respectively. The microstructure of the harder laser-clad coatings comprised heavily dislocated primary dendrites of face-centered cubic (fcc) austenite and eutectic matrix consisting of M_7C_3 carbides randomly distributed in an fcc austenitic phase. Increasing the process speed and reducing the laser power and powder feed rate generally enhanced the microstructure fineness and microhardness and reduced the thickness of the laser-clad coating and the HAZ.

INTRODUCTION

The use of lasers for modification of surface microstructures and material properties has emerged as a field of growing importance to materials science and engineering. Laser-assisted surface engineering provides unique means of developing novel nonequilibrium microstructures possessing improved tribological and corrosion properties. The classification of the different laser processing regimes is based on the energy requirements for laser heating, melting and vaporization corresponding to the particular treatment [1-3], as shown in figure 1. In laser cladding, a high-power laser beam melts a thin layer of the substrate that mixes with the liquid cladding alloy to produce a surface coating exhibiting different metallurgical properties than the substrate.

Laser cladding is influenced by a number of process variables, as well as the interdependence of these variables. The adsorption of the laser energy and the thermal properties of the materials control the thermal history of the clad melt, such as the temperature profile in the clad zone, the time for which the material is molten, and the solidification rates. The overall composition and microstructure of the clad material are determined by the degree of mixing due to convection and diffusion and the cooling rates during liquid-solid and solid-solid state transformations. In view of the high cooling rates, in the range of 10^3 to 10^6 K/s, fine microstructures, increased solid solubility of alloy elements, and nonequilibrium crystalline and amorphous phases can be produced. For example, fine-grained dendritic and hard microstructures were developed by laser cladding Co-Cr-W-C and Ni-Cr-B-Si-C alloys [4]. The increase in solid solubility and the high cooling rates during laser cladding of Fe-Cr-Mn-C alloy produced first fcc M_6C carbide precipitates in a ferritic matrix followed by the formation of hexagonal close-packed (hcp) M_7C_3 carbide precipitates in the ferritic matrix [5]. However, significantly harder microstructures comprising of fine primary dendrites of

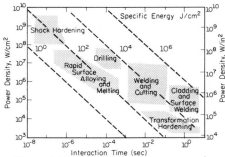

Figure 1 Power density versus interaction time map for laser processing applications.

Table 1 Experimental Ranges of Process Parameters

Laser beam power (P)	2 - 6 kW
Laser beam diameter (d)	2 and 4 mm
Process speed (v)	6.4 - 63.5 mm/s
Interaction time ($t = d/v$)	0.03 - 0.63 s
Powder feed rate (f)	0.04 - 2.0 g/s
Power density ($W = 4P/\pi d^2$)	15.9 - 159.2 kW/cm^2
Specific energy ($E = P/dv$))	2362 - 23622 J/cm^2
Flow rate of Ar gas:	
Main gas	1.59 g/s
Feeder	>0.62 g/s
Delivery nozzle	>0.145 g/s

an fcc austenitic phase and interdendritic eutectic consisting of a network of pseuodohexagonal M_7C_3 carbides rich in Cr randomly dispersed in an fcc austenitic phase were obtained by laser cladding Fe-Cr-W-C quaternary alloy [6].

The potential of laser cladding process for developing hardfacing, wear resistant materials has been demonstrated [7]. Coatings produced by laser cladding a powder mixture consisting of chromium carbides, Ni, Cr, Mo, and Si exhibited wear properties superior to those of plasma sprayed coatings with a similar composition [8]. Fiction and wear properties significantly better than those of Stellite-6 were obtained by laser cladding Fe-Cr-Mn-C alloy on low-carbon steel [9]. The increased wear resistance in the preceding study may be attributed to the hard microstructure of the surface produced due to the inherent rapid solidification and high concentration of key elements. Significant differences in the microstructure fineness, compositional homogeneity, and microhardness of laser-clad Fe-Cr-C-W alloy were obtained by varying the processing conditions [10]. Thus, dendritic, cellular and planar novel microstructures possessing extended solid solutions, metastable crystalline phases, and amorphous solids can be developed, depending on the temperature gradient and solidification rate [11].

It is evident from the foregoing that variations in the laser-cladding conditions may yield markedly different microstructures and properties. The main objective of this article, therefore, is to provide additional insight to the significance of key process parameters on the microstructure, geometry and microhardness of laser-clad coatings.

EXPERIMENTAL PROCEDURES

Thin plates of AISI 1018 steel were polished and used as substrates for laser cladding. The laser-clad coating material was a powder mixture of Fe-Cr-C-W quaternary alloy with a weight ratio of 10:5:1:1. This alloy was selected for investigation because of the relatively limited information about its microstructural and mechanical properties resulting from laser cladding and in order to provide additional confirmation for previously reported results [6,10]. The additions of the Cr and W alloying elements and the high concentration of C are desirable because they provide a strong network of carbide formation, improve the corrosion and oxidation resistance, and promote solid solution strengthening. Laser cladding was performed with a 10 kW, continuous wave CO_2 (AVCO HPL 10) laser using an underfocused $TEM_{01}*$ laser beam. The main components of the apparatus are the laser source, the F7 cassegrain optics, the powder delivery dispenser, and the worktable with three-dimensional control. The independent and dependent laser process parameters and their definitions and experimental ranges are listed in table 1. Characterization of the laser-processed clad coatings was performed by using optical microscopy, scanning electron microscopy (SEM), energy dispersive X-ray spectroscopy (EDS), electron microprobe analysis (EMPA), transmission electron microscopy (TEM), and Vickers microhardness testing.

RESULTS AND DISCUSSION

The laser-clad coating microstructures exhibited primarily dendritic and feathery morphologies. Representative micrographs revealing the effect of the process speed and powder feed rate on the laser-processed microstructure are presented in figure 2. Comparison of figures 2(a) and 2(b) demonstrates that increasing the process speed enhanced significantly the refinement of the dendritic microstructure, evidently due to the higher solidification rates resulting from the shorter interaction time. On the basis of the very small arm spacing of the dendrites, of the order of 1 μm as shown in figure 2(b), the cooling rates were approximately estimated in the range of 10^3 to 10^6 K/s [11]. Although the dendrites produced at a higher powder feed rate (figure 2(c)) are coarser than those shown in figure 2(b), the fineness of the microstructure is still relatively high, indicative of the fairly good mixing of the alloying elements and the melted substrate material. The continuous coating/substrate interfaces reveal a strong metallurgical bond. The corresponding EMPA profiles of the elemental distributions shown in figure 3 provide additional information about the effect of process parameters on the compositional homogeneity and thickness of the laser-clad coatings. As shown in figures 3(a) and 3(b), shorter interaction times yielded thinner and more homogeneous microstructures. According to the results shown in figures 3(b) and 3(c), increasing the feed rate produced a thicker coating exhibiting variations in the elemental distribution near the interface. However, the profiles shown in figure 3(c) are relatively more uniform than those shown in figure 3(a) and, in addition, demonstrate a significant increase in the amounts of Cr and W. The former effect is desirable since incorporation of such elements enhances carbide formation and promotes solid solution strengthening.

  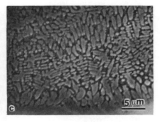

Figure 2 SEM micrographs showing the effect of process speed and feed rate on the laser-clad coating microstructure for P = 3 kW and d = 2 mm: (a) v = 31.75 mm/s, f = 0.04 g/s, (b) v = 63.5 mm/s, f = 0.04 g/s, and (c) v = 63.5 mm/s, f = 0.1 g/s.

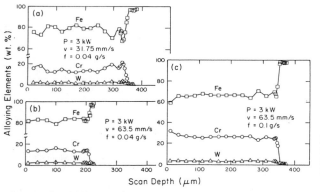

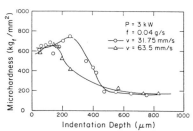

Figure 4 Microhardness profiles revealing the effect of process speed on the laser-clad coating thickness and microhardness (d = 2 mm).

Figure 3 EMPA profiles showing the effect of process speed and feed rate on the distribution of alloying elements through the laser-clad coating thickness (d = 2 mm).

Figure 4 shows microhardness profiles through the laser-clad coating thickness corresponding to the microstructures shown in figures 2(a) and 2(b). The laser-processed coatings exhibit similar average hardnesses of approximately 650 kg$_f$/mm^2, i.e., significantly higher than the substrate hardness (about 160 kg$_f$/mm^2). Although the effect of the process speed on the average coating microhardness appears to be secondary, there is a significant effect on the shape of the hardness profile and thickness of the laser-treated region. The larger scatter in the hardness of the coating produced at a lower process speed is in agreement with the SEM and EMPA results showing a relatively coarser and less homogeneous microstructure (figures 2(a) and 3(a)).

The effect of the process speed on the laser-clad coating geometry can be interpreted by considering the cross section micrographs shown in figure 5. The coatings are dense and exhibit a continuous interface with the substrate. Increasing the process speed reduced significantly the thickness

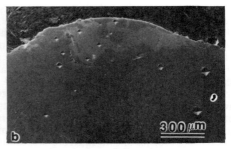

Figure 5 SEM micrographs showing the effect of process speed on the laser-clad coating geometry for P = 3 kW, f = 1.44 g/s, and d = 2 mm: (a) v = 31.75 mm/s, and (b) v = 63.5 mm/s.

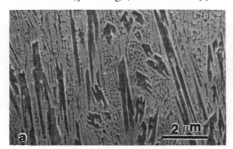
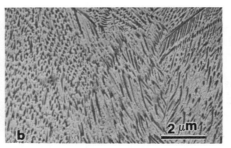

Figure 6 SEM micrographs showing the effect of process speed on the laser-clad coating microstructure for P = 6 kW, f = 0.58 g/s, and d = 4 mm: (a) v = 6.4 mm/s, and (b) v = 12.7 mm/s.

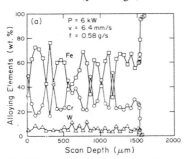
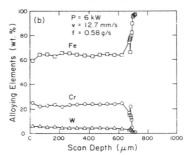

Figure 7 EMPA profiles showing the effect of process speed on the distribution of alloying elements through the laser-clad coating thickness (d = 4 mm).

and width of the laser-processed region. The indentations shown in figure 5(b), qualitatively illustrate the significantly higher hardness of the coating microstructure and the negligible effect of laser heating on the substrate material. That the indentations just below the interface are smaller than those farther away reveal the presence of a thin HAZ. Microscopy studies have revealed the formation of fully restrained martensite at the top of the HAZ and partially transformed martensite and untransformed ferrite at the bottom of the HAZ, where evidently the temperature and quenching rate was not as high as at the interfacial region [10].

The micrographs shown in figure 6 illustrate that at relatively high power levels the laser-clad microstructures comprised dendritic and feathery morphologies. The dependence of the microstructure fineness on the process speed is again apparent and the trend is similar to that observed at lower laser powers. The average arm spacing of the dendrites shown in figures 6(a) and 6(b) is approximately equal to 3.8 and 1.8 μm, respectively. The EMPA elemental profiles shown in figure 7 yield quantitative information about the effect of the process speed on the compositional homo-

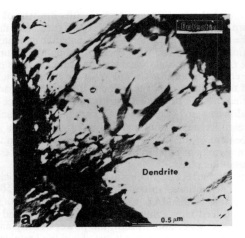

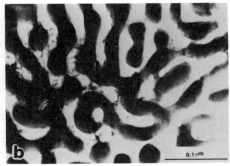

Figure 8 High-magnification bright-field TEM micrographs: (a) primary dendrite, and (b) eutectic matrix ($P = 6$ kW, $v = 12.7$ mm/s, $f = 0.58$ g/s, and $d = 4$ mm).

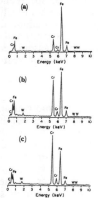

Figure 9 EDS spectra of the primary dendrites and the interdendritic eutectic matrix: (a) primary phase, (b) light contrast eutectic features, and (c) dark contrast eutectic features ($P = 6$ kW, $v = 12.7$ mm/s, $f = 0.58$ g/s, and $d = 4$ mm).

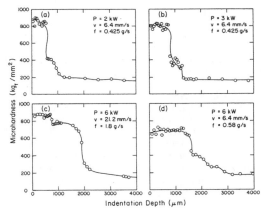

Figure 10 Microhardness profiles revealing the effect of laser power, process speed and feed rate on the laser-clad coating thickness and microhardness ($d = 4$ mm).

geneity and thickness of the laser-clad microstructures. Although thicker coatings were obtained at a relatively low process speed, i.e., long interaction time, figure 7(a) shows that the resulting compositional homogeneity is poor, in agreement with the microscopy observations. However, figure 7(b) shows that the uniformity of the coating composition was greatly enhanced by increasing the process speed. The refined microstructures shown in figures 6(a) and 6(b) possessed average microhardnesses of about 693 and 900 kg$_f$/mm^2, respectively. The significantly higher microhardness of the microstructure shown in figure 6(b) can be attributed to the appreciable concentrations of Cr and W and especially the remarkable grain refinement obtained under these process conditions.

Since the microstructure shown in figure 6(b) was found to exhibit the higher compositional homogeneity and microhardness [10], for the experimental range of process parameters investigated, the metallurgical characteristics of this coating are of particular interest. It was recently reported that the microstructure shown in figure 6(b) comprised primary dendrites possessing heavily dislocated substructures and stacking fault fringes [6]. Additional evidence about the constituents of the former microstructure supporting the previous observations is presented in the bright-field TEM micrographs shown in figure 8. Figure 8(a) shows a submicrometer-thick primary dendrite exhibiting stacking faults and dislocation pairs, and figure 8(b) shows that the eutectic matrix consists of a network of ultra-fine lamellar and wavy type morphologies (dark features). From figure 8(b), the average interlamellar spacing is about 20 nm. The appreciable waviness in the eutectic indicates that the solidification rates must have been significantly high. The EDS spectrum shown in figure 9(a) demonstrates that the primary dendritic phase is rich in Fe and contains significant amounts of Cr. However, figures 9(b) and 9(c) show that the concentrations of Cr and W in the interdendritic eutectic matrix are markedly higher than those of the primary dendrites. These findings are in accord with previous results [10]. It has been reported that the primary dendritic phase was an fcc γ austenite and the eutectic matrix comprised pseuodohexagonal $(Cr, Fe, W)_7C_3$ carbides (dark features in figure 8(b)) randomly distributed in a nonequilibrium austenitic γ phase (bright features in figure 8(b)) [6]. Thus, the microhardness enhancement of the laser-clad microstructure shown in figure 6(b) may be attributed to the formation of networks of hcp ternary M_7C_3 carbides in the eutectic matrix and the heavily dislocated primary austenitic phase.

Figure 10 provides additional information about the effect of process parameters on the thickness and microhardness of the laser-clad coatings. The hardness values in the range of 400 to 500 kg_f/mm^2 are attributed to the martensitic microstructure of the HAZ. Comparison of figures 10(a) and 10(b) shows that for the same process speed and feed rate, a thinner and harder coating and a smaller HAZ was obtained by reducing the laser beam power and, in turn, the power density and specific energy. Figure 10(c) shows that increasing the laser power, process speed and feed rate, led to the formation of a harder and markedly thicker coating in conjunction with a much smaller HAZ. However, figure 10(d) demonstrates that increasing the laser power, while mainatining the process speed and feed rate at low levels, similar to those corresponding to figures 10(a) and 10(b), reduced significantly the coating hardness and yielded a larger HAZ. This can be associated with the low solidification rates and the appreciable dilution of the coating microstructure resulting from the prolonged interaction time and the low feed rate.

CONCLUSIONS

An investigation of the effect of controlling process parameters on the microstructure, geometry and microhardness of laser-clad coatings of Fe-Cr-C-W quaternary alloy was conducted. Rapidly solidified microstructures possessing dendritic and feathery type morphologies were laser processed on low-carbon steel substrates. The small arm spacing of the primary dendrites, obtained under certain process conditions, indicated that the cooling rates in the melt zone were significantly higher than those in conventional heat treatment processes. Increasing the process speed and reducing the laser power and powder feed rate generally enhanced the fineness and hardness of the rapidly solidified microstructures, while at the same time reduced the thicknesses of the coatings and the HAZ.

ACKNOWLEDGMENTS

This research was supported by the Surface Engineering and Tribology Program of the National Science Foundation under Grant No. MSS-8996309. The author gratefully acknowledges the experimental assistance of Mr. K. Nagarathnam.

REFERENCES

1) Moore, P.G. and Weinman, L.S.: SPIE, 1980, 198, 120.
2) Steen, W.M.: Met. Mater., 1985, 1, 730.
3) Mazumder, J.: in Interdisciplinary Issues in Materials Processing and Manufacturing, Samanta, S.K., Komanduri, R., McMeeking, R., Chen, M.M., and Tseng, A., eds., ASME, 1987, 2, 599.
4) Corchia, M., Delogu, P., Nenci, F., Belmondo, A., Corcoruto, S., and Stabielli, W.: Wear, 1987, 119, 137.
5) Singh, J., and Mazumder, J.: Mater. Sci. Technol., 1986, 2, 709.
6) Nagarathnam, K. and Komvopoulos, K.: Metall. Trans. A, 1993, 24A, in press.
7) Gnanamuthu, D.S.:Proc. Conf. Applications of Lasers in Materials Processing, ASM, 1979, 177.
8) Belmondo, A., and Castagna, M.: Thin Solid Films, 1979, 64, 249.
9) Eiholzer, E., Cusano, C., and Mazumder, J.: Proc. Int. Congress on Application of Lasers and Electro-Optics, Laser Inst. Amer., 1984, 44, 159.
10) Komvopoulos, K. and Nagarathnam, K.: J. Eng. Mater. Technol., 1990, 112, 131.
11) Flinn, J.E.: Rapid Solidification Technology for Reduced Consumption of Strategic Materials, Noyes Publications, Park Ridge, NJ, 1985, 42-64.

Materials Science Forum Vols. 163-165 (1994) pp. 423-428

LASER CLADDING IN PRE-MADE GROOVES
GUIDE-LINES FOR GROOVE DESIGN

J.E. Flinkfeldt [1] and Th. F. Pedersen [2]

[1] Division of Materials Processing, Luleå University of Technology, Sweden

[2] Advanced Welding Centre, FORCE Institutes, Broendby, Denmark

ABSTRACT

Post machining of treated components after laser cladding are necessary in most cases. Depending of the quality of the chosen cladding material this job could be laborious and time consuming. By preparing grooves of a certain design on the work piece before laser cladding, the excess material to be removed can be minimised and post machining operations facilitated. The paper describes experimental works with laser cladding in pre-made grooves on cast iron specimens with the aim to investigate the influence of special groove design on the cladding results. The results are discussed in terms of macro surface roughness, efficient layer thickness, residual stress in the cladded layer and base material intermixing in the cladding. The results indicate that laser cladding in grooves is a way to reduce cladding material consumption and to increase the coating rate.

INTRODUCTION

Hard facing of worn-out or new-made machine components can be accomplished by laser cladding. Post machining of the components after the cladding is necessary in most cases and depending on the quality of the chosen cladding material this job could be hard and time consuming. One way to facilitate post machining, and to make the cladding process more favourable, would be to minimise the excess material to be removed. By preparing grooves of a certain design on the work piece before laser cladding this goal could be achieved. This work aims to investigate the influence of special groove design on the cladding results with regard to coating quality and process productivity and to determine the most suitable groove design to obtain:
• Minimised surface roughness after cladding
• Minimised post machining
• Optimised productivity, i.e. covering a given area with sufficient layer thickness and with a
   minimum of passages (tracks)
• Minimised intermixing of base material in the cladding

Some of these objectives are counteracting and hence the choice of cladding parameters must be a compromise to obtain an optimised result.

EXPERIMENTAL PROCEDURES AND RESULTS

Laser cladding was performed with a Rofin-Sinar 6 kW-laser and additive material "Stellite® Nr 6" on cast iron test specimens prepared with grooves (Figure 1). Uncomplicated geometry and easiness of machining were the criteria for choice of the groove design.

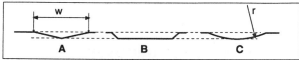

Figure 1   Basic groove design

All grooves were 0,5 mm in depth. Groove shapes A and C were prepared in different widths and with gradually altered centre line distance between the grooves according to table 1.

Groove type	Groove width w [mm]	c/c distances between grooves [mm]			Processing speed [m/min]	Beam spot size [mm]	Power on target [W]	Power density on target [W/mm²]
A I	4	4,0	3,0	2,0	0,5; 0,6; 0,7	4,2	2600	≈200
B I	9,3		–		0,5; 0,6; 0,7	4,2	2600	≈200
C I	4	4,0	3,7	3,2	0,5; 0,6; 0,7	4,2	2600	≈200
A II, C II	5	5,0	4,5	4,0	0,5; 0,6; 0,7	5,2	2900	≈140
A III, C III	6	6,0	5,5	5,0	0,5; 0,6; 0,7	6,2	3900	≈130

Table 1. Processing data

Type AI, BI and CI are considered as pre-test series and of these only the results from BI (together with A II, A III, C II and C III) are shown in this report for comparison of trace overlap and base material intermixing in the cladded layer. Coatings processed at a speed of ≤ 0,6 m/min showed sufficient thickness (≥ zero-level) and the 0,6 m/min specimens were chosen for further examination.

Surface roughness and bottom boundary lines of the claddings were measured on test piece cross-sections in a profile measuring microscope for evaluation of groove geometry influence.

Surface roughness and efficient layer thickness

The measurements of macro surface roughness of the cladding, regarding maximum and minimum trace height profiles, showed increased values by increasing overlap as could be expected. Efficient useful layer thickness below zero level (after grinding) also show increasing values by increased overlap. Efficient useful layer thickness was measured as the minimum distance from the original work piece surface (zero level) to the bottom of the cladding. Values are shown in figure 4 and 5.

To minimise the number of passes means consequently that the trace overlap should be minimised. At zero overlap a useful layer appears, however thin, due to melting of the base material in the boundary zones between the passes. At 0,5 mm overlap the useful layer thickness is significantly increased and with only a limited increase of the surface roughness. Groove shape C appears to give a thicker useful layer and a smoother surface than shape A at the same overlap.

Groove shape B was prepared in only one size. Cladding of this groove was done in both 3 and 4 passes which means that overlap values were given. The overlaps were in these cases approx. 40 - 50%.

Base material intermixing

A quantitative element analysis (EDS) was done using an electron microscope CAM-SCAN S4-80DV and a Link Analytical X-Ray system eXL. Samples from test series A III , C III (centre trace) and B I was chosen to examine the amount of base material (Fe) in the cladding. Fe-content and location of the test points are shown in figure 4 and 5. Around 3% of the Fe-content originates from the cladding material Stellite 6.

Average Fe-contents in the cladding on groove shape A and C, shown in figure 4, are decreasing with increasing trace overlap. At overlap 0,5 -1.0 mm the Fe-content becomes stable. Despite the higher beam power density used in test runs on B type grooves, the Fe-content in the cladding is significantly lower than in A and C, figure 5, and show minimum values in the middle part of the cladding due to the limited contact area between each of the cladding tracks and the base material in that zone. Hardness values in the cladding (500 - 700 Hv) all exceed the original hardness of Stellite 6, which is approximately 400 Hv.

Residual stresses

Residual stresses have been measured on the A-II, A-III, C-II and the C-III series of Stellite 6 claddings by the help of an X-ray diffraction technique [1].

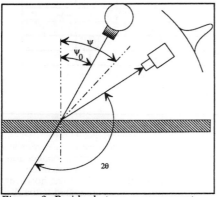

The stresses have been measured with X-ray diffraction of Cr-Kα radiation from the cobalt phase. With laser claddings the biggest problem with residual stress measurements is that the crystalline micro structure has a pronounced texture, because all the dendrites grow in the same direction. Since diffraction has to be made for a number of angles compared to the surface plane, this presents a problem. The intensity of the diffraction line varies very much with the direction, and in some directions it even drops to zero. Therefore, the first step is to scan the angle, Ψ, to determine this variation, and then measure the position, 2θ, of the diffraction line in directions with a reasonable intensity [2].

*Figure 2. Residual stress measurement
conditions*

Material properties of Stellite 6			
Elastic modulus, E	Poisson's ratio, v	Cr-Kα absorption coef.	2θ₀
205 GPa	0,28	1097 cm⁻¹	130°

*Table 2 . The material properties of Stellite 6
used in the residual stress calculations.*

The stress, σ, is then calculated from the formula:

$$\sigma = \frac{-E}{2 \cdot (1 + v)} \cdot \cot(\theta_0) \cdot \frac{\pi}{180} \cdot \frac{\partial(2\theta)}{\partial(\sin^2 \psi)}$$

Before the stresses were measured, the surface was ground to obtain a flat surface. The original surface was still visible at the track overlaps, so about 0.1-0.2 mm has been removed. The ground surface was electropolished with Struers A2 electrolyte at 1A/cm² to remove the layer damaged by the grinding. The stress was then measured in the middle of the track, in a direction perpendicular to the tracks. The stress is measured as an average over an area of approximately 2x4 mm².
The position, 2θ, of the diffraction peak was measured with three different methods:
1) as the centre of the full width at half maximum (FWHM) intensity of the peak,
2) as the centre of gravity (COG) of the peak and, 3) as the top of a parabola fitted to the top 15% of the peak. The stress values obtained with the three methods were averaged to minimise the uncertainty of the measurement.

It was also attempted to measure the stress parallel to the track, because the cracks that appeared in the cladding were all perpendicular to the tracks. This was not possible, however, because of a diffraction peak, either from the substrate or from the interdendritic carbide phase interfered with the peak used for the measurement.

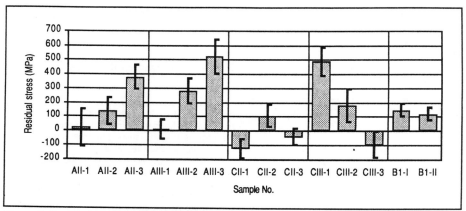

Figure 3, Residual stress in Stellite 6 claddings

For the A-series there seems to be a clear correlation between the layer thickness and the tensile residual stress; The thicker the cladding - the higher the stress. With the C-series the connection is not quite as clear. In the CII-series the useful layer thickness is almost uniform for the three specimens and there is not either a large variation in the stress level. In the CIII-series, however, there seems to be the opposite relation between thickness and stress level as compared to the A-type grooves. Here the stress decreases with increasing layer thickness.

DISCUSSION AND CONCLUSIONS

To achieve a sufficient layer thickness by laser cladding with subsequent passages, conventionally on a <u>flat</u> surface, a certain degree of trace overlap is needed, usually 30-50 %.

A profiled groove of type A and C seems to need a cladding overlap of at least 0,5 mm, corresponding to ≥ 8 - 10 %, to obtain an efficient layer thickness of ≥ 0.4 mm. The results show that with an overlap of 1.0 mm (15-20%) the efficient layer thickness comes close to the original max. depth of the pre-machined groove. As a consequence of decreased overlap the Fe-content in the cladding will rise due to the larger contact area to the work piece for each cladding track. Circular bottom profiles (groove type C) seem to be more advantageous than the V-shaped (type A) regarding layer thickness as well as base material intermixing and residual stresses.

The trials indicate that, if the rather high degree of intermixing of base material in the cladding can be accepted, laser cladding in grooves certainly is a way to reduce cladding material consumption and to increase the coating rate. In these trials cast iron was used as base material, and cladding on steel materials may give less intermixing values due to higher melting point.

REFERENCES

1) Noyan I.C., Cohen J. B.: Residual stress, Springer Verlag 1987, pp 119- .

2) Pedersen T.F. et al.: BRITE-project nr 2178. Laser treatment as a tool for tailoring the surface composition of alloy components for engineering applications. Final report FORCE Institutes 1992.

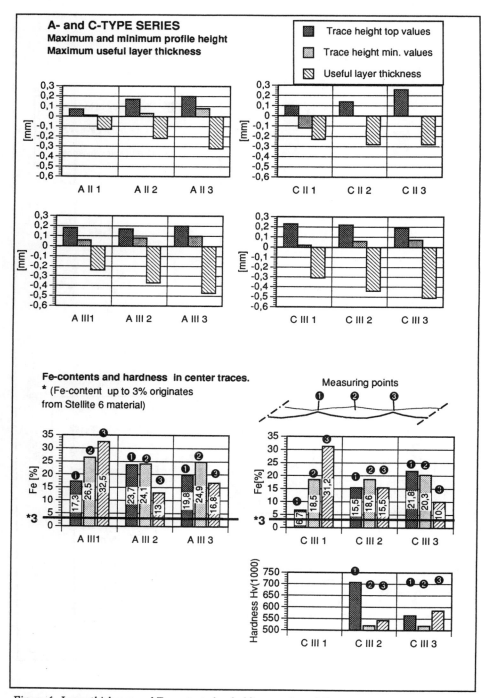

Figure 4. Layer thickness and Fe-contents for claddings on A- and C-type grooves.

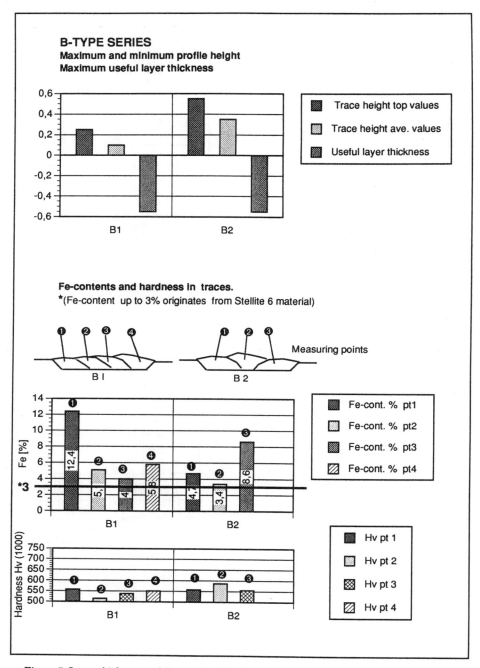

Figure 5. Layer thickness and Fe-contents for cladding on B-type grooves

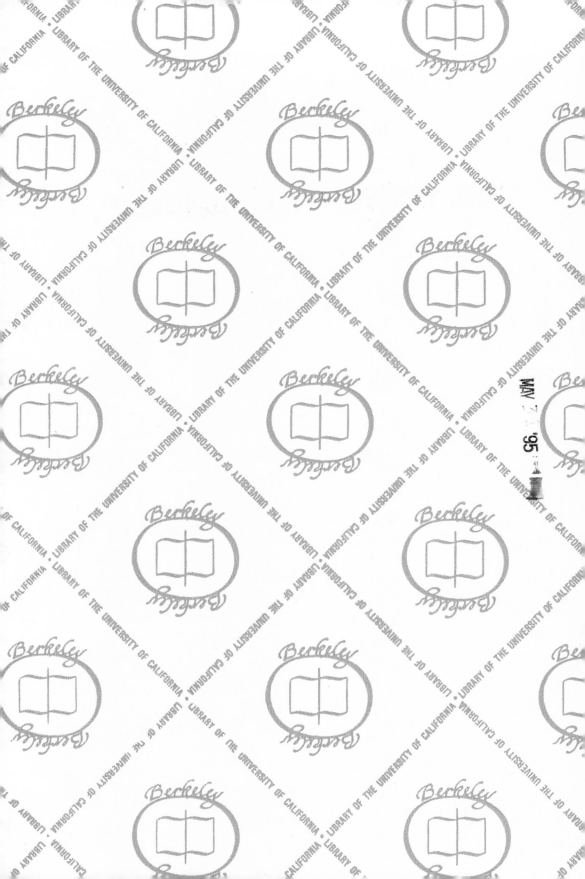